In *The Advancement of Science, and Its Burdens*, Professor Holton carries farther his analysis of how modern science works and what its influences are on our world, with particular emphasis on the role of the thematic elements – those often unconscious presuppositions that guide scientific work to success or failure. Many of the conclusions emerge from the author's continuing study of Albert Einstein's historic contributions, as well as of other scientists' contrasting styles of research. Professor Holton also discusses the often unforeseen consequences of the advancement of modern science – its fruits as well as its burdens.

Some of the specific questions Professor Holton addresses include: What was Einstein's overall scientific program? How did his work shape the imaginations of twentieth-century artists and writers? Are there national differences between styles of scientific research? By what mechanisms is progress in science achieved despite the enormous diversity of individual, often conflicting efforts? What is the belief system of contemporary scientists and do they still need a guiding philosophy of science? What are the uses and dangers of metaphor in frontier research? Is there a sequence of steps by which high-level theories are constructed? What limits may society validly place on the scope of research programs? And finally, in his Jefferson Lecture, Professor Holton asks whether a new organic relationship has developed between "pure" research and technologically directed development, by which science can serve both the search for truth and the national interest.

The advancement of science, and its burdens

The advancement of science, and its burdens

The Jefferson Lecture and other essays

GERALD HOLTON

Mallinckrodt Professor of Physics
and Professor of the History of Science, Harvard University

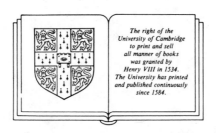

The right of the
University of Cambridge
to print and sell
all manner of books
was granted by
Henry VIII in 1534.
The University has printed
and published continuously
since 1584.

CAMBRIDGE UNIVERSITY PRESS

CAMBRIDGE

LONDON · NEW YORK · NEW ROCHELLE
MELBOURNE · SYDNEY

Published by the Press Syndicate of the University of Cambridge
The Pitt Building, Trumpington Street, Cambridge CB2 1RP
32 East 57th Street, New York, NY 10022, USA
10 Stamford Road, Oakleigh, Melbourne 3166, Australia

First published 1986

Printed in the United States of America

Library of Congress Cataloging-in-Publication Data
Holton, Gerald James.
The advancement of science, and its burdens.
Bibliography: p.
Includes index.
1. Science – Miscellanea.   2. Einstein, Albert,
1879–1955 – Knowledge – Science.   I. Title.
Q173.H734   1986   509'.04   86-9722

British Library Cataloguing in Publication Data
Holton, Gerald
The advancement of science, and its burdens: the Jefferson lecture and other essays.
1. Science – Philosophy
I. Title
501'.8   Q175

ISBN 0 521 25244 X hard covers
ISBN 0 521 27243 2 paperback

CONTENTS

PREFACE

Readers of my previous publications – for example, the books *Thematic origins of scientific thought: Kepler to Einstein* (Harvard University Press, 1973) and *The scientific imagination: Case studies* (Cambridge University Press, 1978) – will not be surprised that in this volume I continue to apply the approach developed in earlier studies to the questions of how modern science is done, and what its influences are on our world. These two questions are directly connected with some of the current controversies in such fields as history of science, philosophy of science, sociology of science, cultural history, science policy, and education.

In order to base the discussion on the foundation provided by a detailed case study, I continue, in the first six chapters (Part I), my researches on the work of Albert Einstein as well as on the influences – in both senses: those upon it during the nascent phase, and those flowing from it after publication. The next part (Chapters 7 through 9) provides an investigation of other styles of research in the advancement of science, contrasting with the approach dominating Part I. And the last set of chapters (Part III) deals more directly with the often unforeseen consequences of the progress of contemporary science, with its fruits as well as its burdens.

On the way through these chapters one could choose to neglect the development of a point of view in favor of focusing on some of the separate questions that become prominent in individual chapters: By what mechanisms is progress in science achieved despite the enormous diversity of individual, often conflicting, efforts? What is the modern scientists' belief system, and do they still need a guiding philosophy of science, whether held consciously or not? What are the uses and dangers of metaphor in scientific research? In education? Is there a sequence of steps by which high-level theories are constructed? What was Einstein's overall scientific program? How did his work shape the culture of science and the imagination of twentieth-century artists and writers? What limits may society validly place on the work of researchers? Has a new relationship developed between "pure" research and mission-directed development, one that allows science to serve better both the pursuit of truth and the national interest?

It is, however, possible to approach the book more holistically, for a main preoccupation underlying all chapters is what one can mean by "understanding" the development of modern science. To this, the historian of science cannot expect one single, simple answer but must proceed in different directions to partial or complementary answers, only the totality of which can have a claim to credibility.

To those in the maturer and exact sciences, such a process may at first not seem congenial. These researchers are used to understanding the current state of their science (rather than its historical development) in another way – in terms of the integrated picture of the physical world obtaining at the moment. To use an analogy: They can imagine themselves as having gained the Olympian height from which they might, in principle, perceive at a glance the whole varied landscape below. From such a unified conception one can hope to deduce, and therefore to "understand," every detail of every phenomenon in the given science. And though the sciences are still a long way from attaining such a goal, the existence of a few scientists with remarkably synoptic understanding gives courage to the rest.

Moreover, the problems encountered in a science usually turn out to have one right answer. Most scientific problems are understood in about the same way everywhere on earth, with different schools of thought in the natural sciences existing relatively rarely and briefly. Third, most experts in a given field share more or less the same epistemology and ideology. And last, the raw material, the data base for any problem in a natural science, is usually relatively certain, because it can be reproduced at will in any suitably equipped laboratory.

But all of this of course does not apply to the study of the history of a science. The very contrary is the case. The history of science has indeed become a vigorous and rapidly rising professional discipline, but unlike the natural sciences, it does not have a well-developed theory. In this respect it seems to me still in a pre-Newtonian stage. Attempts to impose an overarching theory of historical development are premature and misguided.

Yet in my view this state of affairs does not constitute the slightest handicap for understanding the history of science in a profound way. Two aspects of "understanding" are perfectly within our present capabilities: (1) to understand thoroughly many of the individual main *events* in the history of a science (events, not speculative syntheses and other constructs) and (2) to see *connections* between many of these events. An analogy may again help: One can have a comfortable and operational understanding of the geography of one's city if one is familiar with many of the chief intersections and quite a few of the connecting roads between them, without having been in every street or having seen the initial master plan.

By "event" I mean the development, appearance, or publication of a scientific paper, or an influential scientific address, or a specific discovery, or a letter, or a photograph made during the use of laboratory equipment, or a page of a laboratory notebook, and so forth. Each of these has a physical residue that can be studied and that lends itself to the eventual formation of a consensus among competent observers who come to a historic case from different directions. It is in this sense analogous to what an elementary particle physicist calls an event, for example, a trace of tracks in a spark chamber. The task of historians of science, then, is to use these events as the underlying factual base and to proceed inductively from that base.

Readers familiar with my publications will know that in my view an understanding of an event, and ultimately of the connection between events, follows from the proper description of the event, and that this description consists in principle of providing a separate account of each of the main components that generally produce an event. These components – as elaborated in the introduction and first chapter of each of the two books cited above – are in brief the following: the state of "public," shared scientific knowledge within the subject at the time of the chosen event; the trajectory of the state of public, shared scientific knowledge leading up to the time chosen; the state of the "private" scientific knowledge of the particular scientist at the time of the chosen event; the trajectory of the scientific activity of the person under study, up to the chosen time; the prior development of the particular scientist in terms of his or her psychological formation; the sociological setting at the time chosen (e.g., the effect of the educational system on the preparation of the scientist); cultural, ideological, or political elements that may have shaped the thinking of the scientist; the epistemological assumptions and the logical structure of the document under study; and, last but not least, and in my own studies most important, the thematic presuppositions that guided the work of the scientist as well as of his or her followers and opponents.

The exhaustive description of a particular event, one that includes all of these nine components, is neither likely nor necessary. But it is a worthy ideal, as the goal in the eventual accumulation of insight into a particular case. In the meantime, the program proves its worth by helping to make better sense of what otherwise would be a chaotic or arbitrary collection of personalities, documents, and controversies and by guiding our search for coherent patterns characterizing the development of science and its reciprocal interactions with the rest of our culture.

A brief comment seems in order with respect to the last chapter, entitled "The advancement of science, and its burdens: the Jefferson Lecture." It was given as the tenth of these annual lectures, at the invitation of the National Endowment for the Humanities, which is responsible to Congress for the selection of the lecturers. The announced conditions of the award require that the lecture be of some pragmatic use, for the speaker is to bring to bear his experience "upon aspects of contemporary culture, and matters of broad public concern." At the same time, the chosen scholar is also burdened with the announcement by the NEH that the award "is the highest honor the federal government confers for distinguished intellectual achievement in the humanities." To this double-barreled challenge is added the setting. The lecture is given in the first instance in Washington, D.C., before an audience of about fourteen hundred, with the invitation list apparently containing policymakers, from Supreme Court justices down to congressional staffers, as well as administrators and a contingent of scholars.

I cannot hide the fact that, despite all the problems inherent in the task of engaging such a diversified audience under these presuppositions, I felt grateful for the opportunity of speaking at that very time, namely a few months after the inauguration of President Reagan's administration – a time when, with hardly any audible debate, federal support for most major cultural, scientific, and educational programs, built up by all the nation's previous administrations, was being threatened with decimation. My concern over that new policy could of course form only a small part of my explicit presentation; but it was a pleasant reward to find indications later that the lecture had been found of some use even in that world beyond Milton's grove of academe.

I am happy to acknowledge support from the National Endowment for the Humanities, the National Science Foundation, and the Guggenheim Foundation for research on which much of the work yielding these chapters is based. Such support does not, of course, imply that any of these agencies agrees with my conclusions.

I also wish to express my indebtedness to the estate of Albert Einstein for permission to cite from the writings of Einstein. I benefited from discussions on several of the chapters with colleagues too numerous to identify separately. Ms. Kristin Peterson provided editorial help with patience and discernment in assembling the chapters of the book. My special thanks go again to my assistant, Ms. Joan Laws, for seeing these essays through all the stages, from tentative drafts to final publication.

G.H.

PART I

Einstein and the culture of science

1
Thematic presuppositions
and the direction of scientific advance

On June 10, 1933, Albert Einstein delivered the Herbert Spencer Lecture at the University of Oxford. By that time he was a man without a country, passing through that haven as a refugee from Fascism, as so many others, illustrious or unknown, were to do after him. Like them, he retained a warm and thankful memory of the hospitality here. We may assume that he took special care in preparing the lecture. Philipp Frank, his biographer and colleague, called it the "finest formulation of his views on the nature of a physical theory."[1]

The published version[2] has been rarely analyzed or even adequately understood. Now that we have access to so many more of Einstein's published and unpublished documents, his essay turns out to be a very appropriate entry for a study of scientific explanation, both of Einstein's own contribution to the subject and of more recent approaches.

The "eternal antithesis"

Einstein's choice of "the method of theoretical physics" as his topic was by no means casual. In fact, for much of his life he seems to have been almost obsessed by the need to explain what he called his epistemological credo. From about 1911 to the end, he wrote on it again and again, almost as frequently as on physics itself. On occasions great and small, he reverted to his self-appointed task in his remarkably consistent way – with the single-minded patience of a hedgehog, and the glorious stubbornness that characterized him from his boyhood on.

His home-made philosophical system of the practicing scientist, of which he wrote so often, seemed to his philosophical commentators something of a house of cards too, a patchwork of pages from Hume, Kant, Ernst Mach, Henri Poincaré, and many others. Indeed, Einstein himself cheerfully acknowledged once that he might appear "as a type of unscrupulous opportunist," appearing by turns as a realist, idealist, positivist, or even Platonist or Pythagorian. Yet the method he preached and practiced turned out to be remarkably robust. Many of today's physicists, without knowing its origin, have adopted a style of attempting fundamental and daring advances that owes a great deal to Einstein's credo, even as Einstein's dream of finding a unification of the forces of nature has, in its modern form, turned out to be the stuff of which Nobel prizes are made.

In his own day, however, Einstein had good reason to suspect that few physicists and philosophers understood what he was saying about scientific methodology, or even could describe clearly what they themselves were doing. And so, rather like Galileo, he took his epistemological message to the wider public. He opened the formal part of his Herbert Spencer lecture with the famous sentence: "If you want to find out anything from the theoretical physicists about the methods they use, I advise you to stick closely to one principle: don't listen to their words, fix your attention on their deeds."

Here he objects to scientists who speak about the products of their imaginations as if these were "necessary and natural" – not "creations of thought" but "given realities." To expose their mistake, he invites us to pay "special attention to the relation between the content of a theory" on the one hand, and "the totality of empirical facts" on the other. These constitute the two "components of our knowledge," the "rational" and the "empirical"; these two components are "inseparable"; but they stand also, Einstein warns, in "eternal antithesis."

To support this conception, Einstein now gives a very brief sketch of a dichotomy built into Western science. The Greek philosopher-scientists provided the necessary confidence for the achievements of the human intellect by introducing into Western thought the "miracle of the logical system," which, as in Euclid's geometry, "proceeds from step to step with such precision that every single one of its propositions was absolutely indubitable." But "propositions arrived at by purely logical means are completely empty as regards reality"; "through purely logical thinking we can attain no knowledge whatsoever of the empirical world." Einstein tells us that it required the seventeenth-century scientists to show that scientific knowledge "starts from experience and ends with it."

It seems therefore that we are left with a thoroughly dualistic method for doing science: on the one hand, Einstein says, "the structure of the system is the work of reason"; on the other hand, "the empirical contents and their mutual relations must find their representation in the conclusions of the theory." Indeed, virtually all of Einstein's commentators have followed him in stressing this dualism – and leaving it at that. For example, F.S.C. Northrop summarized the main content of Einstein's Oxford lecture in these words: An "analysis of Einstein's conception of science shows that scientific concepts have two sources for their meanings: The one source is empirical. It gives concepts which are particulars, nominalistic in character. The other source is formal, mathematical and theoretical. It gives concepts which are universals, since they derive their meaning by postulation from postulates which are universal propositions." [3]

This is a view of science (even of Einstein's science) of which there are many versions and variants. I would call it a two-dimensional view. It can be defended, up to a point. All philosophies of science agree on the meaningfulness of two types of statements, namely propositions concerning empirical matters that ultimately boil down to meter readings and other public phenomena, and propositions concerning logic and mathematics that ultimately boil down to tautologies. The first of these, the propositions concerning empirical matters of fact, can in principle be rendered in protocol sentences in ordinary language that command the general assent of a scientific community; I like to call these the *phenomenic propositions*. The second type of propositions, meaningful in so far as they are consistent within the system of accepted axioms, can be called *analytic propositions*. As a mnemonic device, and also to do justice to Einstein's warning about the "eternally antithetical" nature of these propositions, one may imagine them as lying on a set of orthogonal axes, representing the two dimensions of a plane within which scientific discourse usually takes place. A scientific statement, in this view, is therefore analogous to an element of area in the plane, and the projections of it onto the axes are the aspects of the statement that can be rendered, respectively, as protocol of observation (e.g., "the needle swings to the left") and as protocol of calculation (e.g., "use vector calculus, not scalars").

Now it is the claim of most modern philosophies of science which trace their roots to empiricism or positivism, that any scientific statement has "meaning" only in so far as it can be shown to have phenomenic and/or analytic components in this plane. And indeed, in the past, this

Procrustean criterion has amputated from science its innate properties, occult principles, and all kinds of tantalizing questions for which the consensual mechanism could not provide sufficiently satisfying answers. A good argument can be made that the silent but general agreement to keep the discourse consciously in the phenomenic–analytic plane where statements can be shared and publicly verified or falsified is a main reason why science has been able to grow so rapidly in modern times. The same approach also characterizes the way science is taught in most classrooms and is "rationalized" in most of the current epistemological discussions.

Problems for the two-dimensional view

Nevertheless, this two-dimensional view has its costs. It overlooks or denies the existence of active mechanisms at work in the day-to-day experience of those who are actually engaged in the pursuit of science; and it is of little help in handling questions every historian of science has to face consciously, even if the working scientist, happily, does not. To illustrate, let me mention two such puzzles. Both have to do with the direction of scientific advance, and both will seem more amenable to solution once the dualistic view is modified.

1. If sound discourse is directed entirely by the dictates of logic and of empirical findings, why is science not one great totalitarian engine, taking everyone relentlessly to the same inevitable goal? The laws of reason, the phenomena of physics, and the human skills to deal with both are presumably distributed equally over much of the globe; and yet the story of, say, the reception of Einstein's theories is strikingly different in Germany and England, in France and the United States. On the level of *personal* choice of a research topic, why were some of Einstein's contemporaries so fatally attracted to ether-drift experiments, whereas he himself, as he put it to his friend de Haas, thought it as silly and doomed to failure as trying to study dreams in order to prove the existence of ghosts? As to skills for navigating in the two-dimensional plane, Einstein and Bohr were rather well matched, as were Schrödinger and Heisenberg. And yet there were fundamental antagonisms in terms of programs, tastes, and beliefs, with occasional passionate outbursts between scientific opponents.

Or, again, how to understand the great variety of different personal styles? The physicist Edwin C. Kemble described his typical mode of work, with some regret, as the building of a heavy cantilevered bridge, each piece painstakingly anchored on a well-secured base. Robert Oppenheimer, on the other hand, one might think of as a spider building a web; individual extensions were achieved by daring leaps, and the resulting structures were intricate and shimmering with beauty, but perhaps a bit fragile. Enrico Fermi, whom many regard as the inventor of teamwork in modern physics, ran his laboratory like a father who had assembled around himself a group of very bright offspring.

And then there is the scientist who moves through his problem-area alone, as the fur trapper did through Indian territory. Bernard DeVoto described it in his book *Across the wide Missouri*. The trapper "not only worked in the wilderness. He also lived there. And he did so from sun to sun by the exercise of total skill." Learning how to read formal signs was of course essential to him, but more important was

> the interpretation of observed circumstances too minute to be
> called signs. A branch floats down a stream – is this natural,
> or the work of animals, or of Indians or trappers? Another
> branch or a bush or even a pebble is out of place – why? . . .
> Buffalo are moving down wind, an elk is in an unlikely place
> or posture, too many magpies are hollering, a wolf's howl is
> off key – what does it mean?

What indeed does all this variety of scientific styles mean? If science *were* two-dimensional, the work in a given field would be governed by a rigid, uniform tradition or paradigm. But the easily documented existence of pluralism at all times points to the fatal flaw in the two-dimensional model.

2. A second question that escapes the simple model, and to which I have devoted a number of case studies in recent years, is this: why are many scientists, particularly in the nascent phase of their work, willing to hold firmly, and sometimes at great risk, to a form of "suspension of disbelief" about the possibility of falsification? Moreover, why do they do so sometimes without having any empirical evidence on their side, or even in the face of disconfirming evidence?

Among countless examples of this sort, Max Planck, responsible for the idea of the quantum but one of the most outspoken opponents of its corpuscular implications, cried out as late as May 1927, "Must we really ascribe to the light quanta a physical reality?" – and this four years after the publication and verification of Arthur H. Compton's findings. On the other hand, when it came to explaining the electron in terms of what Planck called "vibrations of a standing wave in a continuous medium," along the lines proposed by de Broglie and Schrödinger, Planck gladly accepted the idea and added that these principles have "already [been] established on a solid foundation" – and all that before Planck had heard of any experimental evidence along the lines provided by Davisson and Germer.[4]

"I do not doubt at all . . ."

Einstein was even more daring. As I have documented elsewhere,[5] immediately after the publication of his 1905 relativity paper there was published what purported to be an unambiguous experimental disproof of it. The young man remained unperturbed. Later, when the gravitational red shift, predicted by general relativity theory for the spectral lines from stars with large masses, turned out to be very difficult to test, and the experimental results were neither systematic nor of the predicted amount, Einstein again simply waited it out. To Max Born he wrote later that, even in the absence of all three of the originally expected observable consequences of general relativity, his central gravitation equations "would still be convincing," and that in any case he deplored that "human beings are normally deaf to the strongest [favourable] arguments, while they are always inclined to overestimate measuring accuracies."[6]

To be sure, if one looks hard, one can find in Einstein's voluminous writings a small number of statements of the opposite kind. An example of this sort, written shortly after the triumphant announcement of Eddington's results late in 1919, is one sentence in the 1920 edition of Einstein's popular exposition, *Relativity, the special and general theory*: "If the red shift of spectral lines due to the gravitational potential should not exist, then the general theory of relativity will be untenable." Sir Karl Popper, in his recent *Autobiography*, indicates that his own falsifiability criterion owed at its origin much to what he perceived to be Einstein's example, and he cites this specific sentence, which he says he read with profound effect when he was still in his teens.

Those of us who have admired Sir Karl's work can only be grateful that he came upon Einstein's sentence in the 1920 edition that helped set him on his path. In its earlier editions and frequent printings of 1917, 1918, and 1919, Einstein's book had ended very differently. There, Einstein acknowledged that his general relativity theory so far had had only one observable consequence, the precession of the orbit of Mercury, whereas the predicted bending of light and of the red shift of spectral lines owing to the gravitational potential were too small to be then observed. Nevertheless, Einstein drew this conclusion, in a sentence with which he ended his book in its first fifteen printings: "I do not doubt at all that these consequences of the theory will also find their confirmation." [7]

Suspension of disbelief

To illustrate that Einstein is not so different from other scientists when it comes to the willingness to suspend disbelief, it will be worth making an excursion to note how an experimentalist of great skill went about his business in much the same way, but in the privacy of his laboratory. Some time ago I came across laboratory notebooks of R.A. Millikan for 1911–12 that contained the raw data from which he derived his measured value of the basic unit of electric charge, the electron. [8] Millikan's earlier attempts in this direction had been quite vulnerable and had come under bitter attack from a group of research physicists at the University of Vienna, chiefly Felix Ehrenhaft, who believed not in a unitary but in a divisible electron, in subelectrons carrying charges such as one-fifth, one-tenth, or even less of the ordinary electron. Now, in gearing up his response, Millikan had two strong supports for his counterattack. One was his unflagging preconception that there is only one "electrical particle or atom," as he put it, a doctrine he believed to have been proposed first and convincingly by Benjamin Franklin. His other support was the kind of superb skill described in the passage quoted from Bernard DeVoto's book.

Millikan's publication came in the August 1913 issue of the *Physical Review*, and effectively ended the scientific portion of the controversy. It contains data for 58 different oil drops on which he has measured the electric charge. He assures his readers, in italics: "*It is to be remarked, too, that this is not a selected group of drops, but represents all of the drops experimented on during 60 consecutive days.*" Four years later, in his book *The electron*, Millikan repeats this passage, and all the data from the 1913 paper, and he adds for extra emphasis: "These [58] drops represent all of those studied for 60 consecutive days, no single one being omitted." [9]

At the Millikan Archive of the California Institute of Technology, the laboratory notebooks are kept from which the published data were derived. If we put our eye to that keyhole in the service of the ethology of science, we find there were really 140 identifiable runs, made over a period of six months, starting in October 1911. Anyone who has done research work in a laboratory cannot help but be impressed by the way Millikan handles his data, and by the power of a presupposition shrewdly used.

To prepare for the proof from Millikan's laboratory records, let me remind you of the chief point of Millikan's oil drop experiment. In a simplified form that nevertheless retains the scientific essentials as well as its beauty and ingenuity, it is now a standard exercise in the repertoire of school physics. A microscopic oil droplet is timed as it falls through a fixed distance in the view field. It will have some net electric charge to begin with, if only owing to the friction that acted on it when it was initially formed and expelled from the vaporizer. Other electric charges may be picked up from time to time as the droplet encounters ionized molecules in the gas through which it falls. Neither of these charges influences the droplet's motion, so long as it falls freely in the gravitational field. But when an electric field of the right sign and magnitude is suddenly applied, the drop will reverse its course and will rise the more rapidly the larger the electric charge on it. Comparing the times taken for falling and subsequent rising allows one to calculate the net charge owing to friction on the drop, q_{fri}, while comparing the times for alternate risings yields the net charge owing to the encounter with gas ions, q_{ion}.

As one watches the same droplet over a long time, through its many up and down excursions, one can accumulate a large number of values for q_{fri} and q_{ion}. Now the fundamental assumption Millikan makes throughout his work is that q_{fri} as well as q_{ion} are always some integral multiple of a unit charge equal in magnitude to the charge of the electron, e. Conversely, from the full set of data, he can determine the magnitude of e which is common to all of the values obtained for q_{fri} and q_{ion}, both being assumed to be always equal to 1, or 2, or 3, . . . $\times$ e. These assumptions become plausible when the scatter of values for e turns out to be small when computed from either q_{fri} or q_{ion} – and when the mean values of e, so differently based, are nevertheless closely equal for a given droplet.

This is just what happens for the 58 "runs" or droplets discussed in the published (August 1913) paper of Millikan. One of the runs he had made on the Ides of March 1912 and recorded in Millikan's laboratory notebook, is typical. The difference between the values of e, computed on the two different bases, is only about 0.1 percent, and not far from the limits set by the apparatus itself. The page on which both the data and the calculations appear records Millikan's exuberance and pleasure in the lower left corner: "Beauty. *Publish* this surely, *beautiful!*"

Millikan continued immediately to take data on another oil droplet, entering the data on the next page. This time things did not go well. It was now a heavier drop, hence its time of fall was shorter. The numbers of charges it picked up as it went along were not greatly different, and it did not stay in view as long as one would have liked. Now the difference between the average values of *e*, calculated from q_{fri} and q_{ion} respectively, were 1 percent apart, instead of 0.1 percent. So Millikan notes in his private laboratory book on that page: "*Error high* will not use," – and indeed it does not appear among the 58 droplets that made it into the published paper. From Millikan's point of view, it was a failed run, or, in effect, no run at all. The magnitude of the difference in the values of *e* obtained in those two ways was awkwardly large, although not so surprising as to threaten Millikan's fundamental assumptions. Instead of wasting time, he simply went on to a next set of readings, using another droplet.

But the discarded set of observations – and many others like it in the same laboratory notebook – would have appeared very differently if examined from another set of presuppositions. Thus, the discarded entries make sense if one assumes that the smallest charge involved in the oil drop experiment is not *e*, but, say, $\frac{1}{10}$ *e*. In that case, the number of charges on a given droplet would not have been in succession, 11, 13, and 14, as Millikan had to assume, but could have been 109, 129, and 139; and correspondingly, the difference between the (now smaller) elementary charges obtained in the two ways would be of the order of 0.1 percent, instead of Millikan's 1 percent. The "high" error was first of all a judgment stemming from Millikan's presupposition that the smallest charge in nature could not be a fraction of the charge of the electron *e*. To be sure, it was a presupposition supported (although more indirectly) by arguments in many other branches of physics.

Millikan's decisions seem to us now eminently sensible; but the chief point of the story is that, in 1912, Millikan's assumption of the unitary nature of the electric charge was by no means the only one that could be made. On the contrary, a chief reason for his work at the time was to perfect his method and support his claim against the constant onslaught of Felix Ehrenhaft and his associates who, for a couple of years, had been publishing experiments in support of their own, precisely opposite presupposition, namely, in favour of the existence of *sub*electrons.

It is also part of the historical setting that, at the time, Millikan was really just beginning belatedly on his career as a research physicist, whereas Ehrenhaft – at a venerable and much better equipped university – had begun to be widely recognized and rewarded years earlier as a fast-rising star in experimental physics. It was only after losing the argument with Millikan, and probably as a result of it, that he began a rapid decline as a scientist. When Millikan was doing his experiments, the matter was still in the balance. If Ehrenhaft had had access to Millikan's notebook, he would have found precisely those runs most valuable for his purposes which, for Millikan, were "failed."

Conversely, Millikan's own presupposition helped him to identify difficulties of the usual experimental nature which he did not feel were worth following up. For many of those he entered a plausibility argument on the spot (e.g., that the battery voltages must have changed, convection interfered, the stop-watch might be in error). The laboratory notebooks record Millikan's frank comments in such cases. The most revealing of the lot – revealing both of Millikan's insights that dust particles might intrude in the observation chamber, and of the willingness to take risks on behalf of his presupposition – is a marginal note entered for a long run that yielded a value of e far outside the expected limit of error: "$e = 4.98$ which means that this could not have been an oil drop."

Like the trapper in Indian country, he was advancing on dangerous territory, but with a framework of beliefs and assumptions within which judgments are possible. The chief gain was the avoidance of costly interruptions and delays that would have been required to pin down the exact causes of discrepant observations. Obviously, this is not a method we recommend to our beginning students. But obviously also, any discussion of the advance of science that does not recognize the role of the suspension of disbelief at crucial points is not true to the activity.[10]

Toward a third mechanism

Einstein would not have been surprised by Millikan's notebook. Perhaps because of his experience with the reception of his special theory of relativity, he took a dim view of new experiments that, like Ehrenhaft's, made strong claims not explainable in terms of theoretical systems which embrace a greater complex of phenomena. Very early in his career, Einstein had, it seems to me, formed a clear view about the basic structure of nature: at the top there is a small number of eternal, general principles or laws by which nature operates. These are not easy to find – partly because God is subtle, and partly because they do not stop at the boundaries between fields that happen to be occupied by different theories.

Below this upper layer of a few grand laws lies a layer of experimental facts – not the latest news from the laboratory, but hard-won, well-established, aged-in-the-bottle results, many going back to Faraday and Fresnel, and now indubitable. These experiences or key phenomena are the necessary consequences of the visible compliance with the general laws.

But between these two solid levels is the uncertain and shifting region of concepts, theories, and recent findings. They deserve to be looked at, but skeptically; they are man-made, limited, fallible, and if necessary, disposable. Einstein's attitude was perhaps best expressed in a remark reported to me by one of his colleagues in Berlin, the physical chemist Hermann F. Mark: "Einstein once told me in the lab: 'You make experiments and I make theories. Do you know the difference? A theory is something nobody believes except the person who made it, while an experiment is something everybody believes except the person who made it.' "

What, then, must one conclude from the fatal predisposition for the ether on the part of Lorentz, Poincaré, and Abraham; Max Planck's predisposition for the continuum and against discreteness; Robert Millikan's predisposition for a discrete rather than a divisible electron; Einstein's predisposition for a theory that encompasses a wide rather than a narrow range of phenomena – all in the face of clear and sometimes overwhelming difficulties? These cases – which can be matched and extended over and over again – show that some *third mechanism* is at work here, in addition to the phenomenic and analytical. And we can find it right in Einstein's lecture on the method of theoretical physics: The two-dimensional model in it, which first strikes the eye, gives way on closer examination to a more sophisticated and appropriate one. In addition to the two inseparable but antithetical components there is indeed a third – not as clearly articulated here as in some others of Einstein's essays, but present nevertheless. The arguments for it float above the plane bounded by the empirical and logical components of the theory.

Einstein launches on it by reminding his audience, as he often did, that the previously mentioned phenomenic–analytic dichotomy prevents the principles of a theory from being "deduced from experience" by "abstraction" – that is to say, by logical means. "In the logical sense [the fundamental concepts and postulates of physics are] free inventions of the human mind," and in that sense different from the unalterable Kantian categories. He repeats more than once that the "fundamentals of scientific theory" are of "purely fictitious character." [11] As he puts it soon afterwards, in the essay "Physics and reality" (1936),[12] the relation between sense experience and concept "is analogous not to that of soup to beef, but rather to that of check number to overcoat." The essential arbitrariness of reference, Einstein explains in the Spencer Lecture, "is perfectly evident from the fact that one can point to two essentially different foundations" – the general theory of relativity, and Newtonian physics – "both of which correspond with experience to a large extent" – namely, with much of mechanics. The elementary experiences do not provide a logical bridge to the basic concepts and postulates of mechanics. Rather, "the axiomatic basis of theoretical physics . . . must be freely invented."

But if this is true, an obvious and terrifying problem arises, and Einstein spells it out. He writes: How "can we ever hope to find the right way? Nay, more, has this right way an existence outside our illusions? Can we hope to be guided safely by experience at all when there exist theories such as classical mechanics, which do justice to experience to a large extent, but without grasping the matter in a fundamental way?"

We have now left the earlier, confident portion of Einstein's lecture far behind. The question raises itself whether the activities of scientists can ever hope to be cumulative, or whether we must stagger from one fashion, conversion, or revolution to the next, in a kind of perpetual, senseless Brownian motion, without direction or *télos*.

At that point, Einstein states his clear conviction: "I answer with full confidence that there is, in my opinion, a right way, and that we are capable of finding it." Here, Einstein goes suddenly beyond his earlier categories of empirical and logical efficacy and offers us a whole set of selection rules with which, as with a good map and compass, that "right way" may be found. Here, there, everywhere, guiding concepts emerge and beckon from above the previously defined plane to point us on the right path.

The first directing principle Einstein mentions is his belief in the efficacy of formal structures: The "creative principle resides in mathematics" – not, for example, in mechanical models. On the next page, there unfolds itself a veritable hymn to the guiding concept of simplicity. Einstein calls it "the Principle of searching for the mathematically simplest concepts and their connections," and he cheers us on our way with many examples of how effective it has already proved to be: "If I assume a Riemannian metric [in the four-dimensional continuum] and ask what are the *simplest* laws which such a metric can satisfy, I arrive at the relativistic theory of gravitation in empty space. If in that space I assume a vector field or anti-symmetrical tensor field which can be derived from it, and ask what are the simplest laws which such a field can satisfy, I arrive at Maxwell's equations for empty space"; and so on, collecting victories everywhere under the banner of simplicity.

And over there, at the bottom of another page, we find two other guiding concepts in tight embrace: the concept of parsimony, or economy, and that of unification. As science progresses, Einstein tells us, "the logical edifice" is more and more "unified," the "smaller the number [is] of logically independent conceptual elements which are found necessary to support the whole structure." Higher up on that same page, we encounter nothing less than "the noblest aim of all theory," which is "to make these irreducible elements as simple and as few in number as is possible, without having to renounce the adequate representation of any empirical content."

Yet another guiding concept given in Einstein's lecture concerns the *continuum*, the field. From 1905 on, when the introduction of discontinuity in the form of the light quantum forced itself on Einstein as a "heuristic" and therefore not fundamental point of view, he clung to the hope and program to keep the continuum as a fundamental conception, and he defended it with enthusiasm in his correspondence. It was part of what he called his "Maxwellian program" to fashion a unified field theory. Atomistic discreteness and all it entails was not the solution but rather the problem. So here, in his 1933 lecture, he again considers the conception of "the atomic structure of matter and energy" to be the "great stumbling block" for a unified field theory.

One cannot, he thought, settle for this basic duality in nature, giving equal status both to the field and to its antithesis. Of course, neither logic nor experience forbade it. Yet it was almost unthinkable. As he once wrote to his old friend Michele Besso, "I concede . . . that it is quite possible that physics might not, finally, be founded on the concept of field – that is to say, on continuous elements. But then out of my whole castle in the air – including the theory of gravitation and most of current physics – there would remain almost nothing." [13]

We have by no means come to the end of the list of presuppositions which guided Einstein. But it is worth pausing to note how plainly he seemed to have been aware of their operation in his scientific work. In this too he was rare. Sir Isaiah Berlin, in his book *Concepts and categories* [p.159], remarked: "The first step to the understanding of men is the bringing to consciousness of the model or models that dominate and penetrate their thought and action. Like all attempts to make men aware of the categories in which they think, it is a difficult and sometimes painful activity, likely to produce deeply disquieting results." [14] This is generally true; but it was not for Einstein. There are surely at least two reasons for that. It was, after all, Einstein who realized the "arbitrary character" of what had for so long been accepted as "the axiom of the absolute character of time, viz., of simultaneity [which] unrecognizedly was anchored in the unconscious," as he put it in his "Autobiographical notes." "Clearly to recognize this axiom and its arbitrary character really implies already the solution of the problem." [15] (Giving up an explicitly or implicitly held presupposition has indeed often had the characteristic of the great sacrificial act of modern science; we find in the writings of Kepler, Planck, Bohr, and Heisenberg that such an act is a climax of a period that in retrospect is characterized by the word "despair.")

Having recognized and overcome the negative, or enslaving, role of presuppositions, Einstein also saw their positive, emancipating potential. In one of his early essays on epistemology ("Induction and deduction in physics," 1919), he wrote:

> A quick look at the actual development teaches us that the great steps forward in scientific knowledge originated only to a small degree in this [inductive] manner. For if the researcher went about his work without any preconceived opinion, how should he be able at all to select out those facts from the immense abundance of the most complex experience, and just those which are simple enough to permit lawful connections and become evident? [16]

In essay after essay, Einstein tried to draw attention to this point of view, despite – or because of – the fact that he was making very few converts. The Herbert Spencer lecture can be seen as part of that mission. A decade and a half later, in his "Reply to criticisms," we see him continuing in this vein. Thus, he acknowledges that the distinction between "sense impressions" on the one hand, and "mere ideas" on the other, is a basic conceptual tool for which he can adduce no convincing evidence. Yet he needs this distinction. His solution is simply to announce, "we regard the distinction as a category which we use in order that we might the better find our way in the world of immediate sensation." As with other conceptual distinctions for which "there is also no logical-philo-sophical justification," one has to accept it as "the presupposition of every kind of physical thinking," mindful that "the only justification lies in its usefulness. We are here concerned with 'categories' or schemes of thought, the selection of which is, in principle, entirely open to us and whose qualification can only be judged by the degree to which its use contributes to making the totality [*sic*] of the contents of consciousness 'intelligible.' " Finally, he curtly dismisses an implied attack on these "categories" or "free conventions" with the remark that "Thinking with-out the positing of categories and of concepts in general would be as impossible as is breathing in a vacuum." [17]

The thematic dimension

His remarkable self-consciousness concerning his fundamental presup-positions throughout his scientific and epistemological writings allows one to assemble a list of about ten chief presuppositions underlying Ein-stein's theory construction. Examples are symmetry (as long as possible); simplicity; causality (in essentially the Newtonian sense); completeness and exhaustiveness; continuum; and invariance. (We shall elaborate on this point in Chapter 2.)

To these ideas, Einstein was obstinately devoted. Guided by them he would continue his work in a given direction even when tests against experience were difficult or unavailable. Conversely, he refused to accept theories well supported by the phenomena but, as in the case of Bohr's quantum mechanics, based on presuppositions opposite to his own. Much the same can be said of most of the major scientists whom I have studied. Each has his own, sometimes idiosyncratic map of fundamental guiding notions – from Johannes Kepler to our own contemporaries.

With this finding, we must now reexamine the mnemonic device of the two-dimensional plane. I remove its insufficiency by defining a third axis, rising perpendicularly out of it. This is the dimension orthogonal to and not resolvable into the phenomenic or analytic axes. Along it are located those fundamental presuppositions, often stable, many widely shared, that show up in the motivation of the scientist's actual work, as well as in the end-product for which he strives. Decisions between them, insofar as they are consciously made, are judgmental (rather than, as in the phenomenic–analytic plane, capable in principle of algorithmic decidability). Since they are not directly derivable either from observation or from analytic ratiocination, they require a term of their own. I call them *themata* (singular *thema*, from the Greek θέμα, that which is laid down, proposition, primary word).

On this view – and again purely as a mnemonic device – a scientific statement is no longer, as it were, an element of area on the two-dimensional plane but a volume-element, an entity in three-dimensional space, with components along each of the three orthogonal (phenomenic, analytic, and thematic) axes. The projection of the entity down upon the two-dimensional place continues to have the useful roles I stressed earlier; but for our analysis it is also necessary to consider the line element projected onto the third axis, the dimension on which one may imagine the range of themata to be entered. The statements of differing scientists are therefore like two volume-elements that do not completely overlap, but have some differences in their projections.

The scientist is generally not, and need not be, conscious of the themata he uses, but the historian of science can chart the growth of a given thema in the work of an individual scientist over time, and show its power upon his scientific imagination. Thematic analysis, then, is in the first instance the identification of the particular map of themata which, like the lines in a fingerprint, can characterize a scientist or a part of the scientific community at a given time.

Most of the themata are ancient and long lived; many come in opposing dyads or triads that show up most strikingly during a conflict between individuals or groups that base their work on opposing themata. I have been impressed by the small number of thematic couples or triads; perhaps some 50 have sufficed us throughout the history of the physical sciences: and of course I have been interested to see that, cautiously, thematic analysis of the same sort has begun to be brought to bear on significant cases in other fields.[18]

With this conceptual tool we can return to some of the puzzles we mentioned earlier. Let me point out two. If, as Einstein claimed, the principles are indeed free inventions of the human mind, there should be an infinite set of possible axiom systems to which one could leap or cleave. Virtually every one of these would ordinarily be useless for constructing theories. How then could there be any hope of success, except by chance? The answer must be that the license implied in the leap to an axiom system of theoretical physics by the freely inventing mind is the freedom to make such a leap, but not the freedom to make *any leap whatever*. The freedom is narrowly circumscribed by a scientist's particular set of themata that provide constraints shaping the style, direction, and rate of advance of the engagement on novel ground. And insofar as the individual maps of themata overlap, the so-called progress of the scientific community as a group is similarly constrained or directed. Otherwise, the inherently anarchic connotations of "freedom" could indeed disperse the total effort. D. Mendeleev wrote: "Since the scientific world view changes drastically not only from one period to another but also from one person to another, it is an expression of creativity. . . . Each scientist endeavors to translate the world view of the school he belongs to into an indisputable principle of science." However, in practice there is far more coherence than this implies, and we shall presently look more closely at the mechanism responsible for it.

A second puzzle was where the conceptual and even emotional support comes from which, for better or worse, stabilizes the individual scientist's risky speculations and confident suspensions of disbelief during the nascent phase. In case after case, as in the example of Millikan, we see that choices of this sort are made often on thematic grounds. Millikan was devoted to the atomistic view of electricity from the beginning, while his chief opponent, probably under the influence of Ernst Mach and his school, came to look for precisely the opposite evidence, for example, subelectrons that in principle have no lower limit of charge at all. Similarly, Einstein and his opponents such as Kaufmann were divided sharply on the explanatory value of a plenum (ether), and on the range of fundamental laws across the separate branches of physics.

The Ionian enchantment

But of all the problems that invite attention with these tools, the most fruitful is a return visit to that mysterious place, early in Einstein's 1933 lecture, where he speaks of the need to pay "special attention to the relations between the content of the theory and the totality of empirical fact (*Gesamtheit der Erfahrungstatsachen*)." The *totality* of empirical fact! It is a phrase that recurs in his writings, and indicates the sweep of his conscious ambition. But it does even more: it lays bare the most daring of all the themata of science and points to the holistic drive behind "scientific progress."

Einstein explicitly and frankly hoped for a theory that would ultimately be utterly comprehensive and completely unified. This vision drove him on from the special to the general theory, and then to the unified field theory. In one of his letters to his biographer, Carl Seelig, Einstein likened his progress to the construction of an architectonic entity through three stages of development. Each stage is characterized by the adoption of a "limiting principal," a formal condition which restricts the choice of possible theories. For example, in going from special to general relativity theory, Einstein had to accept, from 1912 on, that physical significance attaches not to the differentials of the space–time coordinates themselves, as the strict operationalists would insist, "but only to the Riemannian metric corresponding to them." This entailed Einstein's reluctant sacrifice of the primacy of direct sense perception in constructing a physically significant system; but otherwise he would have had to give up hope of finding unity at the base of physical theory.

The search for one grand architectonic structure is of course an ancient dream. At its worst, it has sometimes produced authoritarian visions which are as empty in science as their equivalent is dangerous in politics. At its best, it has propelled the drive to the various grand syntheses that rise above the more monotonous landscape of analytic science. This has been the case in the last decades in the physical sciences. Today's triumphant purveyors of the promise that all the forces of physics will eventually melt down to one, who in the titles of their publications use the term "The Grand Unification," are in a real sense the successful children of those earliest synthesis-seekers of physical phenomena, the Ionian philosophers.

To be sure, as Sir Isaiah warned in *Concepts and categories*, there is the danger of a trap. He has christened it the "Ionian Fallacy," defined as the search, from Aristotle to Bertrand Russell and our day, for the ultimate constituents of the world in some nonempirical sense. Superficially, the synthesis-seekers of physics, particularly in their monistic exhortations, appear to have fallen into that trap – from Copernicus, who confessed that the chief point of his work was to perceive nothing less than "the structure of the universe and the true symmetry of its parts,"[19] to Einstein's contemporaries such as Max Planck, who exclaimed in 1915 that "physical research cannot rest so long as mechanics and electrodynamics have not been welded together with thermodynamics and heat radiation," [20] to today's theorists who, in their more popular presentations, seem to imitate Thales himself and announce that one entity explains all.

A chief point in my view of science is that scientists, insofar as they are successful, are in practice rescued from the fallacy *by the multiplicity of their themata, a multiplicity which gives them the flexibility that an authoritarian research program built on a single thema would lack.* I shall develop this, but I can also agree quickly that something like an Ionian Enchantment, the commitment to the theme of grand unification, was upon Einstein. Once alerted, we can find it in his work from the very beginning. In his first published paper (1901), he tries to understand the contrary-appearing forces of capillarity and gravitation, and in each of his next papers we find something of the same drive, which he later called "my need to generalize." He examines whether the laws of mechanics provide a sufficient foundation for the general theory of heat, and whether the fluctuation phenomena that turn up in statistical mechanics also explain the basic behaviour of light beams and their interference, the Brownian motion of microscopic particles in fluids, and even the fluctuation of electric charges in conductors. And in his deepest work of those early years, in special relativity theory, the most powerful propellant is Einstein's drive toward unification; his clear motivation is to find a more general point of view which would subsume the seemingly limited and contrary problems and methods of mechanics and of electrodynamics.

Following the same program obstinately to the end of his life, he tried to bring together, as he had put it in 1920, "the gravitational field and the electromagnetic field into a unified edifice," leaving "the whole physics" as a "closed system of thought." In that longing for a unified world picture, a structure that encompasses "the totality of empirical facts," one cannot help hearing the voice of Goethe's Faust, who exclaimed that he longed "to detect the inmost force that binds the world and guides its course" – or, for that matter, Newton himself, who wanted to build a unifying structure so tight that the most minute details would not escape it.

The unified Weltbild as "supreme task"

In its modern form, the Ionian Enchantment, expressing itself in the search for a unifying world picture, is usually traced to Von Humboldt and Schleiermacher, Fichte and Schelling. The influence of the "Nature

Philosophers" on physicists such as Hans Christian Oersted – who in this way was directly led to the first experimental unification of electricity and magnetism – has been amply chronicled. At the end of the nineteenth century, in the Germany of Einstein's youth, the pursuit of a unified world picture as the scientist's highest task had become almost a cult activity. Looking on from his side of the Channel, J. T. Merz exclaimed in 1904 that the lives of the continental thinkers are "devoted to the realization of some great ideal. . . . The English man of science would reply that it is unsafe to trust exclusively to the guidance of a pure idea, that the ideality of German research has frequently been identical with unreality, that in no country has so much time and power been frittered away in following phantoms, and in systematizing empty notions, as in the Land of the Idea." [21]

Einstein himself could not easily have escaped being aware of these drives toward unification, even as a young person. For example, we know that as a boy he was given Ludwig Büchner's widely popular book *Kraft und Stoff* (*Energy and matter*), a book Einstein often recollected having read with great interest. The little volume does talk about energy and matter; but chiefly it is a late-Enlightenment polemic. Büchner comes out explicitly and enthusiastically in favor of an empirical, almost Lucretian scientific materialism, which its author calls a "materialistic world view." Through this world view, the author declares, one can attain "the unity of energy and matter, and thereby banish forever the old dualism." [22]

But the books which Einstein himself credited as having been the most influential on him in his youth were Ernst Mach's *Theory of heat* and *Science of mechanics*. That author was motivated by the same Enlightenment animus, and employed the same language. In the *Science of mechanics*, Mach exclaims: "Science cannot settle for a ready-made world view. It must work toward a future one . . . that will not come to us as a gift. We must earn it! [At the end there beckons] the idea of a unified world view, which is the only one consistent with the economy of a healthy spirit." [23]

Indeed, in the early years of this century, German scientists were thrashing about in a veritable flood of publications that called for the unification or reformation of the "world picture" in the very title of their books or essays. Max Planck and Ernst Mach carried on a bitter battle, publishing essays directly in the *Physikalische Zeitschrift*, with titles such as "The unity of the physical world picture." Friedrich Adler, one of Einstein's close friends, wrote a book with the same title, attacking Planck.

Max von Laue countered with an essay he called "The physical world picture." The applied scientist Aurel Stodola, Einstein's admired older colleague in Zurich, corresponded at length with Einstein on a book which finally appeared under the title *The world view of an engineer*. Similarly titled works were published by other collaborators and friends of Einstein, such as Ludwig Hopf and Philipp Frank.

Perhaps the most revealing document of this sort was the manifesto published in 1912 in the *Physikalische Zeitschrift* on behalf the new *Gesellschaft für positivistische Philosophie*, composed in 1911 at the height of the *Weltbild* battle between Mach and Planck. Its declared aim was nothing less than "to develop a comprehensive *Weltanschauung*," and thereby "to advance toward a noncontradictory, total conception [*Gesamtauffassung*]." The document was signed by, among others, Ernst Mach, Josef Petzold, David Hilbert, Felix Klein, Georg Helm, Albert Einstein (only just becoming more widely known at the time), and that embattled builder of another world view, Sigmund Freud.[24]

It was perhaps the first time that Einstein signed a manifesto of any sort. That it was not a casual act is clear from his subsequent, persistent recurrence to the same theme. His most telling essay was delivered in late 1918, possibly triggered in part by the publication of Oswald Spengler's *Decline of the west*, that polemic against what Spengler called "the scientific world picture of the West." Einstein took the occasion of a presentation he made in honour of Max Planck (in *Motiv des Forschens*) to lay out in detail the method of constructing a valid world picture. He insisted that it was not only possible to form for oneself "a simplified world picture that permits an overview [übersichtliches Bild der Welt]," but that it was the scientist's "supreme task." Specifically, the world view of the theoretical physicist "deserves its proud name *Weltbild*, because the general laws upon which the conceptual structure of theoretical physics is based can assert the claim that they are valid for any natural event whatsoever. . . . The supreme task of the physicist is therefore to seek those most universal elementary laws from which, by pure deduction, the *Weltbild* may be achieved."[25]

There is of course no doubt that Einstein's work during those years constituted great progress toward this self-appointed task. In the developing relativistic *Weltbild*, a huge portion of the world of events and processes was being subsumed in a four-dimensional structure which Minkowski in 1908 named simply *die Welt* – a Parmenidean crystal-universe, in which changes, for example, motions, are largely suspended and, instead, the main themata are those of constancy and invariance, determinism, necessity, and completeness.

Typically, it was Einstein himself who knew best and recorded frequently the limitations of his work. Even as special relativity began to make converts, he announced that the solution was quite incomplete because it applied only to inertial systems and left out entirely the great puzzle of gravitation. Later he worked on removing the obstinate dualities, explaining for example that "measuring rods and clocks would have to be represented as solutions of the basic equation . . . not, as it were, as theoretical self-sufficient entities." This he called a "sin" which "one must not legalize." The removal of the sin was part of the hoped-for perfection of the total program, the achievement of a unified field theory in which "the particles themselves would *everywhere* be describable as singularity-free solutions of the complete field-equations. Only then would the general theory of relativity be a *complete* theory." [26] Therefore, the work of finding those most general elementary laws from which by pure deduction a single consistent, and complete *Weltbild* can be won had to continue.

There has always been a notable polarity in Einstein's thought with respect to the completeness of the world picture he was seeking. On the one hand he insisted from beginning to end that no single event, individually considered, must be allowed to escape from the final grand net. We noted that in the Herbert Spencer lecture of 1933 he is concerned with encompassing the "totality of experience," and declared the supreme goal of theory to be "the adequate representation of any content of experience" (translated in the first English version of the 1933 lecture, as delivered by Einstein, as "the adequate representation of a single datum of experience"). [27] He even goes beyond that; toward the end of his lecture he reiterates his old opposition to the Bohr-Born-Heisenberg view of quantum physics, and declares "I still believe in the possibility of a model of reality, that is to say a theory, which shall represent the events themselves [*die Dinge selbst*] and not merely the probability of their occurrence." Writing three years later (*Physics and reality*, 1936), he insists even more bluntly:

> But now, I ask, does any physicist whosoever really believe that
> we shall never be able to attain insight into these significant
> changes of single systems, their structure, and their causal
> connections, despite the fact that these individual events have
> been brought into such close proximity of experience, thanks

to the marvellous inventions of the Wilson Chamber and the Geiger counter? To believe this is, to be sure, logically possible without contradiction; but it is in such lively opposition to my scientific instinct that I cannot forgo the search for a more complete mode of conception.[28]

Yet even while Einstein seemed anxious not to let a single event escape from the final *Weltbild*, he seems to have been strangely uninterested in nuclear phenomena, that lively branch of physics which began to command great attention precisely in the years Einstein started his own researches. He seems to have thought that these phenomena, in a relatively new and untried field, would not lead to the deeper truths. And one can well argue that he was right; not until the 1930s was there a reasonable theory of nuclear structure, and not until after the big accelerators were built were there adequate conceptions and equipment for the hard tests of the theories of nuclear forces.

Einstein's persistent pursuit of fundamental theory without including nuclear phenomena can be understood as a consequence of a suspension of disbelief of an extraordinary sort. It is ironic that, as it turned out, even while Einstein was trying to unify the two long-range forces (electromagnetism and gravitation), the nucleus was harboring two additional fundamental forces, and moreover that after a period of neglect, the modern unification program, two decades after Einstein's death, began to succeed in joining one of the nuclear (relatively short-range) forces with one of the relatively long-range forces (electromagnetism). In this respect, the labyrinth through which the physicists have been moving appears now to be less symmetrical than Einstein had thought it to be.

For this and similar reasons, few of today's working researchers consciously identify their drive toward the "grand unification" with Einstein's. Their attention is directed to the thematic differences, expressed for example by their willingness to accept a fundamentally probabilistic world. And yet the historian can see the profound continuity. Today, as in Einstein's time, and indeed in that of his predecessors, the deepest aim of fundamental research is still to achieve one logically unified and parsimoniously constructed system of thought that will provide the conceptual comprehension, as complete as humanly possible, of the scientifically accessible sense experiences in their full diversity. This ambition embodies a *télos* of scientific work itself, and it has done so since the rise of science in the Western world. Most scientists, working on small fragments of the total structure, are as unselfconscious about their par-

ticipation in that grand monistic task as they are about, say, their fundamental monotheistic assumption, carried centrally without having to be
avowed believers.[29]

Thematic pluralism and the direction of advance

Difference between some themata, and sharing of others: this formula in
brief seems to me to answer the question why the preoccupation with the
eventual achievement of one unified world picture did not lead physics to
a totalitarian disaster, as an Ionian Fallacy by itself could well have done.
At every step, each of the various world pictures in use was seen as a preliminary version, a premonition of the holy grail. Moreover, each of these
various, hopeful but incomplete world pictures of the moment was not a
seamless, unresolvable entity. Nor was each completely shared within a
given subgroup. Each operated with a whole spectrum of separable themata, with some of the same themata present in portions of the spectrum
in rival world pictures. Indeed, Einstein and Bohr agreed on far more
than they disagreed on. Moreover, most of the themata were not new –
they very rarely are – but adopted from predecessor versions of the *Weltbild*, just as many of them would later be incorporated in subsequent
versions of it. Einstein freely called his project a "Maxwellian program"
in this sense.[30]

It is also for this reason that Einstein saw himself with characteristic
clarity not at all as a revolutionary, as his friends and his enemies so
readily did. He took every opportunity to stress his role as a member of
an evolutionary chain. Even while he was working on relativity theory in
1905, he called it "a modification" of the theory of space and time. Later,
in the face of being acclaimed the revolutionary hero of the new science,
he insisted, as in his King's College (1921) lecture: "We have here no
revolutionary act but the natural development of a line that can be traced
through centuries." Relativity theory, he held, "provided a sort of completion of the mighty intellectual edifice of Maxwell and Lorentz."[31]
Indeed he shared quite explicitly with Maxwell and Lorentz some fundamental presuppositions such as the need to describe reality in terms of
continua (fields), even though he differed completely with respect to
others, such as the role of a plenum.

On this model we can understand why scientists need not hold substantially the same set of beliefs, either in order to communicate mean-
ingfully with one another in agreement or disagreement, or in order

to contribute to cumulative improvement of the state of science. Their beliefs have considerable fine structure; and within that structure there is, on the one hand, generally sufficient stabilizing thematic overlap and agreement, and on the other hand sufficient warrant for intellectual freedom that can express itself in thematic disagreements. Innovations emerging from such a balance, even as "far-reaching changes" as Einstein called the contributions of Maxwell, Faraday, and Hertz, require neither from the individual scientist nor from the scientific community the kind of complete and sudden reorientation implied in such currently fashionable language as revolution, Gestalt switch, discontinuity, incommensurability, conversion, and so on. On the contrary, the innovations are coherent with the model of evolutionary scientific progress, with which Einstein himself explicitly associated his own work, and which emerges also from the actual historical study of his scientific work.

Thus, I believe that generally major scientific advance can be understood in terms of an evolutionary process that involves battles over only a few but by no means all of the recurrent themata. The work of scientists, acting individually or as a group, seen synchronically or diachronically, is not constrained to the phenomenic–analytic plane alone, and hence is an enterprise whose saving pluralism resides in its many internal degrees of freedom. Therefore we can understand why scientific progress is often disorderly, but not catastrophic; why there are many errors and delusions, but not one great fallacy; and how mere human beings, confronting the seemingly endless, interlocking puzzles of the universe, can advance at all – even if not soon, or inevitably, to the Elysium of the single world conception that grasps the totality of phenomena.

2

Einstein's model for constructing a scientific theory

The epistemological imperative

Judging by his publications and letters, Albert Einstein considered it one of his important tasks constantly to express and elaborate his views concerning the philosophy of science. There seem to be two reasons for that. First, Einstein experienced in his own work in the early years, and then again among his "ablest students," how important discussions concerning the aims and methods of the sciences are.[1] Such interest was not merely a matter of intellectual curiosity but, in his opinion, went to the heart of the task of the innovator: epistemology and science, he said, "are dependent on each other. Epistemology without contact with science becomes an empty scheme. Science without epistemology is – insofar as it is thinkable at all – primitive and muddled." At the same time he warned, however, that the scientist cannot permit himself to be too restricted "by the adherence to an epistemological system."[2] He might therefore seem to be more a philosophical opportunist than school philosopher. However, that accusation seemed to bother Einstein as little as did the more serious attacks from so many other quarters upon his science and his other views. Reading his opinions on age-old questions of methodology, we feel we are given a direct report on a deeply felt, personal struggle with ideas.

A second reason why a scientist concerned with the deep problems should not avoid epistemologic considerations was, in Einstein's opinion, that there simply was no other way. In our time, when the scientific foundations are changing rapidly, "the physicist cannot simply surrender to the philosopher the critical contemplation of the theoretical foundations; for he himself knows best and feels most surely where the shoe pinches."[3]

With these motivations, Einstein found himself publishing constantly on the philosophy of science and, significantly, doing so throughout his most creative period of scientific work (e.g., 1914: *Principles of theoretical physics*; 1916: *On Ernst Mach*; 1918: *Motive for doing research*; 1921: *Geometry and experience*; 1933: *On the method of theoretical physics*; 1936: *Physics and reality*, and many others – and of course in his letters to Besso, Solovine, and other friends). With characteristic persistence, not to say obstinacy, he put on himself the task of presenting what he called his "epistemological credo." Moreover, it is striking how consistent he was in his presentation – at least from about 1914 on, after his formative period during which he had gone through a kind of philosophical pilgrimage.

In the last four decades of his life, he therefore was acting not only as a profound scientist, but also as a popularizer, teacher, and philosopher-scientist in the tradition of Henri Poincaré, Ernst Mach, and others of the generation before him. It is obvious that he took this role as a public educator very seriously, and that he tried his utmost to write clearly and at a level where the intelligent layman would understand him. Hence it came about that the man who was best known for his legendary struggles with the most inaccessible and recondite theories in fact was – and to this day remains – one of the most readable and widely read scientists. His essays have been reprinted in the most distant corners of the world. There is also an ever-growing flood of analyses of his ideas and the way they do or do not coincide with long-familiar questions of philosophy. But it may be appropriate, and in the spirit of Einstein's own intentions, to provide a presentation of a key portion of Einstein's epistemological position, using his own words as far as possible.

Writing to Solovine

In all of Einstein's own writings, one message stands out and is returned to repeatedly: a model of scientific thinking, and indeed of thinking in general. That model forms the core of the first pages of his "Autobiographical notes," which I have analyzed elsewhere.[4] But Einstein's most concise and graphic rendition of his model is to be found in a letter he wrote to his friend Maurice Solovine in 1952. I have always thought that for sheer virtuosity of expression and ability to summarize complex thoughts, this letter is unique in Einstein's correspondence. It is therefore well suited for reexamining his credo. It also invites elaborating on his brief explanations by reference to others of his publications on the same subject and pulling together many of his methodological ideas scattered throughout his writings.

Solovine was one of Einstein's oldest friends; they had met in Bern in 1902, had regularly discussed science and philosophy, and had kept up a correspondence after Solovine had moved away. Writing on April 25, 1952, Solovine confesses that he has trouble understanding a point made in one of Einstein's essays. "Would you be so kind," Solovine asks, "as to explain precisely a passage . . . which is not quite clear. You write: The justification (truth content) of the system rests in the proof of usefulness of the resulting theorems on the basis of sense experiences, where the relations of the latter to the former can only be comprehended intuitively. . . ." Solovine indicates his puzzlement and raises questions.

In his reply of May 7, 1952, Einstein starts in his characteristically relaxed, unpompous manner: "Dear Solo! In your letter you give me a spanking on the behind . . . , but," he continues, "you have thoroughly misunderstood me with respect to the epistemological matter. Probably I expressed myself badly." There follows a memorable explanation of the respective roles of sense experience, intuition, and logic in the functioning of the imagination. As we shall see, and as one would expect, Einstein places the emphasis on the sequence of steps in doing science, in making a discovery or formulating a theory, rather than reformulating the results later on to make them acceptable to publishers of scientific journals or philosophers interested in the justification of proposed theories.

It will also be noted that while the context of Solovine's question and Einstein's reply make it clear that Einstein is talking about a model for thinking in *science*, nowhere in what follows does he use the word *science*; and what fragmentary examples he gives (e.g., relation between the concept "dog" and the corresponding experiences) are not drawn from scientific theory. This is entirely in line with his typical refusal to tolerate unnatural and unnecessary boundaries. For he said repeatedly that one is dealing here with a continuum: "Scientific thought is a development of prescientific thought"; "all this applies as much, and in the same manner, to the thinking in daily life as to the more consciously and systematically constructed thinking in the sciences." This point of view was perhaps best caught in his statement that the "whole of science is nothing more than a refinement of everyday thinking." Just for that reason, however, the critical physicist should not restrict his examination of concepts to his own field of expertise, but should consider "critically a

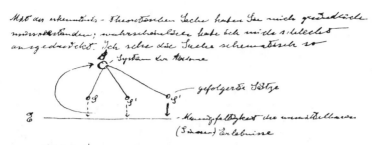

Figure 2.1

much more difficult problem, the problem of analyzing the nature of
everyday thinking" (*Ideas and Opinions* [*I.O.*], pp. 276, 23, 324, 290).
Perhaps for this reason Einstein had placed the question "What, precisely
is 'thinking'?" near the beginning of his "Autobiographical notes" – and
then, during that discussion, referred only rarely to science.

Einstein begins his explanation to Solovine with the sentence: "I see
the matter schematically thus" – and there follows a diagram (not surprisingly, for we know of Einstein's preference for visual thinking). A
sketch of great power and simplicity, it concentrates in a few lines a
wealth of information (Figure 2.1). The diagram indicates an essentially
cyclical process, and Einstein enters on its discussion by laying out the
stage where the process must both begin and end:

"1. The *E* (experiences) are given to us."

This refers to the horizontal line shown at the bottom of the figure,
marked *E* and labeled "Multiplicity [or variety] of immediate (sense) experiences."[5] As usual, these come first in his account [just as he had put
the "reception of sense impressions" as the first item after asking "What
is thinking?" in the "Autobiographical notes"]. And it will have to come
in at the end also, when we return to the level of sense experience to see if

our theory can handle as large a part of the totality of the facts of experience as possible – which is, after all, the final test of a theory.

The thin line marked E is rather deceptive. One might better visualize it as an infinite plane on which the separate and diverse sense experiences or observations that clamor for our attention are laid out, like so many separate points. It does indeed represent the "totality of empirical fact" (*I.O.*, p.271) or "totality of sense experiences." [6] In themselves the points on this plane are bewildering, a universe of elements, a veritable "labyrinth of sense impressions" of which, moreover, we never can be completely sure that they are not "the result of an illusion or hallucination" (*I.O.*, p. 291). In fact, the ultimate aim of science can be defined in this manner: "Science is the attempt to make the chaotic diversity of our sense-experience correspond to a logically uniform [unified] system of thought." The chaotic diversity of "facts" is mastered by erecting a structure of thought on it that points to relations and order: "In this system, single experiences must be correlated with the theoretic structure in such a way that the resulting coordination is unique and convincing" (*I.O.*, p. 323).

An inside: Nobody had to point out to Einstein that sense experiences or "observations" are virtually never pure and unvarnished but theory-dependent. Even the father of positivism, Auguste Comte, had written (*Positive philosophy*, (1829) that without a theory of some sort by which to link phenomena to some principles "it would not only be impossible to combine the isolated observations and draw any useful conclusions, we would not even be able to remember them, and, for the most part, the fact would not be noticed by our eyes." Indeed, sometimes Einstein speaks of "experience" or "facts" in a way very different from what Ernst Mach took to be "elements"; among facts, Einstein in various writings included the impossibility of perpetual motion machines, inertial motion, the constancy of light velocity, and the equality of gravitational and inertial mass.[7] Nevertheless, in their most primitive form the E in Figure 2.1 can be thought of as simple sensory impressions.

The ascent to an axiom system

The diagram in Figure 2.1 now goes on to show what is perhaps Einstein's most insistent conception in epistemology. Rising out of an area just above a portion of the chaos of observables E, there is an arrow-tipped arch reaching to the very top of the whole scheme. It symbolizes

what under various circumstances could be a bold leap, a "widely speculative" attempt, a "groping constructive attempt,"[8] or a desperate proposal, made when one has despaired of finding other roads. There, high
above the infinite plane E, is suspended a well-delimited entity labeled
"A, system of axioms," issuing out of the arrow-tipped arch like a pulse
of light out of the trajectory of a firework.

Einstein writes in explanation:

"2. A are the axioms from which we draw consequences. Psychologically the A are based upon the E. There is, however, no logical path from
the E to A, but only an intuitive (psychological) connection, which is
always 'subject to revocation.' "

Evidently, Einstein holds that in the formulation of ideas – everyday
as well as scientific ones – the process of thinking or discovery does not
follow the classical model of Mill, that is, of erecting a logical ladder by
induction of generalizations from the set of individual observations. That
method is only "appropriate to the youth of science" (*I.O.*, 283). Nor
does Einstein believe as Ernst Mach had counseled, that one should remain as much as possible within the plane of E and confine oneself to
search out the most economic statements of relations among the elements
there; for what that missed, Einstein explained in his "Autobiographical
notes,"[9] was precisely the "essentially constructive and speculative nature
of thought and more especially of scientific thought."

In the schema of Figure 2.1, the arc is just that speculative leap or
constructive groping to A, the axiom or fundamental principles which
in the absence of a logical path have to be postulated, perhaps at first
quite tentatively on the basis of a conjecture, supposition, "inspiration,"
"guess," or "hunch." We are dealing, after all, with the private process of
theory construction or innovation, the phase not open to inspection by
others and indeed perhaps little understood by the originator himself.
But the leap to the top of the schema symbolizes precisely the precious
moment of great energy, the response to the motivation of "wonder" and
of the "passion of comprehension" (*I.O.*, 342) which can come from the
encounter with the chaotic E. Indeed, there is a clear and uncanny parallel
between the process described in Figure 2.1 and the model Einstein proposed to explain the motivation for research. As Einstein puts it there, to
escape from the chaos of the world of experience, the scientist, scholar,
or artist erects a "simplified and lucid image of the world," lifting into it
"the center of gravity of his emotional life."[10]

As one would expect from him, Einstein did not speak of the technique of elevating a supposition or hunch to an axiom or fundamental principle as if it were some hypothetical advice. He had done so himself in his scientific papers and, what is more, confessed it frankly in them. An example of raising a "conjecture" to the status of a "postulate" will be given in Chapter 4. We know that reaching these conjectures, and gathering the courage to raise them to fundamental principles, were not momentary enthusiastic decisions but the results of years of groping. It was in fact forced on Einstein that the kind of fundamental theory he was trying to build could be be attained in no other way.

Two logical discontinuities in J

We have to linger a little more on the implications in Figure 2.1 of the trajectory, arc, ascent, or jump (to which we will now assign a label, J). As Einstein often stressed (cf. analysis of "Autobiographical notes," and many other sources, e.g., *I.O.*, p. 291), there are in fact two sets of logical discontinuities implied in the seemingly smooth curved line. We fashion it by fastening our attention on "certain repeatedly occurring complexes of sense impressions" and "relating to them [*zuordnen*] a concept." The concept is then a kind of "mental knot" or "mental connection" between sense impressions, and is "primary" if close to sense experience. But we select the concept without some logical necessity, really "arbitrarily" in the sense that "considered logically this concept is not identical with the totality of sense impressions referred to; but it is a free creation of the human (or animal) mind" (*I.O.* pp. 291, 293 – "human or animal mind": another unnecessary barrier unceremoniously discarded!).

The same theme of the logical discontinuity in the formation of concepts appears again and again. For example: "All concepts, even those which are closest to experience, are from the point of view of logic freely chosen conventions."[11] And again: "There is no inductive method which could lead to the fundamental concepts of physics. Failure to understand this fact constituted the basic philosophical error of so many investigators of the nineteenth century." Several times Einstein referred to David Hume's attack on induction, showing that "concepts which we must regard as essential, such as, for example, causal connection, cannot be gained from material given to us by the senses" (*I.O.*, pp. 307, 21).

To the same end, Einstein also reminded his readers frequently of the fatal error that had been made for so long in thinking that the basis of Euclidean geometry was logically necessary; this error was caused by forgetting the empirical base and hence the limited experiential context within which all concepts are fashioned. A similar illusion was the great obstacle to formulating the Special Relativity Theory (*I.O.*, pp. 298–299), namely that there exists a universal time applicable to all events in space as a whole, a concept of time long held to be an a priori given, necessary conception, seemingly independent from our sense experience. This error was caused by forgetting that the notion of time itself arises initially in our everyday experience by watching sequences of events happening at one locality, rather than in all of space.

Deprived of any certainty that our concepts have a necessary connection with the corresponding experiences, we begin to see the precariousness of the business of theory construction. But we can do no better. We create new concepts, perhaps suggested at first only tentatively, and gather them together with old concepts whose usefulness has been tested in previous struggles, knowing that neither one nor the other is sacred and unchangeable, neither induced nor in any other way securely abstracted from the plane of experiences below. It may be that this discontinuity is symbolized by the small gap in the drawing between the horizontal line *E* and the arc rising from that region to *A* above.

There is a second logical discontinuity which also enters to make it a "mistake to permit theoretical description to be directly dependent upon acts of empirical assertions." [12] This concerns the relation of concepts to one another when they are used together to make a system of axioms – for example, some postulated laws of nature ["propositions expressing a relationship among primary concepts"(*I.O.*, p. 293)]. Not only each individual concept, but the whole "system of concepts is a creation of man," achieved in a "free play," the justification for which lies only in the pragmatic success of the scheme being built up to give ultimately a "measure of survey over the experience of the senses which we are able to achieve with its aid." [13]

The two-fold discontinuity, then, is a good part of the reason why Einstein repeats, again and again, sentences like this one from 1918: "There is no logical path to these elementary laws; only intuition, supported by being sympathetically in touch with experience [*Einfühlung in die Erfahrung*]" (*I.O.*, p. 226). The repeated insistence was in good part in opposition to the then current form of positivism, which, for example, saw the goal of scientific work to be the economic statement of relations among observables. To this day, Einstein's formulation causes hostility from some philosophical quarters, which insist on exaggerating this particular element in Einstein's total schema. (On the other hand, it should also be said that Einstein's anti-inductivism has encouraged some of the most interesting contributors to philosophy of science in our day.) Another reason for Einstein's dogged insistence on the fundamental dualism between experience and theory, sometimes offered in surprising and inopportune contexts (*I.O.*, p. 356), may have also been caused by a persistent intellectual discomfort. To this great unifier, the existence of an unbridgeable chasm must have presented a challenge of its own.

In no way can Einstein's message on this point be taken to celebrate irrationality, to give primacy to intuition, or the like. Rather, it represents two truths which he had learned, so to speak, on his very own body. One was the liberating warning that just because all theories are "man-made" and "the result of an extremely laborious process of adaptation," they are also "hypothetical, never completely final, always subject to question and doubt." The other message was, precisely against this somber knowledge, to encourage the assertion of ingenuity and innovation, in science as well as outside, if necessary against prevailing dogma. (Einstein quipped: "Would Faraday have discovered the law of electromagnetic induction if he had received a regular college education?") If accused of dragging down, from the Olympian fields, "the fundamental ideas of thought in natural science, and to attempt to reveal their earthly lineage," Einstein would answer that he did so "in order to free these ideas from the taboo attached to them, and thus to achieve greater freedom in the formation of ideas and concepts. It is to the immortal credit of D. Hume and E. Mach that they, above all others, introduced this critical conception" (*I.O.*, pp. 323, 344, 365).[14]

Constraints and freedoms

We might point to other properties of the concepts by means of which axioms are formulated. Although Einstein does not stress it often explicitly, one type of conceptual construction needed to keep the concept from floating away like some arbitrary soap bubble is the definition which we give to every abstract term (point, length, time interval, electric charge). While the definition of any term is logically arbitrary, it is connected with observables by our "operational definition" or "semantical rule" to which we agree to adhere once it has been fashioned. As early as 1916, Einstein wrote, "Concepts have meaning only if we can point to objects to which they refer and to the rules by which they are assigned to these objects."[15]

Good examples of this operational approach to concepts can be found in Einstein's own careful analysis of the mental (mathematical) and physical operations of measuring time in his first relativity paper, or in his description of what is meant by such conceptions as "solid body" or "space" in several of his later essays. Therefore, one might elaborate the diagram in Figure 2.1 by drawing thin, vertical lines between E and A to indicate that such connections are made whenever we choose the convention or "meaning" assigned to a term that is part of the scientific (or any) vocabulary.

The other constraint on our choice of concepts – even though they "have a purely fictitious character," being the "free inventions of the human intellect, which cannot be justified either by the nature of that intellect or in another fashion *a priori*" – lies in Einstein's call for frugality and simplicity. After all, the aim of any good theoretical system is "the greatest possible sparsity of the logically independent elements (basic concepts and axioms)."[16] Any redundancy or elaboration must be avoided, for "it is the grand object of all theory to make these irreducible elements as simple and as few in number as possible." For example, it was, in his view, "an unsatisfactory feature of classical mechanics that in its fundamental laws the same mass appears in two different roles, namely as inertial mass in the laws of motion, and as gravitational mass in the law of gravitation" (*I.O.*, pp. 272, 273, 308). The equivalence of these two interpretations of mass signaled to him a truth which needed to be stated as a basic axiom (in General Relativity Theory), rather than saddling the theory with a proliferation which did not seem to be inherent in the phenomena.

In good part as a result of Einstein's own work and example, physical scientists have indeed succeeded in showing, during this century, that only a very small number of postulated fundamental laws, employing a surprisingly small number of fundamental concepts, are needed to encompass ("explain") at least in principle an ever-growing infinity of separate facts of experience. This does not mean at all that everything is explained, or even in principle already explainable; but still, it is a "wonder"[17] and a motivation for further work. This success also has some curious consequences, to which we shall return later.

One corollary of this method of hypothesizing is that during the period of constructing a theory the innovator must give his proposed "jump" to the axioms a chance to prove itself. Hence in this early and usually private stage of theorizing the researcher has to grant himself a freedom, the right of "suspension of disbelief," a moratorium on premature attempts at falsification (i.e., on attempts to discredit the hypothesized postulate by disproving it).[18] Though the very idea is contrary to the naïve picture of the scientist, it is an essential part of the scientific imagination. In Einstein's case it is connected with his ability to tolerate ambiguities, to keep unresolved problems and polarities long before his mind's eye.

A related corollary is more disturbing, however. Just as there were in principle infinitely many points on the E level at the bottom of the schema, there are in principle infinitely many possible axioms or systems of axioms A at the top. The choice a given scientist makes out of all possibilities cannot be entirely arbitrary, since it would easily involve him in an infinitely long search. How does one in fact make this choice? That is, what guides or constraints do exist which help (or hinder) the innovator in making his particular jump to A rather than to some other A', which another researcher, on the basis of the same E, may prefer to make? Einstein says nothing about it in this letter; but he wrote sufficiently elsewhere to help us deal with the question, as we shall see later.

The logical path

To return now to Einstein's letter to Solovine. He continues:

"3. From A, *by a logical path*, particular assertions are deduced – deductions which may lay claim to being right."

This sentence takes us to the area in the schema where rigorous analytical thinking enters, where the scientific imagination indeed requires the "logical path." "Logical thinking is necessarily deductive" (*I.O.*, p. 307), starting from the hypothetical concepts and axioms which were postulated during the earlier upward swing of the schema. Therefore we are now proceeding downward from the axioms, deriving the necessary consequences or predictions; if A, then $S, S', S'' \ldots$ should follow; or as in the 1905 relativity paper, if the Principle of Relativity and the Principle of the Constancy of the Velocity of Light are assumed in the first place as axiom system A, then there follow necessarily, without any further fundamental assumptions, the transformation equations for space and time coordinates, the relativity of simultaneity, the so-called length contraction and time dilation effects, and, at the end of the 1905 paper, "the properties of the motion of the electron which result from the system of equations and are accessible to experiment. . . . [These] relations are a complete expression for the laws according to which, by the theory here advanced, the electron must move." This wealth of results is a natural consequence of Einstein's powerful deductions. In an excellent paper entitled "Logical economy in Einstein's 'On the electrodynamics of moving bodies,' " Robert B. Williamson [19] has shown the logical consistency and parsimony of Einstein's detailed argument. These features make it even more plausible that the whole work represents the crystallization of years of effort.

Some who have criticized Einstein's remarks as giving too much weight to other logically speculative concepts have tended to overlook the definite role which Einstein did give to the logical phase of the scientific imagination. If he argues for recognizing the necessary inspirational component in the formation of fundamental hypotheses at the level of *A*, he also goes on to say "the structure of the system is the work of reason." This part of the scientist's work, where inference follows inference, requires "much intense, hard thinking," but at least is a task that one can learn in principle "at school." It is only the earlier step, that of establishing the principles in the first place from which deduction can proceed, for which "there is no method capable of being learned and systematically applied. . . . the scientist has to worm these general principles out of nature" (*I.O.*, pp. 272, 282, 221).

Testing against experience

Continuing in his letter to Solovine, Einstein now comes to the fourth and final step that brings us back to the plane from which we started:

"4. The *S* are referred [or related] to the *E* (testing against experience)."

Still anxious to make the necessary distinction between what logic can and cannot be expected to do during the process of theory construction, Einstein adds parenthetically:

Carefully considered, this procedure also belongs to the extralogical (intuitive) sphere, because the relations between concepts appearing in *S* and the experiences *E* are not of a logical nature. [Perhaps this is why Einstein draws the vertical arrow-tipped lines from *S*, *S'*, . . . as *dotted* lines.] This relation of the *S* to the *E* is, however (pragmatically), far less uncertain than the relation of the A to the E. (For example, the concept 'dog' and the corresponding experiences.) If such correspondence could not be attained with great certainty (even if not logically graspable) the logical machinery for the 'comprehensibility of reality' would be completely worthless (example, theology).

The quintessence is the eternally problematic connection between the world of ideas and that of experience (sense-experiences).

The main point of interest in this passage is the first sentence, "The S are referred to the E." Even its simplicity of expression does not hide the difficulties in the content. We are now at the crucial last phase of the schema, and we are looking down, from the predictions and other consequences (S, S' . . .) of the partly hypothesized, partly deduced scheme, to find whether corresponding observations can in fact be found to exist on the plane of experience E. If these are found, we can say that our various predictions have been borne out by observation and that we therefore have a right to regard with more confidence the previous steps that led us to this last one – the jump J from E to A, the postulation of A, and the deduction of the S. We thus have completed the cycle implied in the schema $E \rightarrow J \rightarrow A \rightarrow S \rightarrow E$. For simplicity, I will refer to this schema as Einstein's *EJASE* process of scientific theory construction.

But Einstein knew well that even if the predictions are borne out, one must not be too confident that the theory, the whole structure of conjectures, postulations, and deductions, is necessarily right. This is so for three reasons. First, right predictions can be drawn from wrong axioms. Hence theories that have turned out to be fundamentally in error (Aristotelian theory of elements, phlogiston theory, caloric theory) nevertheless were for a long time thought to be "verified" by the coincidence of deductions and observation.

Second, it is even impossible *in principle* to consider a theory "proven" once and for all, since this would entail subjecting it to an infinity of tests by observation, and not just now but for all future times. There is no such thing as a final verification or confirmation of a theory by experiment or observation. The most one can ever claim is that a theory gains more and more plausibility or usefulness the longer the various predictions derivable from it are found to correspond to the growing area of available sense experience – and the fewer the contradictions.

Third, and most important, Einstein came to realize that except perhaps in the simplest cases, one cannot rely on what someone claims to be "experimental facts" without much probing. The "confirmations" of theories have often turned out to be the result of misinterpretation of the data or a malfunctioning of the experimental apparatus. Einstein had more than once been held up in this theoretical work by claims of experimental scientists that turned out to be wrong. As he was reported to have said in the mid-1920s:

> You must appreciate that observation is a very complicated process. The phenomenon under observations produces certain events in our measuring apparatus. As a result, further processes take place in the apparatus, which eventually and by complicated paths produce sense impressions and help us to fix the effects in our consciousness. Along this whole path – from the phenomenon to its fixation in our consciousness – we must be able to tell how nature functions, know the natural laws at least in practical terms, before we can claim to have observed anything at all.[20]

Criteria for a good theory: I. "external validation"

What then can one expect to be the proper relation between the *S* and the *E* in an adequate theory, at least one of the more ambitious kind that was of interest to Einstein, a theory whose object is the "totality of physical phenomena"? [21] As we saw, in Einstein's sentence "The *S* is referred to the *E*," the phrase "referred to" [*in Beziehung gebracht*] is not at all the same thing as "verified against," as one might expect to read here if the proposed test of a good theory were the test of "verification." But that would be a pre-twentieth-century view which has turned out to be too optimistic an estimate of the solidity of scientific theories.

In fact, Einstein had proposed some years earlier two criteria for a good theory, two tests "according to which it is at all possible to subject a physical theory to a critique." The first test is what Einstein called the criterion of "external validation," and it is "concerned with the validation [*Bewährung*] of the theoretical foundations by means of the material of experience [*Erfahrungsmaterial*] lying at hand." The criterion is simply this: "The theory must not contradict empirical fact." [22]

Note that this is a *principle of disconfirmation* or of *falsification*, and hence much more sophisticated than any injunction to seek "confirmation" by empirical test. It is more generous, because in the absence of disconfirmation one can hold on to the theory – "Once a theoretical idea has been acquired, one does well to hold fast to it until it leads to an untenable conclusion" (*I.O.*, p. 343) – and it is also a sharper demarcation criterion because the presence of believable disconfirmations soon discredits a theory, whereas a continued absence of verification merely delays the final decision.

The disconfirmation criterion does not mean at all, however, that presumed confirmations, or coincidences of S with corresponding elements of E, would be unwelcome. On the contrary; in fact, the purpose of the majority of actual experimental investigations is guided by the hope of finding correspondence of this type by which the plausibility of some previously examined theory would be increased. But for the reasons given above, absence of verification is not conclusive either way, permitting one to be skeptical about or to hold on to the theory until further notice, depending on one's prejudice. But what really decides the matter is stubborn and repeated evidence of disconfirmation.

And it really should be stubborn and repeated evidence. One cannot abandon a theory every time a disconfirmation is reported. That would be extreme experimenticism, not warranted by the delicate and difficult nature of experiments in modern science. One should be as reasonably skeptical about experiments that disconfirm as about those that confirm – and particularly if the experimental disconfirmation of one theory is used to support another which, on *other* grounds, is less appealing.

Criteria for a good theory: II. "inner perfection"

What can this other ground be? What can make a theory more appealing or less so, other than the criterion of "external validation"? The answer is given by Einstein's second criterion for subjecting a theory to a critique. He called it the criterion of "inner perfection," and it concerns itself with choosing the superstructure in the $EJASE$ scheme, namely, the J, A, and S. One must remember that there is no guarantee in a given case that the elements of a theory are unique. It often happens that two quite different theories, with different J, A, and S, arise out of concern with the same material of experience, and moreover give equally good correspondence between their sets of S and the relevant sense experiences. The most famous case is of course that of Ptolemaic theory and the theory of Copernicus in the sixteenth century. While different with respect to their basic axioms, both theories arose out of the need to account for the same regularities and irregularities in E, in the observed motion of celestial

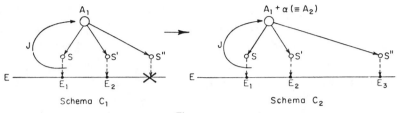

Figure 2.2

bodies, and the predictions derivable from both theories had about the same degree of correspondence with the observables.

Einstein's second criterion was frankly stated in his "Autobiographical notes": "The second point of view is not concerned with the relation to the material of observation but with the premises of the theory itself, with what may briefly but vaguely be characterized as the 'naturalness' or 'logical simplicity' of the premises (of the basic concepts and of the relations between these which are taken as a basis)." [23]

This of course is not an entirely new idea; Einstein acknowledges that it "has played an important role in the selection and evaluations of theories since time immemorial." But in practice, the requirement of naturalness or logical simplicity, or "unity and parsimony" (*I.O.*, p. 23) has never been easy to follow. Einstein is warning here to stay clear of theories that are patched up by ad hoc assumptions introduced just to make the deductions correspond better to the facts of experience as they continue to come in. "For it is often, perhaps even always, possible to adhere to a general theoretical foundation by securing the adaptation of the theory to the facts by means of artificial additional assumptions." Early in his career, Einstein had considered Lorentz's theory of the electron as just such a patchwork in the sense that it avoided factual disconfirmation only by introducing assumptions specially chosen for this very purpose (introduction of length contraction to explain absence of predicted effect in the ether-drift experiment). This practice could be represented by a modification in the diagram of the *EJASE* process, as in Figure 2.2, where schema C_1 is modified to yield schema C_2 by changing A to $(A + \alpha)$, α being the modification in A introduced to deal with the problem of obtaining better correspondence between the deductions S and the facts E.

To be sure, theories are likely to grow in some such manner, as they are applied to new areas of phenomena. And in any case such criteria as "naturalness" or "logical simplicity" or "economy" or "unity and par-

simony" are not easy to defend or even specify, for their "exact formulation . . . meets with great difficulties." It requires from us not a mere
"enumeration of logically independent premises," but "a kind of reciprocal weighing of incommensurable qualities"[24] and, therefore, a judge-
ment into which esthetic considerations and other preferences can enter
prominently.

Einstein was aware of a paradox, in that he tried to deal with great
and complex areas of varied experience, and yet looked "for simplicity
and economy in the basic assumptions. The belief that these two objectives can exist side by side is, in view of the primitive state of our scientific knowledge, a matter of faith. . . . This, in a sense religious, attitude
of a man engaged in scientific work has some influence upon his whole
personality" (*I.O.*, p. 357). Again, writing at about the same time (1950)
elsewhere, he acknowledged the a priori implausibility "that the totality
of all sensory experience can be 'comprehended' on the basis of a conceptual system built on premises of great simplicity. The sceptic will say
that this is a 'miracle creed.' Admittedly so, but it is a miracle creed which
has been borne out to an amazing extent by the development of science"
(*I.O.*, p. 342).

An example of Einstein's commitment to simplicity and naturalness
among the fundamental conceptions of science – an example that haunted
Einstein for much of his scientific life – was his unshakable dislike for
the premises and program of quantum mechanics. For the mathematical
description in quantum mechanics deals in principle with statistical ideas
(e.g., densities in the ensemble of systems), eliminating thereby even the
possibility in principle of describing the detailed behavior of the single
object or system – the very thing that lies closest to our experience, as in
the sensory reports made available by cloud chambers, counters, and the
like. To believe that this program is right "is logically possible without
contradiction; but it is so very contrary to my scientific instinct that I
cannot forgo the search for a more complete conception" (*I.O.*, p. 318;
and also often similarly, e.g., *I.O.*, p. 316; letters to Max Born, discussion with Niels Bohr, etc.). This search for a more complete conception,
he knew, might be doomed. "In the end the choice will be made [by the
profession as a whole] according to which kind of description yields the
formulation of the simplest foundation, logically speaking." But until
the evidence is irresistible, he considered it his right to abstain from "the
view that events in nature are analogous to a game of chance. It is open to
every man to choose the direction of his striving" (*I.O.*, pp. 334–335).

Einstein's use of the colorful words "instinct," "striving," "intuition," or "wonder" was not intended as a calculated provocation of some scientists or philosophers, but it has had this effect nevertheless. To make matters worse, he referred to yet another process of importance in the growth of theories, known to every practicing scientist but difficult to define. For even though he acknowledged that these two criteria of "external validation" and "internal perfection" defied precise description, he held that among the "augurs," those who deal deeply with physical theory, there nevertheless exists at any given time agreement in judging the degree of external validation and inner perfection.[25] Once more, the absence of an airtight definition did not preclude him from putting his bet on the usefulness of a concept, in this case that of consensus in groups within the scientific community.

Going beyond the *précis*

In putting together this account of Einstein's own epistemological views, insofar as possible in his own words, I have been trying to do justice to his meticulous sense of the realities, the lack of guarantees, the tentative, fallible, human aspects of each element in theory construction, and of the "eternal antithesis in our area between the two inseparable components of our knowledge, the empirical and the rational" (*I.O.*, p. 271). The schema that has emerged is miles from the self-confident and axiomatic treatments of scientific methodology which Einstein rightly held to have little resemblance to the actual practice of the working scientist. But let us not make a mistake in the opposite direction. Despite all its disclaimers, Einstein's schema implies nothing less than a description for as solid a process of reasoning as is in fact available to scientists.

Of course, Einstein's letter to Solovine was not a solemn publication but a précis shared between friends, in which Einstein reasserts ideas which he had long held (repeating, for example, almost verbatim passages from his essay of 1919 "Induction and deduction in physics"). But it is very suggestive, and invites one, in the spirit of Einstein's own method, to go further beyond these ideas, and so to see how they can serve to handle other problems of theory construction and the scientific imagination. As in science itself, our belief in a scheme increases if we find that it is not merely ad hoc for covering the area within which it was specifically proposed but is successful beyond it. There are specifically two problems that Einstein's schema can help us with: how scientific theories grow and give way to other theories, and how to understand better the controversies involving fundamentally different theories that claim to deal with the same experimental facts.

The growth of a theory

We noted that the schema in Figure 2.1 is not a static one, but is a process which makes a cycle from E via J, A, S, back to E. But a theory can hardly be created and tested by going through the cycle once. Even the theories by which we orient ourselves in our day-to-day life, and a fortiori the established theories of science which we honor and use as tools that have come down to us from earlier workers and controversies, are all the results of cycles of progressive adaptation, making them more acceptable by using the feedback from one cycle to modify the next. Moreover, this process of modification and growth will continue as new phenomena are found that enlarge the original area of application. Physics is constantly "in a state of evolution. . . . Evolution is proceeding in the direction of increasing simplicity of the logical basis" (*I.O.*, p. 322).

The need to go through many cycles ($C_1 \rightarrow C_2 \rightarrow C_3 \ldots$) of the *EJASE* process is forced on us, if by nothing else, by our human limitations. Neither thought by itself nor sensory experience by itself leads to reliable human knowledge. For concepts can be subjected to analysis which gives us certainty of the kind "by which we are so much impressed in mathematics; but this certainty is purchased at the price of emptiness of content." On the other hand, sense experience cannot be related to the concepts, as we have seen, except by adopting essentially arbitrary definitions (conventions), and hence they cannot claim certainty either. The best we can therefore do is to let whatever trustworthiness there is in our theory construction come out of the interplay of thought and sense experience through many cycles, carried out over time. Theories therefore have to be "thoroughly elaborated" (*I.O.*, pp. 276, 277, 282) and have to evolve – first in the mind of the innovator before publication and then in the community of scientists through discussion or controversy.

For example, in going through the first cycle of the schema, the S of the theory at that stage ($S_1, S_1', S_1'' \ldots$) may show an incomplete correlation with the "facts" in the E plane. Einstein gives an example which, he says, is one of the considerations which "kept me busy from 1907 to 1911": in his early attempts to generalize relativity theory, "the acceleration of a falling body was not independent of its horizontal velocity or

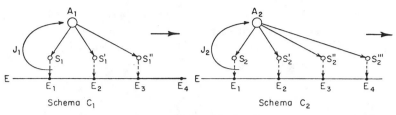

Figure 2.3

the internal energy of the system," contrary to the "old experimental fact" (*I.O.*, p. 287). A discrepancy of this sort forces one to rethink the A, modifying the original axiom system A_1 to become a somewhat different one, A_2.

We recall that Einstein warned that such modification should be made not in a merely ad hoc, brute-force way, but for example by recasting the axiom system into a more generalized form that permits more deductions $S_2, S_2', S_2'' \ldots$ that can be correlated with the E, and if possible from fewer independent concepts. Thus Einstein was able to go from the first principle of restricted relativity theory, that all natural laws must be so conditioned that they are covariant with respect to Lorentz's transformations, to the first principle of general relativity theory, that natural laws are to be formulated in such a way that their form is identical for coordinate systems of any kind of states of motion (*I.O.*, pp. 329–330). In this way Einstein removed his dissatisfactions with the special nature of the original relativity theory, namely that it referred only to systems in uniform motion to which no absolute significance could be attached. The introduction of the principle of equivalence removed the contradiction between the predicted acceleration of a falling body and the observed one, as well as removing an unnecessary duplication (two meanings of mass, as referred to above).

Figure 2.3 represents schematically the progress from the early state of a theory to a later state, from C_1 to C_2 and from C_2 to C_3. Here C_3 could stand for the next step, which Einstein saw needed after his success in fashioning general relativity theory; for he felt that "the theory could not rest permanently satisfied with this success. . . . The idea that there exist two structures of space independent of each other, the metric-gravitational and electromagnetic, was intolerable to the theoretical spirit." Hence, Einstein's persistent attempt to fashion a field theory that corresponds to a "unified structure of space." Again and again, the word "unity" beckons as Einstein's final goal – "seeking, as far as possible, logical unity in the world picture, i.e., paucity in logical elements"; "thus the story goes on until we have arrived at a system of the greatest conceivable unity, and of the greatest poverty of concepts of the logical foundations, which is still compatible with the observations made by our senses" (*I.O.*, pp. 285, 293, 294). Here we glimpse the ultimate goal not only of Einstein but of Mach, Planck, and that whole generation of scientist-philosophers: the unified *Weltbild*. We shall touch on this point again soon in this chapter, and again in Chapter 4.

To give some other examples of the driving force leading from C_1 to C_2 to C_3: Einstein held Newton's mechanics to be "deficient" from the point of view of the requirement of the greatest logical simplicity of the foundation insofar as the choice of the value 2 of the exponent in the inverse square law of gravitation – the very heart of Newton's greatest triumph – was heuristic or ad hoc in the sense that it could be defended only because it worked. And in addition the law of gravitation itself was a separate postulate, not connected with or derivable from other conceptions in mechanics – whereas in general relativity theory it developed as a consequence of the postulates. Similarly, Einstein felt that H. A. Lorentz's synthesis of Newtonian mechanics and Maxwell's field theory contained the "obviously unnatural" mixture of total differential equations (for the equations of motion of particles or points) with partial differential equations (Maxwell's field equations). It led to the need to assume particles of finite dimension, to keep fields at the surface from becoming infinitely large. To Einstein it appeared "certain . . . that in the foundations of any consistent field theory the particle concept must not appear in addition to the field concept" (*I.O.*, p. 306).

One could also visualize the progressive development of scientific theory to take place as the development of the system of concepts at an increasingly higher level of "layers" or strata, each layer having fewer and fewer direct connections with the complexes of sense experiences (*I.O.*, pp. 293–295). In this way, a more phenomenological theory at the early stage of science, for example, the theory of heat before Maxwell, gives way to a more independent set of concepts and axioms that characterizes, for example, kinetic theory and statistical mechanics. Thus the latter eventually allowed one, in the study of Brownian motion, to find the limits of applications of the laws of classical theory, and in addition provided a definite value for the size of atoms and molecules, obtained by several independent methods.

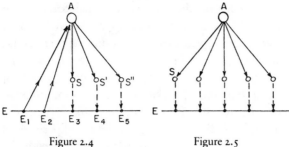

Figure 2.4   Figure 2.5

There is of course a cost in this developmental process. By going cyclically through several stages of theories, each stage is forced to use conceptions more removed from direct experience (e.g., atomism). The distance from the E to the A is larger, the contact with common sense is more and more tenuous. But the fundamental ideas and laws of science attain a more and more unitary character (cf. *I.O.*, p. 303). Eventually all the sciences would attain this final stage.

In the meantime another cost of this process is that the more general the theory becomes, the longer it may have to wait for the correlation of its predictions with the ground of experience. Thus it took the general theory of relativity to 1919 to make proper contact with E. The delay may well test the self-confidence of the theoretician to the utmost. "It may need many years of empirical research to ascertain whether the theoretical principles correspond with reality"; for it may take that long to discover "the necessary array of facts" (*I.O.*, pp. 223–223).

Representations of a finished theory

In the ordinary course of events the development of a theory will take it to a stable canonical form. It enters the textbooks usually as a recast pedagogic scheme which is characterized by a rearrangement to bring out an axiomatic structure and to hide all traces of the speculative phase that motivated and characterized the theory in its early stage. In particular, the textbook tends to hide the J process as if it were an embarrassment. The presentation of the theory at that stage in its life cycle, and of scientific research papers that base themselves on such a theory,[26] are likely to look like Figure 2.4. That is, a few phenomena are cited (E_1, E_2 in Figure 2.4) from which, it is said, the axiom system was induced; and from the latter, predictions are deduced for which corresponding experimental demonstrations $(E_3, E_4 \ldots)$ can in fact be given. Or, as in Figure 2.5, the whole theory is presented as if its starting point is the discovery of an axiom system, from which all follows. That is essentially the style of the *Principia* of Newton and most school books; for example, Newton's laws of motion and of gravitation, at the top, radiate deductions concerning the periodicity of the tides, the shape of planets, and so on, and these in turn are directly confirmed by the experimental evidence below. Or from the fundamental postulates of the kinetic theory, there follow the equations of state of gases, viscosity, diffusion, heat conductivity, radiometric phenomena in gases, and so on; and sure enough, all these can be correlated with their corresponding phenomena over a large range.

Of course, apart from the fact that such a representation no longer reflects the genesis of the theory (which, in any case, is usually of as little interest to most scientists as to most philosophers), a representation such as that in Figure 2.5 brings out the astounding strength of well-developed theories. That is, they give us, to use Einstein's phrase, an "overview" by which the multiplicity of immediate sense experiences of the most diverse kinds are brought into a unified and therefore understandable scheme.

Progress by unification of theories

The next stage in the historic progress of science occurs when a unification of two or more theory systems is forged, as when Galileo joined terrestrial and celestial physics, or when Maxwell produced a synthesis of electricity, magnetism, and optics. Before unification or synthesis, each of the theory systems has its own system of concepts, and while they might be closer to experience than the concepts after unification, they lack unity among their different fundamental postulates within which they are embedded (cf. *I.O.*, p. 302). In terms of our graphic shorthand, what happens is symbolized by Figure 2.6. On the left side, the separate axiom systems for the fields of electricity, magnetism, and optics dominate the respective systems that sit, like separate pyramids, upon their corresponding territories on the *E* plane. After Maxwell's synthesis, the separate axiom systems become merely special cases of a more general one, incorporating Maxwell's equations. The separations between the three areas on the plane of phenomena disappear, the sets of facts pre-

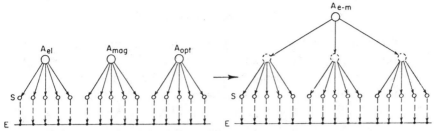

Figure 2.6

viously connected become part of one larger set, and their theoretical description has thereby been simplified (cf. *I.O.*, p. 223).

The same symbolization would be appropriate for many of the great advances of science, for example, the work of P. A. M. Dirac in the late 1920s by which large areas of both physics and chemistry were brought under the control of quantum mechanics. In our time, the current attempts to unify the basic forces of nature is another chapter in this drive to encompass eventually the totality of *E*, all points on the *E* plane, in terms of the fewest possible independent axiom systems. It was of course Einstein's hope through field theory to establish a unified foundation for the whole of physics (cf. *I.O.*, pp. 328–329). Einstein put this aim very clearly:

> Science is the attempt to find a coordination between the chaotic diversity of our sense-experiences and a logically unified system of thought . . . from the very beginning there has always been present the attempt to find a unifying theoretical basis for all these single sciences, consisting of a minimum of concepts and fundamental relationships, from which all the concepts and relationships of the single disciplines might be derived by [a] logical process. This is what we mean by the search for a foundation of the whole of physics. The confident belief that this ultimate goal might be reached is the chief source of the passionate devotion which has always animated the researcher.[27]

It may well be that the ability to carry in mind the entities and relations symbolized in the right-hand portion of Figure 2.6 is what it means to "have a *Weltbild*," to have a mental picture of the physical universe. And even though there is no guarantee of reaching that goal, we have sufficient glimpses of it to inspire the work.

The role of thematic presuppositions

We now must come back to deal with an important point which was left open. The problem can be put as follows: Since the leap from E to A at the beginning of the schema in Figure 2.1 is logically discontinuous and represents the "free play" of imagination, and since from such a leap can result an infinite number of A – virtually all of which will turn out to be useless for the construction of the theory system – how can one ever expect to be successful in this process except by chance? The answer must be that the license implied in the J process is the freedom to *make* a leap, not the freedom to make *any* leap at random. Something must guide or channel J if only because the premises later must pass such tests as those of naturalness and simplicity in order to meet Einstein's second criterion for a good theory.

The chief guide is a constraint that shapes the work of every scientist engaged in a major work on novel ground: the constraint provided by explicit or, more usually, implicit preferences, preconceptions, presuppositions. Einstein himself recognized and commented on this repeatedly: "If the researcher went about his work without any preconceived opinion, how should he be able at all to select out those facts from the immense abundance of the most complex experience, and just those which are simple enough to permit lawful connections to become evident?" [28] By way of example, he discussed the dilemma that, in formulating the laws of mechanics, one has to follow either the "natural tendency of mechanics to assume . . . material points," which necessarily leads to the presupposition of atomism, or else to erect a mechanics of continuous media based on another "fiction," for instance, that "the density and the velocity of matter depend continuously upon the coordinates and time" (*I.O.*, p. 302). These "fictions" – not unrelated to what Frank Kermode in another context has called the "necessary fictions" that are found at the heart of literary works – have of course considerable practical value. For example, they guide the development of mathematical tools (in Einstein's last example, partial differential equations), but they are much more than that. He referred to them as " 'categories' or schemes of thought, the selection of which is, in principle, entirely open to us and whose qualification can only be judged by the degree to which its use contributes to making the totality of the content of consciousness 'intelligible.' "

An example of such a category is the distinction between sense im-

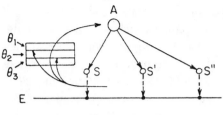

Figure 2.7

pressions and "mere ideas" (Schlipp, p. 673). He warned that "we do not conceive of the 'categories' as unalterable (conditioned by the nature of the understanding [and in this respect "distinct from that of Kant"]) but as (in the logical sense) free conventions. They appear to be a priori only insofar as thinking without the positing of categories and of concepts in general would be as impossible as is breathing in a vacuum." [29]

As I have tried to show in a number of case studies of scientific work, stretching from the time of Kepler to that of Bohr and Einstein and to the frontier work of today, we can recognize the existence of, and at certain stages of scientific thinking, the necessity of postulating and using, precisely such unverifiable, unfalsifiable, and yet not arbitrary conceptions, a class to which I have referred as themata.[30] Different scientists may be attracted to different themata, and allow themselves to be led by them to different degrees.

Among the themata which guided Einstein in theory construction are clearly these: primacy of formal (rather than materialistic) explanation; unity (or unification) and cosmological scale (egalitarian applicability of laws throughout total realm of experience); logical parsimony and necessity; symmetry; simplicity; causality; completeness; continuum; and of course constancy and invariance. It is themata such as these which explain in specific cases why he would unshakably continue his work in a given direction even when testing against experience was difficult or unavailable. It explains equally why Einstein refused to accept theories that were well supported by correlation with phenomena but which were based on thematic presuppositions opposite to his own (as in the case of the quantum mechanics of Niels Bohr's school).

This conception can be built into a modification of Figure 2.1, to exhibit the function of themata in the *EJASE* process. Figure 2.7 shows a number of possible leaps arising from *E* toward *A*, but only one (or a few) survive the filtering action of the themata which a particular innovator has adopted or incorporated into his conceptual processes. For example, the two conjectures which Einstein raised to the status of postulates at the beginning of his 1905 relativity paper are thematically shaped presuppositions; they conform to the prior imposition of the requirements of large scale and egalitarian applicability of laws, invariance, logical parsimony, and primacy of formal explanations.

Confrontation of rival theories

The *EJASE* schema as now completed lends itself very well to represent the situation that arises when two different theories claim to handle about the same experiential material. Thus Einstein often stressed that his relativistic mechanics, at least in the early years, overlapped with Newtonian mechanics with respect to the range of testable experience, even though the theories corresponded to "two essentially different principles" (*I.O.*, p. 273).

In analyzing controversies between rival explanations drawn from the same ground of available experience, it has been evident to me that the difference of choices of themata by the rivals explains a good deal of the details of their theories and of the course of the controversy. In very abbreviated form, Figure 2.8 will help to make this point. On the left appears A_1 as the axiom system reached by the first of two innovators. His themata are symbolized by θ_1. The system of propositions gives rise to deductions S_1, S_1', S_1'', S_1''', as shown. To most of these, there correspond observations (E_1, E_2, E_3) which can be correlated with the deductions. As is usual, some deductions (S_1''') remain, at least for the time being, without such "verifications," although work may be in progress on just such a problem at that point.

The second innovator is represented by the system on the right side of the figure. His axiom system A_2 was reached through preliminary notions passing through the constraint indicated by θ_2, his own set of themata. The system A_2 as a whole is not so utterly different from A_1 on the left side that there is not some overlap between the deductions made by the two innovators. Thus the deductions S_2, S_2' of the second person refer to the same phenomena E_1 and E_2, at least as far as can be determined at the time, as do S_1 and S_1'. In addition, however, A_2 allows deduction S_1'' for which there is no equivalent in the first system and which claims to be correlated with ("borne out by") E_4.

Something like this is how a scientific dispute can continue for some

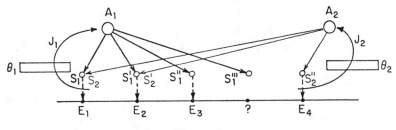

Figure 2.8

time. A_1 or A_2 or both may, in the course of the debate, be progressively modified, with corresponding changes in S_1 and S_2. Eventually, one or the other of the two systems wins out, and this happens usually in one of two ways. The two theory systems, separately, may come to a point of development where there is no essential difference in the number and types of phenomena (experimental evidences) which they can handle. That is, perhaps by some ad hoc adjustment of A_2, it too "can account for" E_3. If this situation persists for some time, a choice is made between the two systems on the basis of the "appeal" of the fundamental presuppositions. This comes down to having the preponderance of the scientific community making a choice on the basis of preferring the system of themata θ_1 or θ_2. Thus, in the early period when Einstein's relativity theory could not be clearly distinguished from Lorentz's or Abraham's by any significant differences in their testable predictions, Max Planck was driven to exclaim in a scientific meeting, when pressed to confess why he believed in Einstein's postulate system rather than its rivals: "I find it more congenial."[31]

An alternative scenario is for one of the two systems to produce more verifiable predictions of observable events than the other, and fewer (or no) uncomfortable disconfirmations. Almost never is the situation so clear that the inability of one theory system to handle a specific experiment produces right away a decision in its disfavor. What is much more likely is that during the period when attempts are made to account for apparent difficulties, the balance of opinion swings toward one of the systems in favor of the other, and the latter slowly fades from view without necessarily ever being "disproved."

The model implied in Einstein's schema can be shown to be useful in further, fruitful extensions, helping us to understand additional details of the working of the scientific imagination. That is a task for another occasion. By having gathered together Einstein's own expressed views, and having tried to correlate them with the schema he himself proposed, we have found a consistent prescription for one way the human mind may puzzle out the order behind the appearances, and for communicating that perception to others in a convincing manner.

Despite this grand ambition we must not think that Einstein in this field, any more than in his science, attempted to impose an absolutistic point of view. He was all too aware of the tentative state of understanding in the field of scientific methodology. The spirit in which he proposed his ideas is well conveyed in a passage in his "Autobiographical notes," immediately after he has begun to give his answer to the question "What, precisely, is 'thinking'?":

> With what right – the reader will ask – does this man operate so carelessly and primitively with ideas in such a problematic realm without making even the least effort to prove anything? My defense: all our thinking is of this nature of a free play with concepts; the justification for this play lies in the measure [*Übersicht*] over the experience of the senses which we are able to achieve with its aid.[32]

As limited human beings confronting the seemingly endless, interlocking puzzles of the universe, we can nevertheless hope to play – as in Newton's metaphor – with pebbles at the shore of a vast ocean. If we do it well, that play can yield the most highly desirable kind of knowledge: a survey (overview, *Übersicht*) of the world of nature that grants us the perception of order guiding the phenomena in their infinite, individual variety, and in their inexhaustible interactions with one another.

3

Einstein's scientific program: the formative years

Albert Einstein's contributions will be studied as long as our civilization exists. But while scientists from their student days on will, on the whole, know of his work indirectly through the textbooks, Einstein's actual words, in his wide-ranging publications and correspondence, will be scrutinized chiefly by the historians of science. One may hope that Einstein would have approved; not only did he publish many essays on historical developments in science,[1] but he was on record, more than once, that a means of writing must be found that conveys the thought processes that lead to discoveries – showing how scientists thought and wrestled with their problems. Moreover, Einstein made the task of the historian easier than did many other scientists, because of the characteristic frankness and consistency in his writings. These traits of his will aid my task to speak about some of the early steps on his path to relativity – steps that made that path a high road and caused the other, more fashionable ones of that day to be seen eventually as blind alleys.

"Recognize the unity of a complex of phenomena"

When viewed in terms of Einstein's early publications,[2] the road that brought him to the threshold of relativity began in apparently quite unimpressive territory. Einstein's first published article, entitled "Consequences of the capillarity phenomena," was sent to the *Annalen der Physik* in December 1900, a half year after Einstein had graduated from the Eidgenössische Technische Hochschule in Zurich and a half year before his getting his first temporary job as a substitute teacher.

At the time, all the excitement in physics lay in a quite different direction. It was just a few years after the discovery of X-rays, radioactivity, the electron. New experimental findings and new theories chased one another rapidly. Einstein was not ready for any of that. As he characterized it later, he was in the middle of years of "groping in the dark." He often remarked that his formal training had been spotty, although on his own he had worked his way through the volumes of classic lectures of Kirchhoff, Helmholtz, Hertz, and Boltzmann's *Gastheorie*, not to speak of Ernst Mach, to whose work Michele Besso had introduced Einstein soon after their first meeting in 1896. Einstein confessed later, "I had no technical knowledge." [3] But he added: "It turned out soon that the general overview [*allgemeine Übersicht*] over physical connections is often more valuable than specialist knowledge and routine."

But capillarity was by no means as dull as it now may seem. Even some five years later, a long paper on capillarity by G. Bakker in the famous volume 17 of the *Annalen der Physik* starts with the panegyric: "The theory of capillarity of Laplace was one of the most beautiful achievements of science"; he goes on to sing the praises of the subject as handled by Gauss, Young, Gibbs, and F. Neumann. Young Niels Bohr's first research, completed in 1906, was on the closely related problem of surface tension of water. For Einstein the capillarity paper was the first of nine publications indicating his deep interest in thermodynamics and, later, statistical mechanics, which he published in the *Annalen der Physik* between 1901 and 1907. [4]

The problem to which Einstein is attending, in this first paper and in the next one, is "the problem of molecular forces." The mechanical work done in a cycle that involves isothermal increases of surfaces of liquids should be zero; but, he says, this is contrary to experience. Therefore, "there is nothing left to do but to assume that the change in surface area is attended by a conversion of work to heat." He promises to proceed from the simplest assumptions [*Annahmen*] about the nature of molecular forces of attraction, and check their consequences in terms of their agreement with experiment. In this work, he says, "I let myself be guided by analogy with gravitational forces." At the end of the paper, he summarizes:

> We can now say that our fundamental assumption has been validated: to each atom there corresponds a molecular force of attraction which is independent of the temperature and independent of the way the atom chemically combines with other atoms. . . . The question whether and in what respect our

forces are related to gravitational forces must be left completely open.

What preoccupies him here, as he writes in a letter to his friend
Marcel Grossmann (April 14, 1901), is "the question concerning the
inner relationship of molecular forces with Newtonian forces at a distance."[5] Now, this is not a problem without ambition! Newton himself,
who had hoped for a relationship between gravitational forces and molecular forces, would have been not uninterested to read this work.

The idea for Einstein's work seems to have been based on Wilhelm
Ostwald's *Allgemeine Chemie*. This is the book first mentioned in all of
Einstein's writings. Indeed, he sent a reprint of the paper to the scientist-
philosopher Ostwald on April 5, 1901, together with a request for a job
in Ostwald's laboratory and the remark that his book had stimulated this
paper. Ostwald, on his side, was not overwhelmed; he did not even reply.
But Einstein's interest in the program of the unification of the forces of
nature was making its tentative first appearance right here. He felt he was
working on important problems, and despite his inability to find a steady
job, he was evidently in good spirits. In the letter to Grossmann he says he
is as merry as a bird, and he adds: "It is a magnificent feeling to recognize
the unity [*Einheitlichkeit*] of a complex of phenomena which to direct
observation appear to be quite separate things." It is only April 1901, but
this is already a familiar Einstein, here searching for bridges between the
phenomena of microphysics and macrophysics, between capillarity and
gravitation.

Einstein's second paper, again, looks not very promising on the surface. It is entitled "Concerning the thermodynamics of potential difference between metals and fully dissociated solutions of their salts, and
concerning a new method to investigate molecular forces." It is dated
April 1902, some two months before he started his job at the patent
office. Over a year has elapsed since his first paper.

He begins with a section entitled "A hypothetical extension of the
second law of the mechanical theory of heat." His earlier method now
has undergone a significant change. In the first paper he had started with
the phenomena, listed reams of experimental data from the literature,
and discussed consequences drawn from them, as was more or less the
rule in papers of the time. Now, in the first section, he postulates a statement of the second law which he recognizes to be outside the limits set by
the available phenomena. By generalizing (*verallgemeinern*) beyond ex-

perience, he proposes to adopt the following statement: "One remains in harmony with experience when one applies the second law to physical mixtures upon whose individual components are acting any arbitrary conservative forces." Moreover, he warns quickly, "We will rely on this hypothesis in what follows even if it does not appear to be absolutely necessary." He has now begun to go for discoveries of postulates frankly beyond the "present experience." The appeal of the *generalization* is taking over, and becomes a directive for research.

The third paper follows, sent in from Bern in June 1902. It is entitled "Kinetic theory of thermal equilibrium and the second law of thermodynamics," the first of three papers in 1902–04 extending Boltzmann's ideas in thermodynamics and statistical mechanics. Einstein introduces an interpretation of statistical probability in which systems run through the various possible states over and over in irregular fashion, unlike thermodynamic descriptions in which equilibrium states once reached are persisted in indefinitely. This is an important bridge to his future work. His power has begun to show now, although it did not attract much attention at the time. Without having read J. Willard Gibbs' *Elementary principles of statistical mechanics*, which appeared in the same year, Einstein is getting some of the same results.

The introductory paragraph of the third paper lays forth the ambition to generalize:

> Despite the success of the kinetic theory of heat in the area of the theory of gases, it has so far not been possible, by the laws of mechanics alone, to provide a sufficient foundation for the general theory of heat. Maxwell's and Boltzmann's theories have come close to this goal. The purpose of the following considerations is to fill the gap. At the same time there will be given an extension of the second postulate which will be of importance for the application of thermodynamics. Moreover, a mathematical expression for entropy will be obtained, from a mechanical point of view.

The first numbered section follows immediately and is entitled "1. Mechanical picture [*mechanisches Bild*] for a physical system." The only footnote in this paper is to Boltzmann's *Gastheorie*, where, incidentally, the important term *mechanisches Bild* is also used in one of the headings.

As if to make sure that we take seriously what is implied by this term, Einstein comes back to it toward the end of his paper. In one sentence he gives the main conclusion: "The second law thus appears as a necessary consequence of the mechanical world picture [*notwendige Folge des mechanischen Weltbildes*]." With this, he has used the term in print for the first time. He was to return to it often later, not least in his two great critiques of the "mechanical world picture" in his "Autobiographical notes" of 1946, and of the whole unsatisfactory development from the *mechanische Programm* of the nineteenth century to our *heutiges Weltbild*, as spelled out in his earlier lecture on "Aether and relativity theory" (1920).

So as of 1902, we see the twenty-three-year-old Einstein entering publicly on the ground where the great fight among German scientists has been in progress between the chief rival world conceptions, one holding to mechanics and the other to electrodynamics as the ground of fundamental explanation. In this 1902 article, he is still concerned with investigating the mechanical world picture as one of the main options. And while he finds it there of use, he ends with a suspicion: "The results are more general than the mechanical representation used to arrive at them." Soon he will see it is in fact far too limited; it cannot handle, for example, Brownian movement. He will also find that its most prominent alternative, the so-called electromagnetic world picture, cannot handle fluctuation phenomena of light. And even the victory of one of the conceptions would, as he later expressed it, leave us with "two types of conceptual elements, on the one hand, material points with forces at a distance between them, and, on the other hand, the continuous fields . . . an intermediate state in physics without a uniform basis for the entirety."[6] He will have to try to build his own *Weltbild*.

The fourth paper, entitled "A theory of the foundations of thermodynamics," is sent off from Bern in January 1903. It is his only paper published that year. In addition to constructing a *Weltbild*, he is also building a career, and of course a family. He and Mileva Marić are married on January 6, 1903, and in the first of his surviving letters to his friend Besso, Einstein writes in that month: "Well now, I am a married man and lead a nice, comfortable life with my wife. She takes excellent care . . . and is always cheerful."[7] He goes on to describe the fourth paper, which he has just sent off

after much reworking and correction. But now it is completely clear and simple, so that I am quite satisfied with it. After postulating the energy principle and the atomic theory, there follow the concepts of temperature and entropy; with the additional aid of the hypothesis that the distribution of states of

isolated systems never go into less probable ones, there follows
also the second law in its most general form namely the impossibility of a perpetuum mobile of the second kind.[8]

Hindsight makes it easy to see that the structure of the argument in
this paper is perhaps its most interesting feature: first, the postulation of
general principles, now with hardly a nod to the detailed phenomena;
then the derivation of logical consequences; and at the end, the test
against experience – in this case, experience of the most general kind.

More than a year later, on April 14, 1904, Einstein writes to his friend
Conrad Habicht: "In a few weeks we are expecting a young one. I now
have found in a very simple manner the relation between the elementary
quanta of matter and the wavelengths of radiation."[9] In fact, he has just
sent off his fifth paper to the *Annalen der Physik*, "On molecular theory
of heat." This paper announces itself modestly, in the first sentence, as
merely an addition to work published the previous year. But the plot has
thickened. In his usual introductory paragraph, he sets out the general
plan. He will derive an expression for the entropy of a system which is

> completely analogous to that found by Boltzmann for ideal
> gases and postulated by Planck for his theory of radiation. . . .
> Then a simple derivation will be given of the second law.
> Thereupon the meaning of a universal constant will be investi-
> gated, which plays an important role in the general molecular
> theory of heat. Finally there follows an application to the
> theory of black body radiation, where a highly interesting rela-
> tion is developed between the universal constant mentioned
> above [he means $R/2N$] (one which is determined by the mag-
> nitude of the elementary quanta of matter and of electricity),
> on the one hand, and the order of magnitude of the wavelength
> of radiation, on the other, without making use of any special
> hypotheses.

Thus Einstein introduces his long-lasting concern with fluctuation
phenomena. He shows that the size of energy fluctuations ε in the system,
and therefore its thermal stability, is determined by a universal constant
(which we would write as $k \equiv R/N$). Having applied it to mechanical
systems and thermal phenomena, he then makes the original and daring
jump to an "application to radiation": In order to determine the universal
constant from the size of observable energy fluctuations, he proposes to
go to the only kind of physical system where, he says, experience allows

one to "suppose" (*vermuten*) that there *are* energy fluctuations – the case of an otherwise empty space filled with thermal radiation. Obviously, by now he has indeed learned to "scent out the path that leads to fundamentals," as he later put it.

There follows now an ingenious argument that shows how much he is at home with borderline cases between two different fields as defined in contemporary physics, not having to choose between one and the other, but rather using both – indeed, "recognizing the unity of a complex of phenomena." He considers a volume v with dimensions of the order of the wavelength λ of the radiation in it ($v \simeq \lambda^3$). In that case, the energy fluctuations will be of the same order as the magnitude of the energy itself, or $\bar{\varepsilon}^2 = \bar{E}^2$. Now he uses the Stefan–Boltzmann law in the form $\bar{E} = cvT^4$ and deduces that $\lambda \simeq 0.4/T$. But "by experience" – Einstein does not mention that it is Wien's law (1893) – we know that indeed λ_{max} is $0.293/T$. So his result has the same general lawfulness with respect to T, and the same order of magnitude for the constant. Einstein concludes that in view of the "great generality of our assumptions," this agreement cannot be an "accident." And with that, the paper ends. (The coincidence of prediction within a factor of about 2 remains characteristically an indication for Einstein that things are going quite well.)

But what a great launching of work, in some eight pages! Fluctuation phenomena will remain one of Einstein's trusted tools, not only in the Brownian motion paper, which is almost ready to come over the horizon, or in the papers on quantum theory of radiation and on Bose–Einstein statistics, but even in some of the smaller papers that have received little notice so far. (There is, for example, one of Einstein's short publications in 1908 on a new electrostatic method for measuring small quantities of electricity. With Conrad and Paul Habicht, friends of the earliest days of the Olympia Academy in Bern, Einstein had been interested in building a device for measuring small voltages by multiplication techniques. In fact, Einstein wrote a patent application for the device, which claimed to measure potential differences down to 5×10^{-4} volt. He even found a "clever mechanic" who attempted to build the thing. But Einstein's real interest shows up in the last paragraph of the short 1908 paper that described the device.[10] In research, for example on radioactivity, an electrostatic method of measurement of highest sensitivity might be useful. But "I was led to this plan by thinking how one might find and measure the spontaneous charge appearing on conductors which should exist analogously with the Brownian motion that is required by the molecular theory of heat.")

The next publication is Einstein's inaugural dissertation on "A new determination of molecular dimensions," dated April 30, 1905.[11] We are now in the miraculous year of 1905, the year that will continue to be a challenge for historians of science for a long time. The work does not yet concern itself with Brownian movement. As Einstein later said, he had only just discovered the existence of that problem:

> Not acquainted with the investigations of Boltzmann and Gibbs which had appeared earlier and actually exhausted the subject, I developed the statistical mechanics and molecular-kinetic theory of thermodynamics which was based on the former. My major aim in this was to find facts which would guarantee as much as possible the existence of atoms of definite, finite size. In the midst of this, I discovered that, according to atomistic theory, there would have to be a movement of suspended microscopic particles open to observation, without knowing that observations concerning the Brownian motion were already long familiar.[12]

The inaugural dissertation paper is concerned with the viscosity and diffusion of liquid mixtures, and the calculation of Avogadro's constant from these. Einstein starts by saying that the size of molecules can be found from the kinetic theory of gases, but not yet from the observable physical phenomena of liquids. He proposes now to bring about a fusion between these portions of micro- and macrophysics.

Einstein regards the solute molecule to behave like a solid body suspended in a solvent, and applies hydrodynamic equations of motion to the large molecules, assuming an approximate homogeneity of the liquid. At the end of the paper, applying kinetic theory to liquid solutes with known diffusion coefficients, he gets for Avogadro's number N the value 2.1×10^{23}, and adds quite simply that N found by this method coincides in order of magnitude with N found by other methods "in a satisfactory way."

Einstein's letters show us that his thesis was really part of an older research interest. He had written to Besso more than two years earlier (March 17, 1903),

> Have you by now calculated the absolute magnitude of ions under the assumption that they are spheres, and so large that the equations of hydrodynamics of viscous liquids are appli-

cable? Since we know the absolute magnitude of the electron [charge], this should be a simple matter. I would have done it myself but I don't have literature or time; you can also use diffusion in order to get information about neutral salt molecules in solution. . . . If you don't understand what I mean, I'll write you more explicitly.[13]

His thesis, though not done for ions, is very close to this proposal.

Exploiting the confrontation of theories

This brings us to the three great papers of 1905, sent off to the *Annalen der Physik* at intervals of less than eight weeks. When I first became interested in the history of physics of this century, it struck me that while these three papers – on the quantum theory of light, on Brownian movement, and on relativity theory – seemed to be in quite different fields, they could be traced in good part to the same general problem, namely fluctuation phenomena. Indeed, in the Archive at the Institute for Advanced Study was a letter Einstein wrote to Max von Laue on January 17, 1952, in which the connections are indicated. Einstein discusses von Laue's textbook on relativity theory, and registers a small objection:

When one goes through your collection of verifications of the Special Relativity Theory, one gets the impression Maxwell's Theory may be unchallengeable. But already in 1905 I knew with certainty that it [Maxwell's Theory] leads to wrong fluctuations in radiation pressure, and hence to an incorrect Brownian movement of a mirror [suspended in] a Planckian radiation cavity. In my opinion one can't get around ascribing to radiation an objective atomistic structure, which of course does not fit into the framework of Maxwell's Theory. Naturally, it is comforting that the Special Relativity Theory in essence rests only upon the constant c, and not on a presupposition of the reality and fundamental character of Maxwell's fields. But unhappily the 50 years which have elapsed since then have not brought us closer to an understanding of the atomistic structure of radiation. On the contrary![14]

Here is made explicit the chief connection between Einstein's work on Brownian motion of suspended particles, the quantum structure of radiation, and his more general reconsideration of what he called later "the electromagnetic foundations of physics" itself.[15]

It also struck me forcibly that the style of the three papers was essentially the same, and fitted with the way Einstein had come to see his task in the course of his earlier work. Contrary to the sequence one finds in many of the best papers of the time, for example H. A. Lorentz's 1904 publication on electromagnetic phenomena, Einstein does not start with a review of a puzzle posed by some new and difficult-to-understand experimental facts, but rather with a statement of his dissatisfaction with what he perceived to be asymmetries or other incongruities that others might dismiss as being of predominantly esthetic nature. He then proposes a principle of great generality. He shows that it helps remove, as one of the deduced consequences, his initial dissatisfaction. And at the end of each paper, he proposes a small number of predictions that should be experimentally verifiable, although the test may be difficult.

The way Einstein starts the paper on the quantum theory of light is typical. He writes: "There exists a deep-going, formal distinction between the theoretical representations which physicists have formed for themselves [note it is always the physicists' notions which are at fault] concerning gases and other ponderable bodies on the one hand, and the Maxwellian theory of electromagnetic processes in so-called empty space on the other hand." The problem Einstein sees is that the physicists consider the energy of light and other purely electromagnetic phenomena as continuous functions of space. On the other hand, the energy in ponderable bodies (he names atoms and electrons) is the sum of discrete entities, therefore not to be considered in arbitrarily small elements or in terms of continua. It was a problem of particle/field duality that Einstein did not solve in this paper or, for that matter, to the end of his life.

Indeed, while he was deeply attached to the continuum approach, he used the occasion of this paper to concentrate on the fact that the continuum approach leads to contradictions with experience when applied to phenomena involving the generation and transformation of light (that is, it leads to what later was termed the "ultraviolet catastrophe.") This meant importing the assumption of a discontinuous distribution of light energy into the realm that previously was thought to be covered by the continuist theory of Maxwell – a heuristic act that characterized theories that he considered transient and, though convenient, not fundamental.[16] Even Walter Nernst, an enthusiast for quantum theory, called it in 1911 still only a rule for calculation, "a very odd rule, one might even say a grotesque one." [17]

The heuristic character of the paper was assured also because, as Einstein said frankly, he would have to proceed "without putting at the foundation any picture about the generation and propagation of radiation." He therefore was forced to make the tested law of blackbody radiation the head of an axiom system, and deduce the consequences, finally coming to the photoelectric effect for which the paper is most remembered (announcing, in a courageous, almost off-hand way in two sentences results that should eventually be obtainable in the laboratory).

Nothing was said here explicitly about either the mechanical or the electromagnetic world view, but the paper carried the clear message that neither one nor the other can deal with the phenomena. The importation into electromagnetic theory of the mechanistic thema of discontinuity produced a mixture with which Einstein was never content, even when others had learned to accept it. Two years later, Einstein referred back to this paper to point out that it showed the "electromagnetic world picture" to be "unsuitable," although he had to add that "for the time being we do not possess a complete world picture corresponding to the relativity principle." [18]

The second great paper of 1905 is usually called the "Brownian movement" paper. In the recently finished paper on the quantum theory of light, he had put at the head his dismay over the apparent clash between two theories, based on continuity and discontinuity, respectively; in the Brownian motion work, he deals with and exploits an analogous clash. As he explained in the second part of that series: "According to classical thermodynamics which differentiates *principally* between heat and other forms of energy, spontaneous fluctuations do not take place; however, they do in the molecular theory of heat. In the following we want to investigate according to which laws those fluctuations must take place." [19] Thus the range of classical thermodynamics should be found to be limited, even in volumes large enough to be observable under the microscope. This leads to the prediction of the mean displacement of small objects as a function of time.

And then comes the third great paper of 1905, his first one on relativity theory. It has been amply discussed, and I shall add something about it in Chapter 4. Here I need underline only a few key points of relevance to the present purpose. His first section sets forth the total plan of the paper, and begins by drawing attention to a formal asymmetry in the calculations for determining the current generated during the relative motion between a magnet and a conductor – a phenomenon physicists had long regarded as adequately understood since Faraday's induction experiments of 1831. Einstein's paper does not invoke explicitly any of the several well-known experimental puzzles with which so many others were struggling, as in the interpretation of the ether-drift experiments – not even when the opportunity arises for him to show in what manner his relativity theory accounts for them.

Once more, two separate theories are confronted: classical mechanics, where Newton's laws are thought to hold and Galilean relativity to obtain, and on the other hand, electrodynamics, in which it was thought that a preferred reference system, the ether, existed. To deal with this dichotomy, Einstein proposes to raise a courageous conjecture (*Vermutung*), which he refers to as the principle of relativity, to an assumption or presupposition (*Voraussetzung*). It amounts to a generalization of Galilean–Newtonian relativity, with the result that when properly phrased, the laws of electrodynamics and optics *and* of mechanics have the same validity with respect to transformation between all inertial systems. In addition, Einstein makes another presupposition, that light in empty space is propagated with a definite velocity that is independent of the state of motion of the emitting body.

These two presuppositions suffice him to produce a theory of both electromagnetic and mechanical phenomena that is most parsimonious with regard to the number of assumptions, free of contradictions, and unburdened of long-standing puzzles. It reveals unexpected connections between previously separate concepts, and is ready to permeate every field of physics; for example, while Lorentz and others applied the transformation equations only to electromagnetic phenomena, Einstein had the ambition from the start to let them apply to all of physics. What has to be given up is the support of long and widely used concepts such as the luminiferous ether, absolute space, and absolute simultaneity – and also the old crutch of constructing theories inductively from the mass of present data and puzzling phenomena.

The postulational method and its progress

Einstein's previous work had prepared him to make the necessary sacrifices when adopting a more ecumenical and independent *Gesichtspunkt* with which to escape the clash existing between two different current theories. As our study has shown, he had learned not to trust any of the existing *Weltbilder*. He was initially reluctant even to present his work as a new theory; and after it had been so christened by Planck (1907) and others, he frequently referred to it in print as the "so-called relativity theory,"[20] and until 1911 avoided using the term "relativity theory" in the title of his papers. It is worth noting here that if he had chosen to give it a name, he would, in my opinion, have preferred to call it something like *Invariantentheorie*; the term appears in his correspondence, and is of course much more appropriate.[21]

But Einstein had developed by this time the modern version of the method by which one can hope to reach the general principles of nature. We saw it at work: It is a bold, postulational method, appropriate for dealing with what Einstein came to call "theories of principle" that have the large ambition of handling the "totality of sense experiences." At the top of the methodological hierarchy, Einstein puts a well-established experimental rule, recast in its most general form, and raises it to a postulate – or, if none is available to serve this function, as in the case of relativity theory, his own *Vermutungen* have to be put into this place. From this postulate system he then draws the logical deductions, and these in turn point to eventual tests against experience. Old puzzles disappear, and new questions come to the fore. As Einstein saw first perhaps when, at the age of twelve, the "holy booklet" of Euclidian geometry came into his hands, with the right axiom system at the top, the right consequences develop by necessity. All else is broken crockery.[22]

Insofar as his contemporaries understood what he was doing – and at least until 1911 most did not – it must have seemed a dangerous high-wire act, the contact with experience being supplied by only a very few carefully selected support ropes anchored in solid ground.[23] It is not surprising that few others cared to take such risks, that the progress of Einstein's style among his contemporaries was initially quite slow. And for all the effort, and at the cost of giving up so many established notions and of paradoxical by-products, what did one really get from Einstein's relativity that was not available in a more congenial way from the work of Lorentz, Poincaré, Abraham, and others? If one did not share Einstein's anguish over the limited scope, asymmetries, and ad hoc hypotheses in the other systems, there was nothing inescapable about accepting Einstein's relativity.

The contemporary reception of relativity shows all this clearly. Among established scientists there was really only one who appreciated Einstein's principle of relativity from the beginning: Max Planck in Berlin. He was at hand when that very unconventional paper was received at the *Annalen der Physik*,[24] and gave it valuable exposure and defense, after its September publication, in talks at the German Physical Society on March 23, 1906, and at the Stuttgart Congress of German Scientists and Physicians on September 19, 1906. And by that time Einstein needed a friend indeed; for there had just appeared, immediately upon publication of his own paper, what seemed to be an inescapable experimental disproof of his 1905 paper in the *Annalen der Physik*, by the eminent experimentalist Walter Kaufmann.[25] Lumping together, as was widely done for some years, the work of Lorentz and Einstein since their predictions coincided nearly enough, it announced in italics: "*The measurement results are not compatible with the Lorentz–Einsteinian fundamental assumption.*"

The effect of Kaufmann's announcement was instant and devastating on Lorentz, and also shook Poincaré.[26] Planck, however, showed in a patient analysis that Kaufmann's measurements were not unequivocally either against Einstein's system or for those of his rivals, for example that of Max Abraham.[27] (The latter, characterized by a rigid, undeformable electron, had for many the advantage of fitting in fully with the electromagnetic world picture.)

The published debate shows that nobody in the audience was converted by Planck to what was being called there "*die Relativtheorie*" (the baby still had only a nickname). When pressed to the wall, Planck had to confess that in the absence of persuasive proof or disproof, the Einsteinian postulate that no absolute translation could be found made the difference: "I find it more appealing [*Mir ist das eigentlich sympathischer*]." As Cornelius Lanczos said later, Planck, although a conservative, was caught by the beauty of Einstein's paper.

Three years later, when Wilhelm Wien was converted to what Planck called the small band of relativists, it was not because some crucial experiments left him no other choice, but mainly on quasi-esthetic grounds. Using words that Einstein must have appreciated, Wien said, "What speaks for it most of all is the inner consistency which makes it possible to lay a foundation having no self-contradictions, one that applies to the totality of physical appearances, although thereby the customary conceptions experience a transformation."[28]

Other adherents to the relativistic point of view had recourse to similar predilections. Thus Hermann Minkowski, in the same September 1908 meeting at Cologne at which he delivered his famous address on "Space and time," also attended a session that touched on the merits of Abraham's rigid-electron theory, then still being held up by many as the chief option, as against the theories of both Einstein and Lorentz. Minkowski exclaimed in the discussion period: "The rigid electron is in my view a monster in relation to Maxwell's Equations, whose innermost harmony is the relativity principle. Going at Maxwell's Equations with the idea of the rigid electron seems to me just like going into a concert after stopping up one's own ears with cotton wool."[29]

Lesser scientists had a great deal more trouble with Einstein's work, of course. No wonder that when Einstein submitted his 1905 relativity paper to the Physics Faculty of the University of Bern in 1907 as his *Habilitationsschrift* to gain the right to teach as *Privadozent*, it was rejected with the grade of "unsatisfactory." The professor of experimental physics returned Einstein's copy with the curt remark, "What you have written here, that I don't understand at all."

I have wondered of course how Einstein himself – a previously unknown young fellow without academic connections or credentials – reacted to the tumult caused in 1906 by Kaufmann's "disproof" and the claim of the rival theories. If he had been a naïve falsificationist, he should have eagerly embraced this evidence of "progress" in science, and gone on to other things. But in fact, for about two years he seems to have paid no attention to the Kaufmann publication that had stunned much more experienced scientists. I could find no reference to Kaufmann's attack in his publications or letters until, urged by Johannes Stark as editor of the *Jahrbuch der Radioaktivität und Elektronik*, he responded in a survey published early in 1908.[30]

In a manner that may well have seemed rather haughty, Einstein allowed that Kaufmann's calculations seemed to be free of blatant error, although he added he would wait for a greater variety of observational material before the experiment could be declared free of systematic errors. But in any case, Einstein had no evident worry about the outcome, because, as he put it, the theories of Abraham and Bucherer, which Kaufmann had claimed to be proving, "have a rather small probability, because their fundamental assumptions concerning the mass of moving electrons are not explainable in terms of theoretical systems which embrace a greater complex of phenomena." Even though the "experimental facts" seemed clearly to favor the theory of his opponents, Einstein found the limited scope and ad hoc character of their theories more significant and objectionable than the apparent disagreement between his own theory and the new results of experimental measurements. It is really a classic demonstration of the justified confidence a rare type of scientist can have in the soundness of what Einstein once called his "scientific instinct," assigning "probabilities" to theories on the basis of judgments made even in the face of seemingly contrary evidence.[31]

"My need to generalize," and its dissatisfactions

Eventually, Einstein's 1905 relativity paper was hailed everywhere as one of the chief historic advances of science. But, again, Einstein himself saw it differently. In his own evaluation he stressed the limitations. Thus in his Nobel Prize speech [32] he did admit that "the special relativity theory resulted in appreciable advances," and he gave a short list of these – significantly, first of all that "it reconciled mechanics and electrodynamics," the respective bases of the opposing *Weltbilder* of that age. Next, he agreed that the relativity theory "reduced the number of logically independent hypotheses. . . . It enforced the need for a clarification of the fundamental concepts in epistemological terms. It united the momentum and energy principle, and demonstrated the like nature of mass and energy." He could have added, as he had done at the 1909 Salzburg conference, that energy and mass appeared now to be equivalent magnitudes, as were heat and mechanical energy. And he could have gone on to show other resulting unifications, resulting from the discarding of old conceptual barriers that had been simply "unbearable" for him.[33]

But then Einstein, in his Nobel Prize talk, adds "Yet, it was not entirely satisfactory." The foremost dissatisfaction, of course, was that the special theory "favors certain states of motion – namely those of the inertial frames. . . . This was actually more difficult to tolerate than the preference for a single state of motion, as in the case of the theory of light with a stationary ether," for now there was not even an imagined reason for preference, namely, the light ether. "A theory which from the outset prefers no state of motion should appear more satisfactory." In his "Autobiographical notes" he added, "That the special theory of relativity is only the first step of a necessary development became completely clear to me only in my efforts to represent gravitation in the framework of this theory." This happened precisely in 1907. Even as the special theory of 1905 was very slowly gathering converts on its way toward the eventual public triumph, Einstein had already put it behind himself, and was hard at work on a generalized theory, driven by what he later called, in the letter of November 4, 1916, to W. de Sitter, "my need to generalize" [*mein Verallgemeinerungsbedürfnis*].

As is well known, Einstein was always far more interested in what remained to be done than in what he had accomplished. There was so much more that kept escaping the task of grand unification that increasingly stood before him as the real goal. It is one of the great ironies that the very extent and depth of the advances Einstein himself helped to launch – including his contributions to quantum theory – eventually made it impossible for the physical phenomena to be gathered in one grand, relativistic *Weltbild* in his own time. As we have seen, writing as early as 1907, he recognized that the task was far from being accomplished; while critical of both the mechanical and the electromagnetic world pictures throughout his life, he continued to long for a unified world picture, the *Einheitlichkeit des Weltbildes*.[34] But the goal was always eluding him, and in the last decade of his life, he looks ahead and asks wistfully "how the theoretical foundation of the future will appear. Will it be a field theory; will it be in essence a statistical theory?"[35]

The question has not been answered yet. Today we are surrounded by various competing, overlapping, largely unarticulated scientific world views. It reminds one of the remark of Edward Gibbon in *The decline and fall of the Roman Empire* (Chapter II, Section I): "The various modes of worship which prevailed in the Roman world were all considered by the people as equally true, by the philosophers as equally false, and by the magistrates as equally useful."

Parts of the legacy

A large portion of any physicist's approach to the study of nature today can be traced directly to the influence of Einstein's vision of an overarching relativistic world picture. No matter how far modern physical scientists regard themselves to have moved away from Einstein's own aims, they share at least three basic components with him.

The first is the hope of, and a method for, achieving progressively more unified theories that provide us with a sense of order by encompassing the immense range and variety of phenomena. As I have stressed, Einstein's postulational method for constructing the deep "theories of principle" came to the fore even in his early paper: On the basis of a few key phenomena and a sensitive interpretation of them, exert the scientific imagination to the utmost to formulate axiom systems of great generality. There follows the sober deduction of consequences, and then the tests against sense experience. If, perhaps after a long series of such cycles, a theoretical structure is attained that seems firm enough in one limited field, the process is further generalized to encompass larger and larger portions on the plane of experience. In Einstein's words,

> Thus the story goes on until we have reached a system of the greatest conceivable unity, and of the greatest paucity of concepts of the logical foundations which is still compatible with the properties of our sense observations. We do not know if this ambition will ever result in a definite system. If asked one's opinion, one is inclined to answer No. While wrestling with the problems, however, one is sustained by the hope that this highest goal can really be attainable to a very high degree.[36]

It is therefore clear that Einstein himself was destined, from the start, to go on from the special to the general theory of relativity, and then beyond it to a unified field theory.

The second component of Einstein's way of building a picture of the world lies in the choice of, and unshakable devotion to, the relatively few themata [37] which characterize a theory with "inner perfection." He knew well that these are not easily defended, or even easy always to specify; their "exact formulation . . . meets with great difficulties. . . . The problem here is not simply one of a kind of enumeration of the logically independent premises . . . but that of a kind of reciprocal weighing of incommensurable qualities," [38] hence a judgment into which presuppositions, esthetic consideration, and other preferences of that sort can and do enter prominently. (Thus John A. Wheeler has recorded that Einstein gave a talk at Princeton in which he made the point that "the laws of physics should be simple." Someone in the audience asked, "But what if they are not simple?" "Then I would not be interested in them.")

Many of these themata appear of course also in the physics of his predecessors and contemporaries; that is what assures the continuity of the total scientific enterprise. But this particular set, and the tenacity with which it was held, characterize Einstein's own style. Thus, adherence to these themata helps explain, in specific cases, why Einstein would obstinately continue his work in a given direction even when testing it against experience was difficult, discouraging, or not possible. It also explains the deep personal anguish he must have felt when other physicists whom he loved – Lorentz, Planck, above all Bohr – based some of their work on key presuppositions that were antithetical to his own. Most physicists today believe that he was not right in giving such loyalty to some of these thematic presuppositions. But on this, too, we do not know what the future holds.[39]

The third part of our Einsteinian legacy concerns what lies beyond the ability of mere humans to perceive regularity and necessity in nature. In later years, Einstein confessed that his progress on the road to relativity had tested not only his physical knowledge but his physical and psychological fortitude as well. It is a fitting comment on the success of his labors that today beginning physics students can enter easily into this once so fought-over field of knowledge. Moreover, they are hardly aware that they have inherited his grand goal, as well as so many conceptual and methodological tools for its pursuit – the goal of pushing toward the attainment of a synoptic and coherent overview of the vast sea of phenomena.

It is, however, again typical of Einstein that, in his role as scientist as well as humanist, he saw how the very progress already made brings us face to face with a deep puzzle. In a famous paragraph, he wrote in 1936: "It is a fact that the totality of sense experience is so constituted as to permit putting them in order by means of thinking – a fact which can only leave us astonished, but which we shall never comprehend. One can say: the eternally incomprehensible thing about the world is its comprehensibility."[40]

After the publication of this paragraph, Einstein received a plaintive letter from one of his oldest and best friends, Maurice Solovine. They had first met in Bern in 1902. Now, a half century later, Solovine had come upon this passage while translating it for a French edition of essays in Einstein's book *Mein Weltbild*. Solovine was worried. How could there be a puzzle about the understandability of our world? Was this not a dangerous notion to allow into science, mankind's most rational activity?

Einstein's reply to Solovine tried to set his mind at ease:

You find it remarkable that the comprehensibility of the world (insofar as we are justified to speak of it at all) seems to me a wonder or eternal secret. Now, *a priori* one should expect a chaotic world that can in no way be grasped by thinking. One could, even *should*, expect that the world turns out to be lawful only insofar as we intervene to provide order. It would be the sort of order like the alphabetic ordering of words of a language.

The kind of order which is provided, for example, by

Newton's Theory of Gravitation is of a quite different character. Even if the axioms of the theory are put forward by human agents, the success of such an enterprise does suppose a high degree of order in the objective world, which one has no justification whatever to expect *a priori*. Here lies the sense of "wonder" which increases ever more with the development of our knowledge. . . .

The nice thing is that we must be content with the acknowledgement of the wonder, without there being a legitimate way beyond it.[41]

At this end of Einstein's century, we can indeed be content with what he has done to increase not only our sense of rational order, but through it also our sense of wonder.

4

Einstein's search for the *Weltbild*

The concern for sharing the meaning of his findings across disciplinary boundaries was clear in Einstein's work. In addition to some 300 papers in the physical sciences, Einstein published widely in other fields, including philosophy of science and social analysis, and he also actively embraced his role as educator, humanist, and moral voice. This diversity has led Einstein's commentators to view him from two quite different perspectives.

One view is the familiar one of the physicist whose accomplishments are at the base of virtually every field of contemporary physical science. But even from this vantage point one must not allow the very magnitude and depth of Einstein's separate scientific contributions (e.g., in statistical mechanics, quantum physics, relativity, and cosmology) to obscure his adherence to a larger ambition in physical science which characterized the generation of scientist-philosophers of his time: the achievement of a coherent, complete, and unified scientific world picture. In his own words:

> The aim of science is, first, the conceptual comprehension
> and connection, as *complete* as possible, of the sense experi-
> ences in their full diversity, and, second, the accomplishment
> of this aim *by the use of a minimum of primary concepts and
> relations* (seeking the greatest possible logical unity in the
> world picture [*Streben nach möglichst logischer Einheitlichkeit
> des Weltbildes*], i.e., logical simplicity of its foundation).[1]
> (Emphases in original.)

The other view of Einstein is that of the person who quietly, and in full awareness of the overwhelming difficulties, hoped to achieve an even larger task – to construct for himself a still more generalized world picture, of such power that it would catch not only physical science but also all other fields of knowledge and behavior, the phenomena of life as such.

Here again, the volume of Einstein's writings supporting this effort is very large. Thus in a book of collected essays, to which his publisher felt justified to give the title *Mein Weltbild*, the five main sections were entitled in translation "How I see the world," "Politics and pacifism," "Germany, 1933," "Jewish problems," and "Scientific contributions."[2]

In Einstein's own life and work, these two searches, for the scientific *Weltbild* and for the more general one which encompassed much more than science, were conducted simultaneously and often converged. Of course, nobody will claim that a hard-and-fast correlation can be established between such elements. As the anthropologist Clifford Geertz has noted,[3] the very concepts "ethos and world view are vague and imprecise; they are a kind of prototheory, forerunners, it is to be hoped, of a more adequate analytical framework." Yet it is not difficult to show that Einstein held the method of gaining knowledge about scientific matters to be also appropriate for thinking one's way through virtually all problems, including those of ethical behavior.[4] For example, his constant fight on behalf of democracy and the sacredness of the individual, in a world hemmed in by totalitarianism and collectivist forces, appears to have been of a piece with his conviction that arbitrary boundaries, classes, or absolutes do not exist in nature anywhere; and his drive to deprovincialize and fuse the different parts of physics through a new synthesis is coherent with his attacks on tribalism or nationalism.

It was therefore rather natural that Einstein was perceived, even by those who understood nothing of his physics, to be a unifier of the most fundamental kind, whose total labors were on behalf of the construction of a world conception that would accommodate science, of course, but also issues of ethics, religion, social institutions, and personal conduct.[5]

While Einstein's double role caused his labors to be seen to belong both to the "cosmos of nature" and to the "cosmos of culture," in Erwin Panofsky's useful distinction,[6] it is true that Einstein himself was frequently puzzled by the public perception of him, as if he did not fully appreciate the wide impact of the demonstration he provided for "how the efforts of a life hang together," to use Einstein's own phrase.[7] When he arrived for his first visit to the United States in April 1921 – to help raise funds for a new university in Jerusalem, to help reopen relations with scholars after the recent war, and to respond to the interest in his work in physics – he was startled to see the large and enthusiastic crowds that had come to meet the ship in New York. A news conference was improvised,[8] and one of the questions was, inevitably, what he thought to be the cause of the mass enthusiasm. Einstein's answer was that "it seemed psychopathological." When the journalist proposed that the wide attention came about because Einstein's theory "seemed to give a new description of the universe . . . the subject of the most fascinating speculations of the mind," Einstein replied that this was possible, but that the essence of his theory was chiefly "the logical simplicity with which it explained apparently conflicting facts in the operation of natural law." It freed science of the burden of "many general assumptions of a complicated nature." But why should this interest the man in the street?

If the journalist replied, it was not recorded. But he might well have answered as follows. The heart of relativity, as Einstein noted, is indeed the discovery of far greater simplicity at the foundations than had been suspected, resulting in a unification of previously separate notions. But precisely for that reason the theory exposes a more rational and harmonious universe than could be known before. That itself, and even a glimpse of it at a distance, was a prospect immensely pleasing and encouraging to the human spirit – then, as now.

Competing world pictures

In preparation for tracing Einstein's attempt to construct his scientific *Weltbild*, we must consider at least briefly the state of science when he began his researches. At the time he wrote his first paper at the age of 21 in 1900, most physicists considered the condition of science to be very good. Some, such as Kelvin and Poincaré, discerned clouds; but the predominant feeling was that of satisfaction, even enthusiasm, with the momentum and direction in which the work was going.

There were in fact four types of success. First, there were great victories in theoretical physics that were now being exploited. Maxwell's theories were well established, yielding a synthesis of electricity, magnetism, and optics. The ether, universally thought to be the medium that carried electromagnetic waves such as light, was a subject of great research interest. As J. J. Thomson said as late as 1909, "The ether is not a fantastic creation of the speculative philosopher; it is as essential to us as the air we breathe. . . . The study of this all-pervading substance is perhaps the most fascinating and important duty of the physicists."[9]

Second, new experimental phenomena had recently presented themselves like so many gold mines – X-rays in 1895, radioactivity in 1896. The Curies, Rutherford, and many others were discovering things unimagined a few years earlier. In 1897, Thomson had found the long-sought electron; with it, the ancient question of what matter really consists of seemed ready for speedy resolution. A. A. Michelson was an artist in the design of equipment of such precision that one could hope to discover whole areas of physics research in the newly accessible decimal places of data. Yet another authority of that sort was Walter Kaufmann of Göttingen, who was measuring how beams of high-speed electrons were being deflected by electric and magnetic fields. New findings chased one another at a dizzy pace. The prolific Ernest Rutherford wrote to his mother in 1902: "I have to keep going, as there are always people on my track. I have to publish my present work as rapidly as possible in order to keep up the race." After publishing his comprehensive survey *Radioactivity* in 1904, Rutherford had to bring out a fully revised second edition the very next year, apologizing in the preface that the flood of new researches had made it necessary.

Third, there was now also an ambitious program for physics: to find relations between the phenomena in mechanics, electricity and magnetism, the theory of matter, and the ether. And fourth, with these somewhat overlapping victories of theory, experiments, and program, there went a preferred method of research. That method, in brief, was to focus in a given field on discrepancies between the best current theory and the new experimental results, and then modify the otherwise satisfactory theory to remove the discrepancies. When, as often happened, there was no other way, the theory was modified by an ad hoc amendment, invented for that purpose in the hope that subsequently a more elegant version of the theory could be formulated. This was on the whole an inductive process, in close touch with experience, and appropriate for building "constructive theories" (to use Einstein's later terminology), rather than the more fundamental "principle theories."[10]

The most astute craftsman of that period was the great Dutch theoretical physicist H. A. Lorentz, with whom Einstein eventually came to form a very close friendship (even though Lorentz never accepted Einstein's relativity fully). An exemplar of the best work of that time is a paper Lorentz published in 1904, in which the subject matter overlaps considerably with Einstein's paper of the following year, although Einstein had not read it at that time. Let us look briefly at the style and sequence of arguments in Lorentz's paper.[11]

He is concerned with determining the influence exerted on electric and optical phenomena – for example, on the measured speed c for the propagation of light – owing to the motion through that great ether sea by laboratory systems, such as the earth itself when it moves with a speed v while such measurements are made. In the early pages of his 1904 paper, Lorentz begins by noting a deficiency of current theory: it admitted of a comparatively simple understanding of phenomena that involved the ratio of speeds v/c – but not so of v^2/c^2, an experimental region attainable with newer, high-precision methods such as that opened by Michelson's interferometer. Michelson's failure to find the small but definite, expected effect of the earth's motion on the measured speed of light had led Lorentz and Fitzgerald to propose removing the discrepancy between theory and experiment by adopting the special hypothesis that the dimensions of all bodies, including the measuring equipment, are affected by their motion through the ether in just such a way as to annul the expected measurement – by what Lorentz himself later called "a fortuitous compensation of opposing effects."

In the next paragraph, Lorentz cites recent experiments by Rayleigh and Brace which, if the Lorentz–Fitzgerald contraction really existed, should have shown the existence of double refraction. But these physicists have obtained a negative result. To make matters worse, Trouton and Noble, in an experiment in which the theory of the electron as then understood would have required a rotation of a charged condensor owing to its motion through the ether, have also failed to find that effect. Moreover, the French physicist Henri Poincaré has objected to the method of dealing with Michelson's negative result by the introduction of a convenient new hypothesis, because the same need may occur each time new facts will be brought to light. Lorentz agrees that this course of inventing special hypotheses for each new experimental result is somewhat artificial, and he promises to take a more fundamental approach in this paper.

In fact, however, he still must introduce explicitly or implicitly a whole stream of special assumptions. Moreover, the results are not what he hoped for. As one puts it now, Lorentz did not attain exact covariance. In his modified theory, Maxwell's equations are not completely invariant, even at low speeds. Still, the thrust of his work is promising. In the last part of the paper he compares his theory with new measurements, this time Kaufmann's on the path of electron beams in electric or magnetic fields, and Lorentz expresses his pleasure that those data indicate "we may expect a satisfactory agreement with my formulae."

The very least one must say about this approach to advancing the state of physics is that it was really quite adequate for forging ahead step by step in this difficult world without looking for basic revisions; and in the following year Poincaré showed how one could improve Lorentz's theory while continuing to hold on to the idea of the ether. As a matter of fact, if it had not been for Einstein and his demonstration that there is another more fundamental way to think about nature, one can well imagine that scientists could have continued to live with such a physics for a long time.

But at a deeper level of thinking about the physics of the time, there was a flaw that concerned scientists, and none more than Einstein. In fact, the temple of science, being built through the victories of theory, experiment, program, and method, was resting on challengeable foundations. Decades earlier, a stark disagreement about the ultimate objective of this whole work had surfaced, a thematic disagreement about the very direction in which the arrow of explanation was pointing. To summarize very briefly an episode in intellectual history which has been discussed by a number of scholars,[12] by 1900 there were major, very different conceptions of what the final structure would look like.

In the German physics literature of the time, the chief contenders were called the "*mechanische Weltbild*" and the "*elektromagnetische Weltbild*," respectively. The first of these was a descendant of the venerable mechanical world conception, symbolized by the grand clock, or Newtonian machine. In this view, all phenomena would be understood if they were modeled on mechanics. It accepted such concepts as inertia as axiomatic, and ether as a mechanical medium of which perhaps electrons and other components of matter are made. As Newton wrote in the preface to the first edition of the *Principia* – flushed with his successes in explaining the motion of the planets, the moon, and the sea by the action of forces on material bodies – "I wish we could derive the rest of the phenomena of Nature by the same kind of reasoning from mechanical principles." Newton had hoped that the same approach would explain chemical and other properties of matter, perhaps even sense perception. For him and his disciples, the arrow of scientific explanation was launched from mechanics. As late as 1894, Heinrich Hertz noted: "All physicists are unanimous that the task of physics is to trace the natural phenomena back to the simplest laws of mechanics."

The second approach, very lively at the time Einstein began his work, was the electromagnetic world picture, which held that explanation must proceed in the opposite direction, namely from electricity to mechanics. The inertia of ordinary objects and of the electron, for example, was now regarded not as a tool of explanation but rather as the puzzle to be explained – perhaps by thinking of matter, in which electric charges are always dispersed, as interacting with the ether through which it moves, giving rise to the inertial effect by a kind of self-induction. Speaking in 1901, Walter Kaufmann explained with enthusiasm:

> Instead of all the fruitless attempts to explain electric phenomena mechanically, can we not try conversely to reduce mechanics to electrical processes? . . . When all material atoms are conglomerates of electrons [as J. J. Thomson had proposed], their inertia follows naturally. [Moreover] to explain gravity, we only have to assume that the attraction between unlike charges is somewhat greater than the repulsion between like charges.[13]

A third, less widely elaborated point of view was identified with Wilhelm Ostwald and, to some degree, with Ernst Mach. Here, the fundamental tool of explanation was energy, and the approach to phenomena was as unhypothetical as possible. This world picture, identifiable by the term *energetics-phenomenology*, cautioned against going too far from direct sensations and observations, and saw the physicist's task as chiefly a correlation of sense experiences which yields Keplerian empirical laws, to serve as the economical expressions of observed regularities. This world conception was to a large degree the product of the victories of monism over its metaphysical opponents in the second half of the nineteenth century, and hence represented ambitions that went far beyond the purely scientific *Weltbild*.

Einstein's critique

We will leave to another occasion a study of the origins and role of the powerful notion of scientific *Weltbild*, particularly among German scientists who attempted to construct an image of reality in which, to quote Max Weber, "events are not just there and happen, but they have meaning, and happen because of that meaning." For the purpose of this paper, it is more appropriate to record some of Einstein's own explicit and characteristic critiques of the world pictures reigning at the time he began his work.

In his essay "Ether and relativity theory" (1920),[14] Einstein gave his view of the historic development of nineteenth-century physics that had failed to produce a "mechanical model of the ether" which would yield a "satisfactory mechanical interpretation of Maxwell's laws of the electromagnetic field." Therefore the physicists of the time gave up their "long-ing for a closed theory" in which the fundamental conceptions came exclusively from mechanics (e.g., mass density, velocity, deformation, forces of compression), and "by and by became used to admitting also electric and magnetic field intensities as fundamental conceptions, along-side the mechanical ones." In this manner, "the purely mechanistic view of nature was given up."

But this was not done without paying a price: "This development led, however, to a dualism at the very foundations which, in the long run, was unbearable. To escape from it, one now tried the reverse tack, reducing the mechanical fundamental concepts to electrical ones," the more so as the experiments on high-velocity electron beams "shook the confidence in the strict validity of the mechanical equations of Newton." That pro-gram too remained unsuccessful. Hertz's theory ascribed to matter and ether mechanical and electrical states that had no coherence with each other. Lorentz's theory removed from matter the electromagnetic quali-ties and from ether the mechanical ones (except for the ether's immo-bility, a mechanical conception that was to be removed by special rela-tivity theory); but this left physics with the conception of a preferred coordinate system, the one at rest with respect to the ether, whereas the phenomena do not exhibit such a preference. "Such asymmetry in the theoretical edifice, to which there corresponds no asymmetry in experi-ence, is for the theoretician unbearable." We glimpse here some of the motivational drive for the construction of a theory of a particular sort that permits an escape from alternative theories that are "unbearable," even if not necessarily disconfirmed empirically.

Writing his "Autobiographical notes" about twenty-seven years later, in 1947, Einstein returns to a critique of the "*mechanische Weltbild*" of physics before relativity theory,[15] above all of its "dogmatic rigidity." The attitude that God had created the Newtonian laws of motion together with the necessary masses and forces could take the scientists just so far. The attempts to base electromagnetism on this theoretical structure were doomed to failure.

But before going into details, Einstein interrupts his critique and asks on what basis one can criticize any physical theory at all. As we noted in Chapter 2, in reply he puts forth two criteria. The first is the requirement that the theory must not be falsified by the empirical facts. He calls this a principle of "external validation." But a second criterion is also necessary, in part because one can almost always adjust a failing theory ad hoc, "by means of artificial additional assumptions" that do not go to the heart of the matter. He calls the second criterion one of "inner perfection," while admitting he can only "vaguely characterize" it as a requirement for "naturalness" or "logical simplicity" of the premises.

With these criteria in mind, Einstein returns to his critique. The mechanistic theories failed on the first criterion; for example, they lacked sound mechanical models to explain optical effects. But more important still were the failures with respect to the second criterion by which quasi-esthetic considerations come into play. On that score, Einstein lists four specific dissatisfactions.

It was particularly offensive (*besonders hässlich*) that the inertial systems in Newtonian mechanics are not only infinite in number but also, each of them, specially distinguished over all accelerated systems. Second, in the definition of force or potential energy there was no inherent necessity, but rather a great deal of latitude. Third, there was an internal asymmetry in the theory, for the concept mass appears in Newton's law of motion and in the law of gravitational force, but not in expressions of, say, the electric force. And lastly, Einstein found it "unnatural" that there are two kinds of energy, potential (field dependent) and kinetic (dependent on motion of particles).

Although on that occasion Einstein gave his detailed objections to the mechanistic world picture, he did not let the electromagnetic world picture escape lightly. He recalled that Lorentz's theory of the electron, based on Maxwell's equations, did not explain the stability of the electric charge constituting the particle. Some nonelectromagnetic force had to be imagined to hold the electron together against its own tendency to disperse explosively. Einstein might have added that as early as his own first paper on the quantum theory of light in 1905, he had made clear his dissatisfaction with the purely electromagnetic viewpoint: the phenomena involving the generation and transformation of light (e.g., the photoelectric effect) forced one to import the assumption of discontinuous, finite quanta of light energy – and hence of a mechanistic thema – into the realm thought to be covered by the continuist electromagnetic theory of Maxwell.[16] Two years later,[17] he reminded his readers that he had "shown that our present-day electromagnetic *Weltbild* is not suited" to explain the phenomena.

Einstein's first relativity paper

When we now look at Einstein's first few publications, we are struck at once by how little interest he shows in the most widely celebrated new victories of the time, or in the startling experiments, or in the research program, or in the reigning method, or in the current world pictures of physics. Instead, he sets out to do physics in his own way.

In Chapter 3 we had noted two things of importance here. One is an early glimpse of his motivation, as expressed in his letter[18] to Grossmann, "It is a magnificent feeling to recognize the unity [*Einheitlichkeit*] of a complex of phenomena that to direct observation appear to be quite separate things." If one carefully studies his papers prior to the relativity theory of 1905, which cover very different fields, one sees again and again *this drive to find common ground between apparently different, well developed fields*. In a letter to the astronomer W. de Sitter, he later called it "my need to generalize" (*mein Verallgemeinerungsbedürfnis*). To generalize, and thereby to unify – that is to be his hallmark.

In his third paper,[19] he phrased his main conclusion as follows: "The second law thus appears as a necessary consequence of the mechanical world picture" – using the phrase *Weltbild* in print for the first time. By 1907 Einstein will be ready to indicate publicly his hope of developing his own, relativistic world picture, or as he puts it, "a complete *Weltbild* which corresponds to the relativity principle."[20]

But we are getting ahead of our story. Let us first look at some crucial early passages of Einstein's 1905 paper in which he transformed the notions of space and time.[21] This paper, now among the most renowned in the history of science, actually did not deal with any of the problems of greatest concern to the physicists of the time; there is not a word about the theory of matter, the nature of the electron, or the properties of the ether. Even in the space of a few paragraphs one can perceive the novelty of his mind, and can see that the true subject matter is the unification, the simplification, the rationalization of the physical world.

The title in translation is "On the electrodynamics of moving bodies" – relativity theory was a name given by others. His first sentence is a curious complaint: "It is known that Maxwell's electrodynamics, as usually understood at the present time, when applied to moving bodies, leads to asymmetries which do not appear to be inherent in the phenomena." Einstein is referring not to the content of the theory but to the form, not to a discrepancy between established theory and some new experiments, as Lorentz had done, but to an old expectation in the minds of physicists of asymmetries, whereas the phenomena do not have those asymmetries.

Specifically, when a current is induced in a conductor while a magnet is in motion with respect to it, the observed current is found to be the same whether the conductor is held stationary and the magnet moved, or the conductor moved and the magnet held stationary. It is only the relative velocity that counts. But the equations physicists used to calculate the current, derived from Maxwell's theory, looked very different for these two arrangements. To Einstein, this indicates the need to take a more general point of view that should allow one to regard the two cases as the same in theory, since they result in the same effect in practice.

But by that early point in the paper, Einstein already may have lost the attention of most of his audience, because he was discussing an experiment, done over seventy years earlier by Faraday, that every physicist knew and few suspected of having further research interest. Of course there were two different equations for calculating the induced current; but they worked, each in its own way. No major physicist was concerned about such "asymmetries" or had voiced this quasi-esthetic dismay before, and there seemed little to be gained by that still-almost-unknown author's idiosyncratic longing for symmetry.

It was only later, largely as a result of Einstein's work, that modern physics found symmetry, invariance, and covariance, three completely intertwined concepts, to be among the most basic tools of thought. (The theory of relativity can be characterized as a theory of symmetry, prescribing the covariance of the laws of nature with respect to the group of space-time transformations that distinguish different frames of reference from which the laws may be described.) I think it was lucky for Einstein and for us that Max Planck, the only physicist of stature to appreciate Einstein's relativity from the beginning, happened to be on hand as an editor of the *Annalen der Physik* when Einstein submitted his paper there for publication.

The next few lines on Einstein's first page continue to be surprising. In proper translation:

> Examples of a similar sort [they are unidentified], together
> with the unsuccessful attempts to ascertain a motion of the
> earth relative to the "light medium" [lumping together all the
> experiments known at the time, anonymously], lead to the
> conjecture that to the concept of absolute rest there correspond
> no properties of the phenomena, not only in mechanics [as long
> known] but also in electrodynamics. . . .

It would take more space than we have to unravel this half sentence fully. The main points are these. The Faraday electromagnetic induction experiment had shown that only *relative* motion, not absolute motion, determines the observed phenomena. The many ether drift experiments – there were now seven, most of them optical ones – which had attracted the fascinated attention of physicists for two decades showed that the motion of the earth with respect to the ether could not be found. Hence, Einstein proposed that the concept of absolute rest or absolute motion has no operational meaning here too. Leaving aside for a moment electricity, magnetism, and optics, and turning to Newtonian mechanics, we recall that there also, much to the regret of Newton himself, absolute motion could not be detected. Ernst Mach, in the *Science of mechanics* that had so impressed Einstein at the age of eighteen, had devoted his most devastating passages to what Mach called the "conceptual monstrosity of absolute space . . . a thought-thing which cannot be pointed to in experience."

At this point we see Einstein's drive toward generalization and unification in action: Mechanics has long accepted the fact that the concept of absolute motion has no meaning. Why not extend this to the other fields of physics? Indeed, Einstein goes on, the examples from electrodynamics and optics suggest "that the same laws [if we phrased them properly] will hold for the phenomena of electrodynamics and of optics, so long as the frames of reference are those in which the equations of mechanics hold good. . . ." Such frames of reference are called inertial systems; they are systems in motion with any constant velocity with respect to the fixed stars, which includes to a sufficient degree of accuracy all actual laboratories. To put it differently: since nobody has been able to find the absolute motion of an inertial system by any test involving electrodynamic and optical phenomena, and since the current theories of electrodynamics therefore have to be adjusted in an embarrassing, ad hoc manner to explain this inability, we should perhaps regard this development as a sign that in the end electrodynamics and optics are not so different from mechanics. Einstein seems to be saying: Why should God have made a world with an asymmetry between the subfields of physics? After all, the apparent boundaries between these subfields were set not by nature but by the historic process of the development of scientific ideas.

To be sure, all this is frankly confessed to be only a "conjecture." If Einstein were developing a "constructive theory," one built up inductively from phenomena, as so many others were doing, he would have had to stop at that point. But as he said later in his "Autobiographical notes," "By and by I despaired of the possibility of discovering the true laws by means of constructive efforts based on known facts. The longer and the more despairingly I tried, the more I came to the conviction that only the discovery of a universal formal principle could lead us to assured results." [22] Here we have reached that part of Einstein's method of theory construction which is the very opposite of most of his contemporaries. Unlike their inductive arguments based on current experiments, he proposes a great leap to universal principles, far beyond the level of the phenomena that first engaged his attention. Without apology he says, "We will raise this conjecture, the purport of which shall hereafter be called the 'principle [not theory] of relativity,' to the status of a postulate. `. . ." Thus, Galilean–Newtonian relativity, on which mechanics had long depended, will be extended. The laws of electrodynamics and therefore of light propagation will be restated so that they too, by their very formulation, give the same results when used in any inertial system. The transformation equations, formerly applied by Lorentz and others to electromagnetic phenomena, are to be used in all parts of physics. [23] Previously separate concepts will be joined; the ancient wall between mechanics and the rest of physics has been breached; and the question of whether the mechanistic, the electrodynamic, or the energetics world picture is preeminent is axiomatically asserted to be finally meaningless. The subfields of physics are now on an equal footing; there is no reason why one of them should explain the others.

But I must not linger, because I am only halfway through the sentence. Einstein adds, without even a comma, that he will also introduce another postulate, seemingly out of thin air, for now he does not even refer to anonymous phenomena. That other postulate, which he says is "apparently irreconcilable" with the first postulate (but only apparently, if one pays proper attention), is that light is always propagated in empty space with a definite velocity that is independent of the state of motion of the emitting body. Those great experimenters who had spent years trying to find how the measured light velocity might change owing to the earth's motion are being told, in this half-sentence, that their maddening failure was only to be expected as a matter of course. It is as if Einstein had followed Goethe's advice to Zelter: "The greatest art in theoretical and practical life consists in changing the *problem* into a *postulate*; that way one succeeds." [24]

Assuming these two postulates, [25] Einstein announces (and later demonstrates) that one can construct deductively a simple and consistent theoretical scheme for dealing with the phenomena of electrodynamics, most parsimonious in its assumptions, ready to permeate every branch of physics, and relieved of long-standing puzzles. Indeed, "the introduction of a 'luminiferous ether' will prove to be superfluous." The beloved ether, the flower of nineteenth-century physics! For many, this unceremonious dismissal was not only unbelievable, it was unforgivable. Einstein does not even bother to show explicitly that, without introducing any further fundamental postulate or assumption, the Lorentz–Fitzgerald contraction, introduced as an ad hoc device in the previous decade, follows as a simple deduction, and that a host of problems that beset electrodynamics has simply disappeared in his formulation. They are not even solved; they just turn out to be nonproblems, the cost of having had the wrong point of view.

Having stated his audacious postulational method,[26] Einstein suddenly changes his tune. In preparation for the new definition of simultaneity, he adopts an instrumental approach as he goes over elementary kinematics in careful detail. If we want to describe the motion of a material point, we must give the value of its coordinates as a function of time. Therefore, "we must understand clearly what one means by 'time.' We have to take into account the fact that all our judgements in which time plays a part are always judgements of *simultaneous events*." To elucidate this, he writes what one of his commentators called the simplest of all sentences in the *Annalen der Physik*: "If for instance I say: 'that train arrives here at seven o'clock,' I mean something like this: 'the pointing of the small hand of my watch to seven and the arrival of the train are simultaneous events.' "

He is in fact saying three important things. First, the time of an event is given by the reading on a clock that is fixed at the place of the event. Time is therefore localized, given separately at each point in space, rather than distributed throughout space in some disembodied way as absolute time by itself was thought to be. Time is an operational concept – as Einstein may have first glimpsed in his reading of Hume's *Treatise of human nature* and Mach's *Science of mechanics*.[27] Einstein is also saying that identifying the occurrence of an event joins time and space measurements; hence he talks about the coincidence of the hand of the watch and the front of the train, coming together at one point and one instant. (Minkowski will later call it the intersection of two World lines in space-time.) And third, he stresses the word "event" (*Ereignis*), which appears eleven times on that page and the next; he introduces a more neutral way of thinking about phenomena, which moves physics away from the old conceptual tools and the controversies that had come to adhere to them. He is discussing coincidences taking place at measuring equipment, not the structure of matter or the pressure of the ether on electrons.

One consequence of that "event"-way of thinking was that events thought to happen at the same instant at different places, that is, simultaneously, when observed in one coordinate system, turned out not to be necessarily simultaneous when observed from another. Hence there was no absolute simultaneity. Another consequence was that this *instrumental* aspect of Einstein's thought strongly appealed to the positivists. In fact, Einstein soon found his name on their banners. But this was their oversimplification. As we have seen, his was a *dual* method in the construction of the deep "principle theories"; for it included also that other element, the courageous postulation beyond induction, calling upon what Einstein frankly referred to as his "intuition," "scientific instinct," or "creative act" – the Platonic leap from a few uncannily chosen, well-established experimental facts to conjecture to postulate. These two aspects of his method acted like the keel and the sail of a boat. Each by itself would not have sufficed to carry him safely forward; together, they defined a style for advancing which helped shape the work in science in this century.

Some lessons

What were some of the lessons that Einstein's early work taught? First, that Einstein was not interested in easy victories, and dared to take great intellectual risks. At that time in the history of physics, almost any bright person working on what was perceived to be the frontier problems, such as the properties of X-rays or radioactivity or the electron, was likely to produce something novel. Instead, Einstein put to himself much harder questions. As he later remarked to his assistant Ernst Straus, "What really interests me is whether God had any choice in the creation of the world." This meant dispensing with everything that lacked the stamp of necessity. It meant suspecting and removing the barriers with which others had become comfortable – precisely the style of other figures that have played the same kind of cultural role: Copernicus giving up differences in the state of motion of the earth and the other planets, Galileo and Newton synthesizing terrestrial and celestial physics, Darwin stressing the continuity of *Homo sapiens* with other life forms, and Freud the psychological continuity of the child and the adult person and of the conscious and unconscious mind.

A second lesson is that this young outsider took his ideas seriously. As noted in Chapter 3, this was tested quickly and in a striking way, for immediately upon the publication of Einstein's 1905 relativity paper there appeared in the *Annalen der Physik* an experimental test by the eminent experimentalist Walter Kaufmann, showing that his results appeared to contradict Einstein's theory. If Einstein had been a naïve believer in the strategy of falsification, he might have accepted this disproof from a source of highest reputation, and gone on to other things; of course he did not.

The certainty with which this young man felt he was unpuzzling the design of the Creator becomes the more remarkable if one follows the story of Kaufmann's experiment further: It took ten years, to 1916, for it to be fully realized that, most surprisingly, Kaufmann's apparatus had been inadequate; apparently there was a leak in the vacuum system, which changed the effective fields available for deflecting the electron beam. By that time, the matter had been settled on other grounds.

The outcome of the search

The third lesson naturally concerns the fate of Einstein's primary ambition. We recall that it was, in his words, to fashion "as complete as possible" a scientific world picture, "seeking the greatest possible logical unity in the world picture."[28] The words *complete* and *unified* (*vollständig* and *einheitlich*) are among the most frequent ones to appear in Einstein's many essays in which he explains his view of science, as well as in his scientific correspondence. Again and again he declared his interest in the construction of that unified world conception, "*das einheitliche Weltbild*," which would find the necessary base of all natural phenomena and at the same time unify the separate, compartmentalized sciences.[29] As early as 1918 he confessed that he saw it as a task confronting every person: "Man seeks to form for himself . . . a simplified world picture that permits an overview [*übersichtliches Bild der Welt*]." Specifically, the world view of the theoretical physicist

> deserves its proud name *Weltbild*, because the general laws upon which the conceptual structure of theoretical physics is based can assert the claim that they are valid for any natural event whatsoever. . . . The supreme task of the physicist is therefore to seek those most universal elementary laws from which, by pure deduction, the *Weltbild* may be achieved.[30]

There is of course no doubt that the special relativity theory of 1905 and its further development during the next few years constituted great progress toward this "supreme task." The Newtonian world and the Maxwellian one were now on the same level, with the relativity principle the tool for deprovincializing and uniting them into a more general entity, in which previously separate subfields found their natural place. As we noted in Chapter 3, in his 1923 lecture in Sweden in lieu of a Nobel Prize address, which he devoted primarily to relativity theory, Einstein gave a list of the "appreciable advances" resulting from the special relativity theory,[31] and significantly put first that "it reconciled mechanics and electrodynamics," the respective bases of the opposing *Weltbilder* of that age. Next, he agreed that the theory "reduced the number of logically independent hypotheses. . . . It enforced the need for a clarification of the fundamental concepts in epistemological terms. It united the momentum and energy principle, and demonstrated the like nature of mass and energy."

He could have continued with a long list of unifications achieved. Within electrodynamics, electrical and magnetic phenomena could now be considered essentially the same thing, viewed from different reference systems. The old conceptions of space and time were now shorn of their absolute character and became subsets of space-time. With the disappearance of the notion of simultaneity of distant events as an absolute, all phenomena now were to be conceived of as propagated by continuous functions in space. In the developing relativistic *Weltbild*, a huge portion of the world of events and processes was subsumed in a four-dimensional structure[32] which Minkowski in 1908 christened "*die Welt*" – a static representation of all past and future events in which the main themata are those of constancy and invariance, determinism, necessity, and completeness.

Above all, the relativity theory acted like a stern filter. Instead of the requirement to adhere only to mechanistic or electromagnetic or energeticist fundamental concepts, as demanded by the previous *Weltbilder*, relativity acted as a selection rule on the form and type of laws of nature itself. As Einstein himself put it: "The whole content of the special theory of relativity is included in the postulate: The laws of nature are invariant with respect to the Lorentz transformations. The importance of this requirement lies in the fact that it limits the possible natural laws in a definite manner."[33] Reading the physics literature of the time, one can watch how, after an initial period of neglecting Einstein's point of view, the majority of German physicists became aware of the limitations of the previously competing world pictures. After the publication of the first serious textbook on relativity theory, by Max von Laue in 1911, the outcome was not in doubt, even if the average physicist was satisfied with using Einstein's conceptual tools for solving relatively narrow problems rather than adopting explicitly his grand program.

Eventually, Einstein's 1905 relativity paper was of course hailed everywhere as one of the chief historic advances of science. But typically, Einstein himself knew, and from an early point on recorded frequently, the *limitations* of his work. The initial solution was quite incomplete because it applied only to inertial systems, and left out the great puzzle of gravitation.

Then, too, there were obstinate questions: What is really the nature of the quantum of light? What did the equality of an object's inertial and gravitational mass portend? And do we have to be left with

> two kinds of physical things, i.e., one, measuring rods and clocks, two, all other things, e.g., the electromagnetic field, a material point, etc. This, in a certain sense, is inconsistent; strictly speaking, measuring rods and clocks would have to be represented as solutions of the basic equation . . . not, as it were, as theoretical self-sufficient entities.

This he called a "sin" which "one must not legalize." [34] The removal of the sin, however, would have to await the perfection of the total program, the achievement of a unified field theory in which "the particles themselves would *everywhere* be described as singularity-free solutions of the complete field-equations. Only then would the general theory of relativity be a *complete* theory." [35] Therefore, the work of finding those most general elementary laws from which by pure deduction a single, consistent, and complete *Weltbild* can be won, had to continue.

This striving – constant, and in the face of decades of disappointment, even heroic – characterized Einstein from beginning to end. As we saw, it showed up even in his first published paper and in his letter of April 14, 1901, to Marcel Grossmann, although Einstein himself surely could not have been aware then where his longing "to recognize the unity" would take him. In retrospect, it is tempting to believe that Einstein's life-program was set early, and that his special relativity theory had to give way to the search for the "generalized" (*verallgemeinerte*) relativity theory, and then to an effort at formulating the unified field theory. For "the idea that there exist two structures of space independent of each other, namely the metric-gravitational and the electromagnetic, was intolerable to the theoretical spirit"; hence, "we are driven to the conviction that both sorts of field must correspond to a unified structure of space."[36]

The generalization of special relativity yielded the demand that the equations expressing the laws of nature must be covariant with respect to all continuous transformations of the coordinates, and this provided an even narrower filter for the laws of nature:[37] "The principle of general relativity imposes exceedingly strong restrictions on the theoretical possibilities. Without this restrictive principle it would be practically impossible for anybody to hit on the gravitational equations." Moreover, Einstein did not "see any reason to assume that the heuristic significance of the principle of general relativity is restricted to gravitation and that the rest of physics can be dealt with separately on the basis of special relativity, with the hope that later on the whole may be fitted consistently into a general relativistic scheme." He was repeating here, near the end of his life, the dream that had kept him at work for many decades – to bring together, as he put it in 1920, "the gravitational field and the electromagnetic field, into a unified edifice," leaving "the whole physics" as a "closed system of thought."[38]

Threats to Einstein's unified world picture

Yet even as Einstein was hailed from about 1920 on as the very exemplar of what the human mind is capable of in science, he became more and more aware of what remained to be done than of what he had accomplished. Occasionally, Einstein thought he saw the end of his road on the horizon, close enough to reach it; but such periods lasted only briefly. His letters and essays document his growing realization that the program of unification itself might after all remain unreachable. By 1931, in an essay celebrating the centenary of Maxwell's birth,[39] he acknowledged the profound change in the "conception of the nature of the physically real" had brought us to a certain impasse. Before Maxwell, the representation of the processes of nature was based on the conception of

material points whose changes consist only in motions that are subject to total differential equations. After Maxwell, one conceived of the physically real as represented by continuous fields, not mechanically explicable, subject to partial differential equations. But while this change in the conception of the real is

the deepest and most fruitful that physics has experienced since Newton, one has to admit that the full realization of the programmatic idea has by no means been carried out so far.
Rather, the successful physical systems which have been constructed since then represent compromises between these two programs, which, despite great progress in certain particulars, and just because of the character of compromise, carry the stamp of the provisional and logically incomplete.

The relativity theories, he added, were unable to avoid the separate introduction of material points and total differential equations. And the new quantum mechanics moved fundamentally away from both the Newtonian and the Maxwellian program; "for the quantities which appear in its laws demand that they describe not the physically real itself, but only the probabilities of the occurrence of the physically real to which we direct attention." Yet, he concluded, somehow one would have to return to the attempt to realize the Maxwellian program, which he defined as "description of the physically real by means of fields which satisfy partial differential equations without singularities."

A decade later [40] he had to confess that the "field theory approach had still not managed to supply a foundation for the whole of physics." At the turn of the century, he explained, the very progress made at that time in understanding groups of new phenomena served, if anything, to

> move the establishment of a unified foundation for physics into
> the far distance, and this crisis at the foundation has been even
> aggravated by the subsequent developments. The development
> is characterized by two theoretical systems essentially indepen-
> dent of each other: the theory of relativity and the quantum
> theory. The two systems do not exactly contradict each other;
> but they seem little adapted to fusion into one unified theory.

As his collaborators all agreed, Einstein never subsided in discouragement, but right after each setback would launch with optimism into another attack that seemed promising. Yet from time to time in his late years, he would write to his friends about his dismay, as in this letter to Max von Laue in 1952:[41]

> Now you will more easily understand why I have seized on
> the quixotic-seeming attempt to generalize the gravitation equa-
> tions. If one cannot trust Maxwell's equations, and because of
> the general relativity principle is limited to a representation
> through field- and differential equations, and if one has de-

spaired of reaching a deep explanation of the theory in an
anschaulich-constructive way, then there seems to be no effort
of another kind open. At the same time one has the well
founded suspicion that one will not be able to hold on to the
continuum. But then one sees oneself displaced into a hopeless
conceptual vacuum. My attempts to interpret physically the
generalized theory of gravitation have been completely without
success. . . .

Einstein saw that the completion of the task he had set for himself and
for those who would follow was endangered by four threats. While they
are more or less closely related to one another, each is characterized by its
own thematic content.

1. *Completeness and exhaustiveness of description.* As we saw in Einstein's statement quoted at the very beginning of this chapter, completeness of description played a fundamental role in his *Weltbild*. Elsewhere
he spoke of what "appears to me to be the programmatic aim of all
physics: the complete description of any individual real situation, as it
supposedly exists irrespective of any act of observation and substantia-
tion." [42] On the last page of one of his last essays, he declared again "the
goal" was "a theory which describes *exhaustively* physical reality, including four-dimensional space, by a field." [43]

Of course, on this point quantum mechanics, or more properly the
Copenhagen interpretation of quantum mechanics, took exactly the opposite stance in holding that the state of a system can be specified not
directly but only indirectly, by a statement of the statistics of the results
of measurements attainable on the system. As is well known, Einstein
did not rebel against the view that quantum mechanics is valid and the
predicted probabilities statistically correct. Rather, the question he asked
was whether the "incompleteness" of quantum-theoretical descriptions,
guaranteed by its attention to ensembles of systems rather than describing
individual events, reflected a fundamental law of nature or merely the
incompleteness of the theory.

To him, the answer was plain: "The incompleteness of the representation leads necessarily to the statistical nature (incompleteness) [*Unvoll-
ständigkeit*] of the laws." [44] Then Einstein exclaims plaintively:

But now, I ask, does any physicist whosoever really believe that
we shall never be able to attain insight into these significant
changes of single systems, their structure, and their causal con-

nections, despite the fact that these individual events have been brought into such close proximity of experience, thanks to the marvelous inventions of the Wilson Chamber and the Geiger counter? To believe this is, to be sure, logically possible without contradiction; but it is in such lively opposition to my scientific instinct that I cannot forgo the search for a more complete mode of conception.

Einstein goes on:

There is no doubt that quantum mechanics has seized hold of a great deal of truth, and that it will be a touchstone for a future theoretical foundation that will have to be able to deduce quantum mechanics as a limiting case – just as electrostatics is deducible from the Maxwellian equations of the electromagnetic field, or thermodynamics from classical mechanics. However, I believe that in the search for this foundation, quantum mechanics cannot serve as a *starting point*, just as one could not find from thermodynamics, or statistical mechanics, the foundations of mechanics. (Emphasis in original.)

In short, in the yet-to-be-completed world picture, explanation would have to be anchored in a foundation that allows complete and exhaustive description of the individual events – despite the contrary pressure from the overwhelming majority of the scientific community.

2. *Causal determinism.* A closely related aspect of the threat from quantum mechanics was the "in-principle" unpredictability or indeterminism that Einstein had already objected to in his earliest work.

This point is best seen in historical perspective. Already in his first major presentation before an international scientific audience, in September 1909 at Salzburg,[45] he indicated his distaste with the lack of symmetry between the undulatory theory of light and the kinetic molecular theory. In the latter, for each molecular collision or similar process, the inverse elementary process can exist. Not so in the emission and absorption of light: only the emission of expanding wave fronts is an elementary process, while the collapse of a spherical wave on a point region (while mathematically possible) requires a very large number of emitting elements. A "Newtonian" theory of emission and absorption would not have this asymmetry.

But it, too, would not solve all problems. When, in 1916, fully a decade before the rise of quantum mechanics and its claims, Einstein again considers "the still so obscure process" of emission and absorption of radiation,[46] he objects to the discovery of indeterminism in his own paper, and implies that it will not be a basic element in the further development of the theory:

> Radiation of spherical waves does not exist. In the elementary process of spontaneous emission, the molecule suffers a recoil of magnitude hv/c in a direction that, in the present state of the theory, is determined only by "chance" [*Zufall*]. . . . The weakness of the theory lies, on the one hand, in the fact that it does not bring us any closer to a connection with the wave theory, and, on the other hand, in the fact that it leaves the time and direction of the elementary processes to "chance," nevertheless I have full confidence in the reliability of the course taken.

Already one hears the overtones of Einstein's many later avowals that it would be intolerable to find, at the foundations of the world picture, a dice-playing God, rather than what Newton had called the "God of Order." It became more and more clear, as Einstein freely admitted, that

> It is probably out of the question that any future knowledge can compel physics again to relinquish a statistical theoretical foundation in favor of a deterministic one. . . . At the present [1940s], we possess no deterministic theory, one directly describing the events themselves and in consonance with the facts of experience.[47]

And closing with one of the strongest statements he allowed himself:

> Many, among them myself, cannot believe that we must renounce once and for all the direct representation of physical reality in space and time, and that we must conceive of the events in nature on the model of a game of dice. It is open to everyone to choose the direction of his striving; and everyone may also draw comfort from Lessing's fine saying, that the search for truth is more precious than its secure possession.

In short, he would rather forgo accepting an otherwise successful *Weltbild* that is fundamentally indeterministic, and instead continue the search, if necessary, indefinitely.

In time, the conflict between the new quantum mechanics and general relativity became more and more obvious, for the former really denies the observability of the basic events which the latter deals with.[48] On the other hand, it cannot be said that today quantum mechanics itself is in its final form. Its problems prompted P. A. M. Dirac to say:[49]

> Some further changes will be needed, just about as drastic as the changes which one made in passing from Bohr's orbits to a quantum mechanics. Some day a new relativistic quantum mechanics will be discovered in which we don't have these infinities appearing at all. It might very well be that the new quantum mechanics will have determinism in the way that Einstein wanted.

3. The independence of the external, objective world. After his early Machist and positivistic phase, Einstein grew, without at first being fully conscious of it, into a rational realist. One can even date when Einstein himself began actively to assert his allegiance. It seems to have been shortly before 1930.[50] Possibly reinforced by Max Planck, Einstein now held the physical laws to describe a reality in space and time that is independent of the observer or theorist. As he put it in one of several early formulations,[51] "The belief in an external world independent of the perceiving subject is at the basis of all natural science," even though the physically real, of which our sense experiences furnish only indirect testimony, "can be grasped by us only in a speculative manner."

When he came to write his "Autobiographical notes" at the age of sixty-seven, Einstein even projected back upon his childhood the discovery of this state of affairs. He described that after his conversion from the "religious paradise of youth" at the age of about twelve, he made his

> first attempt to free myself from the chains of the "merely personal." . . . Out yonder there was this huge world, which exists independently of us human beings and which stands before us like the great, external riddle, at least partially accessible to our inspection and thinking. The contemplation of this world beckoned like a liberation. . . . The mental grasp of this extra-personal world within the frame of the given possibilities swam as highest aim half consciously and half unconsciously before my mind's eye. . . . The road to this paradise was not as comfortable and alluring as the road to the religious paradise; but it has proved itself as trustworthy, and I have never regretted having chosen it.[52]

One can imagine the strong intellectual and psychological pressures on Einstein, when the positivism of the Copenhagen school denied the claim of describing physical reality and declared itself satisfied with the probabilities of occurrences to which we direct attention – occurrences that in principle are linked to the observing process, at that – thus giving up any attempt to make a representation of "what is actually present or goes on in space and time."[53] The very meaning of relativity itself consisted in the discovery of the independence of the laws of nature from the point of view of particular observers. Einstein's repeated, insistent, and obstinate return to these points in his correspondence with Born, Schrödinger, and others testifies to the depth of feelings aroused by the world picture rising out of the new quantum mechanics.

4. *Evolution, not revolution.* Today, when every serious student can understand the elements of relativity, it is hard to reconstruct the overwhelming difficulties Einstein's ideas caused in the first decades of this century. Even Max Planck, a conservative person in thought and expression, nevertheless was moved to enthuse:

> This new way of thinking about time makes extraordinary demands on the physicist's ability to abstract, and on his imaginative faculty. It well surpasses in daring everything that has been achieved in speculative scientific research, even in the theory of knowledge. . . . This revolution in the physical *Weltanschauung*, brought about by the relativity principle, is to be compared in scope and depth only with that caused by the introduction of the Copernican system of the world.[54]

The friends of Einstein's theory called him a great revolutionary in physics and human thought generally – and so did his opponents. Einstein himself took every opportunity to disavow that label. He saw himself, in this as in so much else, essentially as a continuist, and had specific ideas on the way scientific theory developed by evolution. In his famous letter to Conrad Habicht,[55] written in the spring of 1905 to describe what he was then working on, Einstein referred to the paper on the electrodynamics of moving bodies (relativity) with the simple remark that he was using "a modification of the theory [*Lehre*] of space and time." In his small book of 1917,[56] he writes "The most beautiful fate of a physical theory is to point the way to the establishment of a more inclusive theory, in which it lives on as a limiting case."

Indeed, over time he developed a theory of levels, or "stratification of the scientific system."[57] In the striving for a logical unity, the research progressively leads the theory from the "first layer" to a "secondary system" and on to higher levels, each level characterized by being more parsimonious in concepts and relations, particularly the concepts that are directly connected with the complexes of ready sense experience. "So one continues until we have arrived at a system of the greatest conceivable unity, and of the greatest conceptual paucity of the logical foundations that is compatible with the nature of what is given to our senses." There is of course no guarantee that "this greatest of all aims can really be attained to a very high degree," the more so as all we can use are our own, "freely formed concepts." But in this gradual way one can hope to make further and further progress in the construction of the unified system that does beckon. In the course of this pursuit, many discontinuities of a conceptual kind may have to be introduced – and not least the postulation of axioms that cannot be logically connected with the experiential base. Individuals such as Maxwell, Faraday, and Hertz introduced "far-reaching changes."[58] But Einstein did not see them as revolutionary breaks with the past.

Probably just because the word "revolutionary" was so often applied to him in the period following the November 1919 announcements of the experimental test of general relativity, Einstein went out of his way to counteract this tendency.[59] On his arrival in New York in April 1921, he was quoted as having insisted:

> There has been a false opinion widely spread among the general public that the theory of relativity is to be taken as differing radically from the previous developments in physics from the time of Galileo and Newton, that it is violently opposed to their deductions. The contrary is true. Without the discoveries of every one of the great men of physics, those who laid down the preceding laws, relativity would have been impossible to conceive and there would have been no basis for it. Psychologically it is impossible to come to such a theory at once, without the work which must be done before. The men who have laid the foundation of physics on which I have been able to construct my theory are Galileo, Newton, Maxwell, and Lorentz.[60]

Some days later he had himself introduced by M. I. Pupin at Columbia University as the discoverer of a theory which is "an evolution, not a revolution of the science of dynamics."[61] A few weeks later still, in his address at King's College, London, he again stressed that relativity theory "provided a sort of completion to the mighty intellectual edifice of Maxwell and Lorentz. . . ." After that, Einstein continued to dismiss talk of revolutions in modern science, for example with the caustic remark that such writings give "the impression that every five minutes there is a revolution in science, somewhat like the *coups d'état* in some of the smaller unstable republics."[62] Indeed, how else but in an evolutionary way could one hope to approach a stable world picture that would reflect the once-given world in all its parts?

As it has turned out, Einstein's exhortation to seek a unified *Weltbild* is more coherent with the activities of many of today's best theoreticians than has been the case for the previous few decades. In a real sense, contemporary physicists, who use "Grand Unification Program" quite simply as a technical term to identify their current version of that ancient quest, are recognizably following Einstein's general goal. This is not to say that threats to Einstein's own solution for the *Weltbild* have been overcome. Today, completeness, exhaustiveness of description, and causal determinism are not put forward as necessary boundary conditions of the developing scheme; the postulation of an independent, external, objective world, or of the evolutionary model rather than a revolutionary one for the growth of science; hardly ever enters explicitly into the publications of physicists (in good part because they are rather less interested in the kind of epistemological questions that Einstein's generation regarded as intimately tied to science itself). General relativity and quantum mechanics are still far from having "fused"; quantum mechanics and classical mechanics are fundamentally not assimilated; and there are more such fissures that stand in the way of an early synthesis. Above all, there is a historic irony in the fact that quite apart from the claims for attention raised by fields to which Einstein did not contribute – such as high-energy physics – the very extent and depth of the advances Einstein himself helped to launch, including his contributions to quantum theory, eventually made it impossible for the physical phenomena to be all gathered in one grand relativistic *Weltbild* of the sort he longed for.

But Einstein's uncompleted work will appear in any future, more inclusive theory as a limiting case; hence it will live on even by his own severe criteria.[63] So will his spirited hope in the possibility of ultimate success, which animates today's version of the old unification program.

Lorentz's appreciation[64] of Einstein's general relativity theory applies, for new reasons, to the current stage in the evolution of the unitary scientific *Weltbild*: it "has the very highest degree of aesthetic merit; every lover of the beautiful must wish it to be true." Einstein did not live to see it come true in his time, and it may not come to be so in ours. If that is to be called failure, it has to be the kind of noble failure that was also the fate of Newton.

5

Einstein and the shaping of our imagination

In addition to his role as builder of a new view of the physical universe, and as contributor to many branches of physical science, Einstein came to influence twentieth-century culture in ways no other scientist did. His ideas, or views attributed to him, reverberate to this day in fields as distant from his own direct scientific contributions as psychology, linguistics, the analysis of modern art, and the study of the impact of science and technology on ethics.

It is therefore the more important to remember how long it took, by present standards, for his seminal, early scientific work to be understood even by his fellow physicists. Six years elapsed after the first publication of the special theory of relativity before it established itself sufficiently to merit a textbook (Max von Laue's *Das Relativitätsprinzip*), and for some years after that the theory continued to be confused by most scientists with the electrodynamics of H. A. Lorentz. Einstein's ideas on quantum physics, published from 1905 on, were also generally neglected or discounted for years. R. A. Millikan, on accepting his Nobel Prize for 1923, confessed that the validity of Einstein's "bold, not to say reckless" explanation of the photoelectric effect forced itself on him slowly, "after ten years of testing . . . [and] contrary to my own expectation." The transcripts of the questions asked in scientific meetings in the decade after 1905 contain many passages that demonstrate to the historian of science the large intellectual effort required at the time to enter fully into the meaning of the new physics.

Today, virtually every student who wishes to do so can learn at least the elements of relativity or quantum physics before leaving high school, and the imprint of Einstein's work on the different areas of physical science is so large and varied that a scientist who tries to trace it would be hard put to it to know where to start. A modern dictionary of scientific terms contains thirty-five entries bearing his name, from "einstein: A unit of light energy used in photochemistry" and "Einstein–Bose statistics" to "Einstein tensor" and "Einstein viscosity equation."[1] It is ironic that now, several decades after his death, there is in many branches of the physical sciences more awareness of his generative role than would have been credited during the last decade or two of his life. His ideas became essential for laying out conceptual paths for contemporary work in astronomy or cosmology, for unifying gravitation with the quantum field theory of gauge fields, or even for understanding new observations that were not possible in his time but were predicted by him (as in his 1936 paper on the optical lens formed by gravitational fields).

Apart from changing science itself, Einstein has reached into the daily life of virtually every person on the globe in direct or indirect ways through the incorporation of his ideas on physics into a vast range of technical devices and processes. I need cite only some of the most obvious ones. Every photoelectric cell can be considered one of his intellectual grandchildren. Hence, we are in his debt whenever photo emission or absorption is used, in the home or on the job, to capture an image by means of a television camera, or to project the optical soundtrack of a motion picture, or to set the page of a book or newspaper by photocomposition, or to make a telephone call over a modern fiber cable, or (eventually) to replace the oil-fired heater by an array of photovoltaic cells. In each case, if a law required a label on the appliance giving its intellectual content or pedigree, such a display would list prominently: "Einstein, *Annalen der Physik*, 17 (1905), pp. 132–148; 20 (1906), pp. 199–206," and so forth.

One would find an entry of this sort also on the laser, whose beam was probably used to lay out the highway on which one travels to the office or to site the office building itself ("Einstein, *Physikalische Zeitschrift*, 18 (1917), pp. 121–128," etc.). Or again, the same kind of answer comes if one lists key ideas that helped to make possible modern electric machinery, such as power generators, or precision clocks that allow the course of planes and ships to be charted. Einstein appears also, if one looks for the ancestry of the ideas in quantum and statistical physics by which solid-state devices operate, from calculators and computers to the transistor radio and the ignition system – and perhaps even when one takes one's vitamin pill or other pharmaceutical drug, for it is likely that its commercial production involved diffusion processes, first explained in Einstein's papers on Brownian movement and statistical mechanics ("Einstein, *Annalen der Physik*, 17 (1905), pp. 549–560," etc.).

As Edward M. Purcell remarked in his lecture at the Einstein Centennial Symposium at Princeton in 1979, since the magnetism set up by electric currents is a strictly relativistic effect, derivable from Coulomb's law of electrostatics and the kinematics of relativity, and nothing more, it requires no elaboration to discuss "special relativity in engineering": "This is the way the world *is*. And it does not really take gigavolts or nanoseconds to demonstrate it; stepping on the starter will do it!" It is not too much to say that even in our most common experiences, that unworldly theoretician's publications help to explain what happens to us all day – from the moment we open our eyes on the light of the morning, since the act of seeing is initiated by a photochemical reaction ["Einstein, *Annalen der Physik*, 37 (1912), pp. 832–838; 38 (1912), pp. 881–884," etc.].

The proverbial man in the street is quite blissfully ignorant of all that, and has preferred to remain so, even while expecting fully that, mysteriously yet automatically, a stream of practical, benign "spin-offs" continues from the pursuit of pure science. But the philosopher, the writer, the artist, and many others outside the scientific laboratories could not help but be caught up to some extent by the wave that spread beyond science and technology, at first slowly, then with astonishing intensity. As the best scientists were coming to understand what Einstein had done, the trumpets began to sound. When in London on November 6, 1919 the result of the British eclipse expedition was revealed to bear out one of the predictions of general relativity theory, the discussion of implications rose to fever pitch among scholars and laymen, beginning with declarations such as that in *The Times* of London (November 8, 1919): the theory had served "to overthrow the certainty of ages, and to require a new philosophy, a philosophy that will sweep away nearly all that has hitherto been accepted as the axiomatic basis of physical thought." It became evident that, as Newton had "demanded the muse" after the *Principia*, now it would be Einstein's turn.

In fact, Einstein did his best to defuse the euphoria and excess of attention that engulfed and puzzled him from that time on. He saw himself essentially as a continuist and attempted to keep the discussion limited to work done and yet to be done in science. He did not get much help, however. Thus, the physicist J. J. Thomson reported that the Archbishop of Canterbury, Randall Davidson, had been told by Lord Haldane "that relativity was going to have a great effect upon theology, and that it was his duty as Head of the English Church to make himself acquainted with it. . . . The Archbishop, who is the most conscientious of men, has procured several books on the subject and has been trying to read them, and they have driven him to what it is not too much to say is a state of *intellectual desperation.*" On Einstein's first visit to England in June 1921, the Archbishop of Canterbury therefore sought him out to ask what effect relativity would have on religion. Einstein replied briefly and to the point: "None. Relativity is a purely scientific matter and has nothing to do with religion." [2] But of course this did not dispose of the question. Later that year, even the scientific journal *Nature* felt it necessary to print opposing articles on whether "Einstein's space-time is the death-knell of materialism." [3]

Although the crest of the flood and the worst excesses have now passed, debates of this sort continue. More constructively, since modern philosophy is concerned in good part with the nature of space and time, causality, and other conceptions to which relativity and quantum physics have contributed, Einstein has had to be dealt with in the pages of philosophers, from Henri Bergson and A. N. Whitehead to the latest issues of the professional journals. As John Passmore observed correctly, it appeared in this century that "physics fell heir to the responsibility of metaphysics." [4] Some philosophers and philosopher-scientists have confessed that Einstein's work started them off on their speculations in the first place, thus giving some direction to their very careers. One example is P. W. Bridgman, who disclosed that the effort to clarify in his mind the issues in relativistic electrodynamics, when first asked to teach that course, drew him to the task of writing the influential book *The logic of modern physics* (1927).

Philosophy was no doubt destined to be the most obvious and often the earliest and most appropriate field, outside science itself, that the radiation from Einstein's work would reach. But soon there were others, even though the connections made or asserted were not always valid. From Einstein's wide-ranging output, relativity was invoked most frequently. Cultural anthropology, in Claude Lévi-Strauss's phrase, had evolved the doctrine of cultural relativism "out of a deep feeling of respect toward other cultures than our own"; but this doctrine often invited confusion with physical relativity. Much that has been written on "ethical relativity" and on "relativism" is based on a seductive play with words. And painters and art critics have helped to keep alive the rumor of a supposed genetic connection of visual arts with Einstein's 1905 publication.

Here again, Einstein protested when he could and, as so often, without effect. One art historian submitted to him a draft of an essay entitled

"Cubism and the theory of relativity," which argued for such a connection – for example, that in both fields "attention was paid to relationships, and allowance was made for the simultaneity of several views."[5] Politely but firmly, Einstein tried to put him straight, and he explained the difference between physical relativity and vulgar relativism so succinctly as to invite an extensive quotation:

> The essence of the theory of relativity has been incorrectly understood in it [your paper], granted that this error is suggested by the attempts at popularization of the theory. For the description of a given state of facts one uses almost always only one system of coordinates. The theory says only that the general laws are such that their form does not depend on the choice of the system of coordinates. This logical demand, however, has nothing to do with how the single, specific case is represented. A multiplicity of systems of coordinates is not needed for its representation. It is completely sufficient to describe the whole mathematically in relation to one system of coordinates.
>
> This is quite different in the case of Picasso's painting, as I do not have to elaborate any further. Whether, in this case, the representation is felt as artistic unity depends, of course, upon the artistic antecedents of the viewer. This new artistic "language" has nothing in common with the Theory of Relativity.[6]

Einstein might well have added here, as he did elsewhere, that the existence of a multiplicity of frames, each one as good as the next for solving some problems in mechanics, went back to the seventeenth century (Galilean relativity). As to the superposition of different aspects of an object on a canvas, that had been done for a long time; thus Canaletto drew various parts of a set of buildings from different places and merged them in a combined view in the painting (for example, *Campo S. S. Giovanni e Paolo*), the view becoming thereby an impossible *veduta*.

It was therefore doubly wrong to invoke Einstein as authority in support of the widespread misunderstanding that physical relativity meant that all frameworks, points of view, narrators, fragments of plot, or thematic elements are created equal, that each of the polyphonic reports and contrasting perceptions is as valid or expedient as any other, and that all of these, when piled together or juxtaposed, *Rashomon*-like, somehow constitute the real truth. If anything, twentieth-century relativistic physics has taught the contrary: that under certain conditions we can extract from different reports, or even from the report originating in one frame properly identified, all the laws of physics, each applicable in any framework, each having therefore an invariant meaning, one that does not depend on the accident of which frame one inhabits. It is for this reason that, by comparison with classical physics, modern relativity is simple, universal, and, one may even say, "absolute." The cliché became, erroneously, that "everything is relative," whereas the whole point is that out of the vast flux one can distill the very opposite: "some things are invariant."

The cost of the terminological confusion has been so great that a brief elaboration on this point will be relevant. Partly because he saw himself as a continuist rather than as an iconoclast, Einstein was reluctant to present this new work as a new *theory*. The term "relativity theory," which made the confusions in the long run more likely, was provided by Max Planck. As noted in Chapter 3, in correspondence Einstein seemed happier with the term *Invariantentheorie*, which is of course much more true to its method and aim. How much nonsense we might have been spared if Einstein had adopted that term, even with all its shortcomings! To a correspondent who suggested just such a change, Einstein replied (letter to E. Zschimmer, September 30, 1921): "Now to the name relativity theory. I admit that it is unfortunate, and has given occasion to philosophical misunderstandings. . . . The description you proposed would perhaps be better; but I believe it would cause confusion to change the generally accepted name after all this time."

To come back to Einstein's careful disavowal of a substantive genetic link between modern art and relativity: far from abandoning the quest for it, his correspondent forged onward enthusiastically and published three such essays instead of one. Newton did not always fare better at the hands of eighteenth-century divines and literati who thought they were following in his footsteps. Poets rush in where scientists fear to tread. And why not, if the apparent promises are so great? In April 1921, at the height of what Einstein on his first journey to the United States all too easily diagnosed as a pathological mass reaction, William Carlos Williams published a poem entitled "St. Francis Einstein of the Daffodils,"[7] containing such lines as "April Einstein / . . . has come among the daffodils / shouting / that flowers and men / were created / relatively equal. . . ." Declaring simply that "relativity applies to everything"[8] and that "Relativity gives us the clue. So, again, mathematics comes to the rescue of the arts," Williams felt encouraged to adopt a new variable measure for his poems – calling it "a *relatively* stable foot, not a rigid one"[9] – that proved of considerable influence on other poets.

Williams was of course not alone. Robert Frost, Archibald MacLeish, E. E. Cummings, Ezra Pound, T. S. Eliot, and some of their disciples (and outside the English-speaking world, others such as Thomas Mann and Hermann Broch) referred directly to Einstein or to his work. Some were repelled by the vision thought to be opened by the new science, but there were at least as many who seemed to be in sympathy with Jean-Paul Sartre's remark that "the theory of relativity applies in full to the universe of fiction."[10] Perhaps the most well-known of the attempts to harness relativity and literature to common purpose is Lawrence Durrell's entertaining set of novels, *The Alexandria quartet*, of which its author says by way of preface: "Modern literature offers us no Unities, so I have turned to science and am trying to complete a four-decker novel whose form is based on the relativity proposition. Three sides of space and one of time constitute the soup-mix recipe of a continuum."[11] The intention is to use the properties of space and time as determining models for the structure of the book. Durrell says "the first three parts . . . are to be deployed spatially . . . and are not linked in a serial form. . . . The fourth part alone will represent time and be a true sequel."

For that alone one would not have had to wait for Einstein. But more seems to be hoped for; that, and the level of understanding, is indicated by the sayings of Pursewarden recorded in Durrell's novel. Pursewarden – meant to be one of the foremost writers in the English language, his death mask destined to be placed near those of Keats and Blake – is quoted as saying, "In the Space and Time marriage we have the greatest Boy meets Girl story of the age. To our great-grandchildren this will be as poetical a union as the ancient Greek marriage of Cupid and Psyche seems to us." Moreover, "the Relativity proposition was directly responsible for abstract painting, atonal music, and formless . . . literature."[12]

Throughout the novel it is evident that Durrell has taken the trouble to read up on relativity, although chiefly out of impressionistic popularizations such as *The mysterious universe* by James Jeans. Durrell readily confessed that "none of these attempts has been very successful."[13] There is something touching and, from the point of view of an intellectual historian, even a bit tragic about the attempt. In his study *A key to modern British poetry*, Durrell revealed his valid concern to show that as a result of "the far-reaching changes in man's ideas" about the outer and inner universe, "language has undergone a change in order to keep in line with cosmological inquiry (of which it forms a part)."[14] Yet on page after page the author demonstrates that he has been misled by the simplifications of H. V. Routh and Jeans; he believes that Rutherford and Soddy suggested that the "ultimate laws of nature were not simply causal at all," that "Einstein's theory joined up subject and object," that "so far as phenomena are concerned . . . the uniformity of nature disappears," and so forth.[15] The terrible but clarifying remark of Wolfgang Pauli comes to mind, who said about a theory that seemed to him doomed: "It is not even wrong."

If I have spelled out some of the misunderstandings by which Einstein's work, for better or worse, has been thought to have found its way into contemporary culture, the examples of incorrect interpretation prepare us to appreciate that much more the correct ones. I should confess that my own favorite example of the successful transmutation of scientifically based conceptions in the writer's imagination is a novel, and a controversial one. William Faulkner's *The sound and the fury* is more like an earthquake than a book. Immediately on publication in 1929 it caused universal scandal; for example, not until Judge Curtis Bok's decision in 1949 was this, among Faulkner's other novels, allowed to be sold in Philadelphia. On the surface it seems unlikely that this book – even a friendly reviewer characterized it as "designedly a silo of compressed sin" – has any resonance with the ideas of modern physics, by intent or otherwise. At the time he poured himself into the book, Faulkner was still almost unknown, largely self-taught, eking out a meager living as a carpenter, hunter, and coal carrier on the night shift of a power station, his desk the upturned wheelbarrow on which he would write while kneeling on the floor. Yet, even there, he was not isolated if he read even a small part of the flood of articles in newspapers, periodicals, and popular books in the 1920s dealing with the heady concepts of relativity theory – such as the time dilation experienced by a clock traveling through space, the necessity to recognize the meaninglessness of absolute time and space – and the recent quantum physics, with its denial of the comforts of classical causality. Particularly in America, Einstein was quoted down to the level of local evening papers and *Popular Mechanics*, resulting in wide circulation of such haunting epigrams as his remark, made in exasperation to Max Born (1926), that "God does not throw dice." Could any of this have reached Faulkner?

In the second of the four chapters of *The sound and the fury* we follow Quentin Compson of Jefferson, Mississippi, as he lives through a day in

June 1910. It is the end of his freshman year at college and the culmination of a short life wrenched by the degeneration and guilt, the fixations and tribal racism of his whole haunted family – from his father Jason, drinking himself to death, to his idiot brother Benjamin, whose forty-acre pasture has been sold to send Quentin to college. The only resource of human affection he has known came from black laborers and servants, although they have been kept in the centuries-old state of terror, ignorance, and obeisance. But the Compsons are doomed. As the day unfolds, Quentin moves toward the suicide he knows he will commit at midnight.

It is all too easy to discover theological and Freudian motives woven into the text, and one must not without provocation drag an author for cross-examination into the physics laboratory when he has already suffered through interrogations at the altar and on the couch. But Faulkner asked for it. Let me select here from a much more extensive body of evidence in the novel itself.

Quentin's last day on earth is a struggle against the flow of time. He attempts to stem the flow, first by deliberately breaking the cover glass of the pocket watch passed down to him from grandfather and father, then twisting off the hands of the watch, and then launching on seemingly random travel, by streetcar and on foot, across the whole city. His odyssey brings him to the shop of an ominous, cyclopean watch repairer. Quentin forbids him to tell him the time but asks if any of the watches in the shop window "are right." The answer he gets is "No." But wherever he then turns, all day and into the night, he encounters chimes, bells ringing the quarter hours, a factory whistle, a clock in the Unitarian steeple, the long, mournful sound of the train tracing its trajectory in space and time, "dying away, as though it were running through another month." Even his stomach is a kind of space-time metronome. "The business of eating inside of you space to space and time confused stomach says noon brain says eat o'clock All right I wonder what time it is what of it." Throughout, Quentin carries the blinded watch with him, the watch that never knew how to tell real time and cannot even tell relative time. But it is not dead: "I took out my watch and listened to it clicking away, not knowing it could not even lie." [16] And in the streetcar, the clicking away of time is audible to him only while the car has come to a stop.

Quentin has taken a physics course that freshman year and uses it to calculate how heavy the weights must be that he buys to help drown himself. It is, he says wryly to himself, "the only application of Harvard," and as he reflects on it: "The displacement of water is equal to the something of something. Reducto absurdum of all human experience, and two six-pound flat-irons weigh more than one tailor's goose. What a sinful waste, Dilsey would say. Benjy knew it when Damuddy died. He cried."

As midnight approaches, before he is ready to put his "hand on the light switch" for the last time,[17] Quentin is overcome by torment, caused by the shamed memory of his incestuous love for his sister Candace, by her loss, and by his own sense of loss even of the meaningfulness of that double betrayal. In anguish he remembers his father's terrible prediction after he had made his confession:

> You cannot bear to think that some day it will no longer hurt
> you like this now were getting at it. . . you wont do it under
> these conditions it will be a gamble and the strange thing is
> that man who is conceived by accident and whose every breath
> is a fresh cast with dice already loaded against him will not
> face that final main which he knows beforehand he has
> assuredly to face without essaying expedients. . . . that would
> not deceive a child until some day in very disgust he risks every-
> thing on a single blind turn of a card no man ever does that
> under the first fury of despair or remorse or bereavement he
> does it only when he has realized that even the despair or
> remorse or bereavement is not particularly important to the
> dark diceman. . . . it is hard believing to think that a love or a
> sorrow is a bond purchased without design and which matures
> willynilly and is recalled without warning to be replaced by
> whatever issue the gods happen to be floating at the time.

This was not the God Newton had given to his time – Newton, of whom, just two centuries before Faulkner's soaring outcry, the poet James Thomson had sung in 1729 that "the heavens are all his own, from the wide role of whirling vortices, and circling spheres, to their great simplicity restored." Nor, of course, was it Einstein's God, a God whose laws of nature are both the testimony of His presence in the universe and the proof of its saving rationality. But this, it seems to me, defines the dilemma precisely. If the poet neither settles for the relief of half-understood analogies nor can advance to an honest understanding of the rational structure of that modern world picture, and if he is sufficiently sensitive to this impotency, he must rage against what there is left him: time and space are then without meaning; so is the journey through them; so is grief itself, when the very gods are playing games of chance, and all the sound and the fury signify nothing. And this leads to recognizing the way out of the dilemma, at least for a few. At best, as in the case of Faulkner, this rage itself creates the energy needed for a grand fusion of the literary imagination with perhaps only dimly perceived scientific ideas. There are writers and artists of such inherent power that the ideas of science they may be using are dissolved, like all other externals, and rearranged in their own glowing alchemical cauldron.

It should not, after all, surprise us; it has always happened this way. Dante and Milton did not use the cosmological ideas of their time as tools to demarcate the allowed outline or content of their imaginative constructs. Those college students of ours who, year after year, write us dutifully more or less the same essay, explaining the structure of the *Divine comedy* or *Paradise lost* by means of astronomy, geography, and the theory of optical phenomena – they may get the small points right, but they miss the big one, which is that the good poet is a poet surely because he can transcend rather than triangulate. In Faulkner, in Eliot's *The Waste land*, in Woolf's *The waves*, in Mann's *Magic mountain* it is futile to judge whether the traces of modern physics are good physics or bad, for these trace elements have been used in the making of a new alloy. It is one way of understanding Faulkner's remark on accepting his Nobel Prize in 1950: the task was "to make out of the material of the human spirit something which was not there before." [18] And insofar as an author *fails* to produce the feat of recrystallization, I suspect this lack would not be cured by more lessons on Minkowski's space-time, or Heisenberg's indeterminacy principle, or even thermodynamics, although such lessons could occasionally have a prophylactic effect that might not be without value.

Here we suddenly remember that, of course, the very same thing is true for scientists themselves. The most creative ones, almost by definition, do not build their constructs patiently by assembling blocks that have been precast by others and certified as sound. On the contrary, they too melt down the ready-made materials of science at hand and recast them in a way that their contemporaries tend to think is outrageous. That is why Einstein's own work took so long to be appreciated even by his best fellow physicists, as I noted earlier. His physics looked to them like alchemy, not because they did not understand it at all, but because, in one sense, they understood it all too well. From their thematic perspective, Einstein's was anathema. Declaring, by simple postulation rather than by proof, Galilean relativity to be extended from mechanics to optics and all other branches of physics; dismissing the ether, the playground of most nineteenth-century physicists, in a peremptory half-sentence; depriving time intervals of inherent meaning; and other such outrages, all delivered in a casual, confident way in the first, short paper on relativity – those were violent and "illegitimate" distortions of science to almost every physicist. As for Einstein's new ideas on the quantum physics of light emission, Max Planck felt so embarrassed by it when he had to write Einstein a letter of recommendation seven years later that he asked that this work be overlooked in judging the otherwise promising young man.

Moreover, the process of transformation characterizes not only science itself and the flow of ideas from high science to high literature. It also works across the boundaries in other ways. The most obvious example is Einstein's importation into his early physics of an epistemology that he himself thought, with some enthusiasm, to be based on Ernst Mach's kind of positivism. Mach had begun to find him out on this point even while Einstein was still signing his letters to Mach as "Your devoted Student."

It seems clear to me that without this process of transformation, willing or unwilling, of ideas from science and from philosophy, physics would not have come into its twentieth-century form. (A similar statement may well be made for the cases of Copernicus, Kepler, Galileo, and Newton.) The case of Einstein suggests, therefore, that the accomplishments of the major innovator – and not only in science – depend on his ability to persevere in four ways: by giving his loyalty primarily to his own belief system rather than to the current faith; by perceiving and exploiting the man-made nature and plasticity of human conceptions; by demonstrating eventually that the new unity he has promised does become lucid and convincing to lesser mortals active in his field – that he has it all "wrong" in the right way; and, in those rare cases, by even issuing ideas that lend themselves, quite apart from misuse and oversimplification, to further adaptation and transformation in the imagination of similarly exalted spirits who live on the other side of disciplinary boundaries.

It remains to deal with one more, somewhat different mechanism by which Einstein's imprint came to be felt far beyond his own field of primary attention: the power of his personal intervention on behalf of causes ranging from the establishment of a homeland for a persecuted people to his untiring efforts, over four decades, for peace and international security. In retrospect we can see that he had the skill, at strategic periods of history, to lend his ideas and prestige to the necessary work of a Chaim Weizmann or a Bertrand Russell. Even the most famous of these personal interventions, the call on President Roosevelt in 1939 to initiate a study of whether the laws of nature allow anyone to produce an atomic weapon, was of that sort, although it has perhaps been misunderstood more widely than anything else Einstein did. He was, after all, correct in his perception that the Germans, who were pushing the world into a war, had all the skill and intention needed to start production of such a weapon if it was feasible. In fact, they had a head start, and but for some remarkable blunders, they might have fulfilled the justified fears, with incalculable consequences on the course of civilization.

To highlight these personal interactions, I select one as more or less paradigmatic of the considerable effect Einstein had even in some brief or seemingly casual discussion with the right person. In this case it was the Swiss psychologist Jean Piaget. Piaget's work entered its most important phase with the publication in 1946 of *The child's conception of time*. The book begins with a plain acknowledgment:

> This work was promoted by a number of questions kindly suggested by Albert Einstein more than fifteen years ago [1928, at a meeting in Davos]. . . . Is our intuitive grasp of time primitive or derived? Is it identical with our intuitive grasp of velocity? What if any bearing do these questions have on the genesis and development of the child's conception of time? Every year since then we have made a point of looking into these questions. . . . The results [concerning time] are presented in this volume; those bearing on the child's conception of motion and speed are reserved for a later work.[19]

Throughout his later writings, Piaget remarks on this debt: "It was the author of the theory of relativity who suggested to us our work,"[20] or "Einstein once suggested we study the question from the psychological viewpoint and try to discover if there existed an intuition of speed independent of time."[21] In addition, Piaget refers explicitly to notions of relativity and other aspects of Einstein's work.[22]

Looking back at the variety of ways in which Einstein came to impress the imagination of his time and ours, we can discern some rough categories, spread out, as it were, in a spectrum from left to right. At the center portion, corresponding to the largest intensity, one finds the widespread but unfocused and mostly uninformed fascination, manifested in a variety of ways, from enthusiastic mass gatherings to glimpse the man, to the outpouring of popularizations with good intentions, to responses that betray the vague discomfort aroused by the ideas. A good example of the last is an editorial entitled "A mystic universe" in the *New York Times* of January 28, 1928 (p. 14): "The new physics comes perilously close to proving what most of us cannot believe. . . . Not even the old and much simpler Newtonian physics was comprehensible to the man on the street. To understand the new physics is apparently given only to the highest flight of mathematicians. . . . We cannot grasp it by sequential thinking. We can only hope for dim enlightenment." The editorial writer then notes that the ever-changing scene in physics does offer some "comfort":

> Earnest people who have considered it their duty to keep abreast of science by readapting their lives to the new physics may now safely wait until the results of the new discoveries have been fully tested out by time, harmonized and sifted down to a formula that will hold for a fair term of years. It would be a pity to develop an electronic marriage morality and find that the universe is after all ether, or to develop a wave code for fathers and children only to have it turn out that the family is determined not by waves but by particles. Arduous enough is the task of trying to understand the new physics, but there is no harm in trying. Reshaping life in accordance with the new physics is no use at all. Much better to wait for the new physics to reshape our lives for us as the Newtonian science did.

Similarly, in Tom Stoppard's play *Jumpers* a philosopher is heard to ask: "If one can no longer believe that a twelve-inch ruler is always a foot long, how can one be sure of relatively less certain propositions?"

Near this position, as we said, are the enthusiastic misapplications, usually achieved by an illicit shortcut of meaning from, say, the true statement that the operational definition of length is "framework" dependent, to the invalid deduction that mental phenomena in a human observer have thereby been introduced into the very definitions of physical science. (To be sure, the layman has not always been served by the explanation on this point given by the scientists themselves; for example, in such pronouncements as "the object of research is no longer nature in itself but rather nature exposed to man's questioning, and to this extent man here also meets himself." [23]) The irony here is that the first lessons we learned from relativity physics were that short circuits in signification must be avoided, for they were just what burdened down classical physics, and that attention must be paid as never before to the meaning of the terms we use.

When we now glance further toward the left, or blue, end of the spectrum, the expressions of resignation or futility become more explicit. Indeed, among some of the most serious intellectuals there seems, on this point, a sense of despair. By the very nature of their deep motivation they must feel most alienated from a universe whose scientific description they can hardly hope to understand except in a superficial way. The much-admired humanistic scholar Lionel Trilling spoke for many when he stated the dilemma frankly and eloquently:

> The operative conceptions [of science] are alien to the mass of
> educated persons. They generate no cosmic speculation, they
> do not engage emotion or challenge imagination. Our poets
> are indifferent to them. . . .
>
> This exclusion of most of us from the mode of thought
> which is habitually said to be the characteristic achievement of
> the modern age is bound to be experienced as a wound given
> to our intellectual self-esteem. About this humiliation we all
> agree to be silent, but can we doubt that it has its conse
> quences, that it introduced into the life of mind a significant
> element of dubiety and alienation which must be taken into
> account in any estimate that is made of the present fortunes of
> mind? [24]

Einstein, who had intended originally to become a science teacher, came to understand this syndrome, and the obligation it put on him. He devoted a good deal of time to popularization of his own. His avowed aim was to simplify short of distortion. In addition to a large number of essays and lectures, he wrote, and repeatedly updated, a short book on relativity that he promised in the very title to be *gemeinverständlich*.[25] It is, however, overly condensed for most nonscientific beginners. Later, Einstein collaborated with Leopold Infeld in a second attempt to reach out to the population at large by means of a book-length treatment of modern physics. As the preface acknowledged, the authors no longer attempted "a systematic course in elementary facts and theories." Rather, they aimed at a historical account of how the ideas of relativity and quanta entered science, "to give some idea of the eternal struggle of the inventive human mind for a fuller understanding of the laws governing physical phenomena."[26] In fact, there is to this day no generally agreed source, the *reading* of which by itself will bring a large fraction of an adult nonscientific audience to a sound enough understanding of these ideas, even for those who truly want to attain it and are willing to pay close attention. I believe it is a fact of great consequence that it takes a much larger effort, and one starting earlier than most people undertake. To make matters worse, so little has been found out about how scientific literacy is achieved or resisted that little blame can be spun off on the would-be students, young or old.

Going now further along the spectrum in the same direction, we encounter outright hostility and opposition to Einstein's work, either on scientific or on ideological grounds. Almost all scientists, even those initially quite reluctant, became eventually at least reconciled, save (to this day) for Einstein's famous refusal to regard the statistical interpretation as fundamental.

On the other hand, the opposition to Einstein's work on grounds other than scientific has had a longer history. Thus, a number of studies now exist that show the lengths to which various totalitarian groups, for various reasons, felt compelled to go in their attacks.

Turning now to the other, more "positive" half of the spectrum, we see there the gradual acceptance and elaboration of Einstein's work within the corpus of physical science; its penetration into technology (largely unmarked) and into the more thoughtful philosophies of science; Einstein's effect through his personal intervention, causing some historic redirections of research; and its passage into the scientific world picture of our time, as it tries to achieve a unification that eluded Einstein. And beyond that, at the end of the spectrum, where the number of cases is small but the color deep and vibrant, we perceive the examples of creative transformation beyond science. Those are the works of the few who found that scientific ideas, or rather *metaphors* embodying such ideas, released in them a fruitful response with an authenticity of its own, far removed from textbook physics.

This last is the oldest and surely still the most puzzling interplay between science and the rest of culture. Evidently, the mediation occurs through a sharing of an analogy or metaphor – irresistible, despite the dangers inherent in the obvious differences or barriers. We know that such a process exists, because any major work of science itself, in its nascent phase, is connected analogically rather than fully logically, both with the historic past in that science and with its supporting data. The scientist's proposal may fit the facts of nature as a glove fits a hand, but the glove does not uniquely imply the hand, nor the hand the glove.

Einstein spoke insistently over the decades about the need to recognize the existence of such a discontinuity, one that in his early scientific papers asserted itself first in his audacious method of postulation. In essay after essay, he tried to make the same point, even though it had little effect on the then reigning positivism. Typical are the phrases in his Herbert Spencer Lecture of 1933.[27] The rational and empirical components of human knowledge stand in "eternal antithesis," for "propositions arrived at by purely logical means are completely empty as regards reality." In this sense, the "fundamentals of scientific theory," being initially free inventions of the human mind, are of "purely fictional character." The phenomenic-analytic dichotomy makes it inherently impossible to claim that the principles of a theory are "deduced from experience" by "abstraction," that is, by logically complete claims of argument. As he put it soon afterwards, the relation between sense experience and concept "is analogous not to that of soup to beef, but rather that of check number to overcoat."[28]

If this holds for the creative act in science itself, we should hardly be surprised to find the claim to be extended to more humanistic enterprises. The test, in both instances, is of course whether the freely invented check token produces a suitable overcoat. The existence of both splendid scientific theories and splendid products of the humanistic imagination shows that despite all their other differences, they share the ability to build on fundamentals of a "purely fictional character." And even the respective fundamentals, despite all their differences, can share a common origin. That is to say, at a given time the cultural pool contains a variety of themata and metaphors, and some of these have a plasticity and applicability far beyond their initial provenance. The innovator, whether a scientist or not, necessarily dips into this pool for his fundamental notions, and in turn may occasionally deposit into it new or modified themata and metaphors of general power.

Examples of such science-shaped metaphors, each of these by no means a "fact" of the external world, yet revealing immense explanatory energy, are easy to find: Newton's "innate force of matter (*vis insita*)" and the Newtonian clockwork universe; Faraday's space-filling electric and magnetic lines of force; Niels Bohr's examples of complementarity in physics and in daily life; Mendeleev's neat tableau setting for the families of elements, and Rutherford's long parent-daughter-granddaughter chains of decaying atoms; Minkowski's space-time "World," of which our perceptible space and time are like shadows playing on the wall of Plato's cave; and of course the imaginative scenes Einstein referred to – the traveler along the light beam, the calm experimenter in the freely falling elevator, the dark, dice-playing God, the closed but unbounded cosmos, the Holy Grail of complete unification of all forces of nature. So it continues in science.

The allegorical use of such conceptions may, as we have noted, help to shape works of authenticity outside the sciences. And the process works both ways. Thus Niels Bohr acknowledged that his reading in Kierkegaard and William James helped him to the imaginative leap embodied in his physics, Einstein stressed the influence on his early scientific thinking of the philosophical tracts of that period, and Heisenberg noted the stimulus of Plato's *Timeaus*, read in his school years. No matter if such "extraneous" elements are eventually suppressed or forgotten, or even have to be overcome; at an early point they can encourage the mind's struggle. The philosopher José Ortega y Gasset was one of those who struggled with such ideas. In 1921–2, evidently caught up by the rise of the new physics, he wrote on "The Historical Significance of the Theory of Einstein." [29] There he noted correctly that the most relevant issue was not that the triumph of the theory

> will influence the spirit of mankind by imposing on it the
> adoption of a definite route. . . . What is really interesting is
> the inverse proposition: the spirit of man has set out, of its own
> accord, upon a definite route, and it has therefore been possible
> for the theory of relativity to be born and to triumph. The
> more subtle and technical ideas are, the more remote they seem
> from the ordinary preoccupations of men, the more authenti-
> cally they denote the profound variations produced in the
> historical mind of humanity. [30]

I conclude that in pursuing the documentable cases of "impact" of one person or field on another, we have been led to a more mysterious fact, namely, that there exists a mutual adaptation and resonance of the innovative mind with portions of the total set of metaphors current at a given time.

6

Physics in America, and Einstein's decision to immigrate

One might think that when a nation undergoes a revolutionary upheaval against its own political system and even against reason itself, the sons and daughters of Urania, the muse representing science, would be among the last to be touched. For it is generally agreed that research into the behavior of nature is as free from local political overtones as intellectual work can be, and that the achievements of a nation's major scientists – more than those of its poets or statesmen – are embedded in an international and intercultural system of recognition that is guaranteed by the consensual nature of scientific proof itself.

During the rise of some authoritarian regimes, such hopes for leniency in the treatment of scientists were indeed fulfilled. But during the ascent of fascism in Germany (and later in Austria), the world witnessed the enthusiastic persecution of scientists from the very start of the political upheaval. For Albert Einstein the clouds began to gather on the horizon even earlier.[1] Largely because of the outpouring of international acclaim after the November 1919 announcement of experimental support for the predictions of general relativity theory, Einstein came under vicious attack by both political and scientific extremists in Germany. The German ambassador in London even felt constrained in 1920 to warn his Foreign Office privately in a report that "Professor Einstein is just at this time for Germany a cultural factor of first rank. . . . We should not drive such a man out of Germany with whom we can carry on real culture propaganda (*Kulturpropaganda*)."[2] In 1922, following the political assassination of the foreign minister, Walter Rathenau, the news spread that Einstein also was on the list of intended victims. Einstein felt it wise to make a long journey to the Far East and to Palestine, writing from Japan on December 20, 1922, that he had "greatly welcomed" the opportunity for a lengthy absence from Germany so he could "escape the increasing danger." [3]

Ten years later, the power of the state was placed in the hands of Hitler, and the long pent-up floodwaters of hatred finally broke through the dam, washing equally over everyone. Almost overnight, Jewish scientists were dismissed from their posts at the universities and stricken from the rolls of honorific institutions, with virtually no audible protest being raised by their colleagues.

Of the lucky ones who escaped, about one hundred physicists found refuge and a new productive life in the United States between 1933 and 1941.[4] It has often been remarked that their flight turned the personal tragedy, and the tragedy for Germany itself, into an unexpected boon for the intellectual and artistic life of their host country. More than that, in the physical sciences, so the story goes, the influx of refugees from Germany, and later from fascist tyranny in Italy and Austria, provided the necessary critical infusion of high talent that helped to turn the United States rather suddenly into the world's preeminent country for the pursuit of frontier research. Indeed, when it became known in 1933 that Einstein was moving to America, rather than to any of the other countries offering him a haven, the prominent French physicist Paul Langevin was quoted as having announced that the United States would now become "the center of the natural sciences." [5] Lord Rutherford, among others, expressed himself similarly.

These perceptions were not wholly mistaken. But they hid a more interesting and more complex truth that, through the work of a number of scholars, has recently become more evident.[6] It is a truth that I wish to illustrate in this chapter and that may be put succinctly in the following way. At least with respect to the physicists who escaped Hitler's henchmen and fled to the United States, a remarkable symbiosis occurred. While the United States gave European physicists a new life, they in turn provided a new source of energy and a new style of research. This symbiosis would have been impossible without the prior development of a high level of scientific accomplishment in the host country. Einstein did not come to a scientific backwater. On the contrary, he chose to come to the United States chiefly because he was impressed with the achievements already made there (what Robert Oppenheimer later called, with simple understatement, "a rather sturdy indigenous effort" [7]), with the quality of the colleagues, with the conditions of work, and with the bright promise for the future of science in the country. In short, the

United States was, in 1933, a country of natural choice for a physicist whose first loyalty was the pursuit of science.

First contact with America

Albert Einstein's search for a country of refuge and his eventual decision to settle in the United States form a good lens with which to study the migration of physicists to the United States during the 1930s. When Hitler came to power, Einstein was fifty-four years old and intensely occupied with his work in general relativity theory and cosmology. As it happened, in January 1933 he was away from Berlin on a visit to the United States. He vowed that he would not return to his positions at the university in Berlin and the Kaiser Wilhelm Gesellschaft as long as the Nazis were in charge. Suddenly, he was a man without a home, spending the first uncertain months in Belgium and England. His apartment in Berlin and his summer cottage had been raided and sealed, and he had renounced his German citizenship.

In September 1933 he found himself in England, shortly before having to journey back to the United States to spend a few months at the California Institute of Technology (where R. A. Millikan had arranged for Einstein's periodic visits). Einstein did not know that these were to be his last few weeks in Europe. It was by no means clear where he might settle or which of the many options he would choose. One attractive possibility was England. In the preamble to his Herbert Spencer lecture at Oxford in June, Einstein had clearly expressed the hope that this would be the beginning of a closer association.[8] A bill was then pending in the House of Commons to give him the status of a naturalized citizen. Frederick A. Lindemann at Oxford was hard at work arranging for an appointment there.

But offers came to Einstein from many other directions, and in a certain spirit of absent-mindedness, he seems to have accepted quite a few of them. Chairs were waiting for Einstein, or were being arranged for him, in Belgium, Spain, and France, at the Hebrew University in Jerusalem and the newly formed Princeton Institute for Advanced Study. To Langevin, who begged him to consider a post being created for him at the Collège de France, he wrote with characteristic perception, "I find myself in an embarrassing situation, exactly the opposite of that of my compatriots who were chased out of Germany."[9]

In an interview Einstein gave on September 11, 1933, to a reporter from the *Daily Express*, he provided a further glimpse of his unsettled state of mind at the time. Einstein told of Millikan's proposal that he make his home at the California Institute of Technology in Pasadena, then significantly added: "They have there the finest observatory in the world. That is a temptation. But although I try to be universal in thought, I am European by instinct and inclination. I shall want to return here." He never did.

The first indication of the trail that would take him, later in 1933, once and for all to the United States can be found in Einstein's correspondence 20 years earlier. Einstein wrote on October 14, 1913, to George Ellery Hale, the astronomer at Mount Wilson Observatory. Working in Zurich on his first version of general relativity theory, Einstein was at that time by no means a world celebrity. (On the contrary, appended to Einstein's inquiry is a plea from one of his colleagues, Julius Maurer, whom Hale knew, asking the privilege of "a friendly reply to Mr. Professor Dr. Einstein, my honorable colleague at the Polytechnical school.") Einstein was asking the American astronomer's advice on whether one might observe the bending of light from stars near the rim of the sun, when observed against the background of the sun (without an eclipse). Although Einstein's project was not realistic, he was right to consult Hale, whose "rich experience in these matters" Einstein said he valued. Hale was only one of a whole galaxy of American scientists who had demonstrated their experimental prowess. The work of Henry Rowland, Albert Michelson, Theodore Lyman, and R. A. Millikan (not to mention the research of Benjamin Franklin and Joseph Henry) was known to every physicist.

Nor should one overlook the early signs of excellence in theoretical contributions of the Americans. Einstein's own work became the focus of theoretical studies soon after his seminal 1905 publications – for example, by G. N. Lewis at MIT, working alone as well as with his student R. C. Tolman and in collaboration with Edwin Bidwell Wilson; and by H. A. Bumstead at Yale. As Stanley Goldberg has pointed out in his case study of the American response to Einstein's relativity theory, this work by Americans showed a serious understanding of relativity long before the same could be said of some more prominent European scientists, and it exhibited in addition a characteristic "brashness or boldness" of spirit.[10] For example, the Americans accepted the principles of relativity theory as *experimentally proven* (which Einstein himself, aware of the postulational content of his theory, did not claim), and most of them accepted the need to abandon the ether, which many French and British scientists did not do for a long time. The pragmatism of these American theorists, which seems to me part of the antimetaphysical approach that characterized the American style and which later so upset the social scientists arriving from Europe, was an additional early indicator of the vigorous growth of science in America during the early decades of this century.

Einstein himself had noticed this much during his first visit to the United States in the spring of 1921. He had seen a number of universities and was impressed by the promise of the young Americans there, with their unselfconscious manners and their uninhibited urge to do research. As Philipp Frank reported Einstein to have said about the trip, "much is to be expected from American youth: a pipe as yet unsmoked, young and fresh."[11]

When he returned from that visit, Einstein published an essay, "My first impression of the U.S.A.," in which he made six perceptive, rather Tocquevillian points:

1. Contrary to the widespread stereotype, there is in the United States, not a preoccupation with materialistic things, but an "idealistic outlook"; "knowledge and justice are ranked above wealth and power by a large section." (The tumultuous welcome that Einstein was forced to suffer made this point obvious to him.)

2. The superiority of the United States "in matters of technology and organization" has consequences at the everyday level; objects are more solid, houses more practically designed.

3. What "strikes a visitor is the joyous, positive attitude to life." The American is "friendly, self-confident, optimistic – and without envy. The European finds intercourse with Americans easy and agreeable." The American lives for the future: "life for him is always becoming, never being." (Einstein chose not to remark on the occasional evidence of xenophobia.)

4. The American is less an individualist than the European is; he lays "more emphasis on the *we* than the *I*." Therefore, there is more uniformity of outlook on life and in moral and esthetic ideas. But, therefore, one can also find more cooperation and division of labor, essential factors in America's economic superiority.

5. The well-to-do in America have impressive social consciences, shown, for example, in the energy they throw into works of charity.

6. Last but not least, "I have warm admiration for the achievements of American institutes of scientific research. We are unjust in attempting to

ascribe the increasing superiority of American research work exclusively to superior wealth; devotion, patience, and the spirit of comradeship, and a talent for cooperation play an important part in its success."[12]

The Germany to which Einstein had returned in June 1921 offered a bleak contrast. He found the campaign against him progressing more viciously than ever. The very same attitudes that had assured him a good reception in the United States seemed to outrage his opponents in Germany. They saw him as a pacifist, an internationalist who had visited the former "enemy country" less than three years after the end of World War I, a "formalistic" theoretician whose work challenged common sense, a nonconformist, a stubborn and vocal defender of human rights, skeptical of the religious establishment, a Zionist, and a Jew. Einstein himself must have seen in which direction history was lurching: Philipp Frank recalls Einstein telling him in 1921 that he would not likely remain in Germany longer than another ten years.[13] The prediction was close to the mark.

Building the scientific potential

Another indication of the "rather sturdy indigenous effort" and the rapid growth of physics in the United States, even before the refugees arrived in force in the 1930s, is the fact that fully thirteen hundred new Ph.D.s in physics were awarded in the United States in that difficult decade. As if in preparation for this growth, the previous decade had seen a lively exchange across the Atlantic, in both directions. More European post-doctoral physicists chose to go to the United States than anywhere else; conversely, many young Americans went to European centers for a year or two, not as untutored beginners but, as I. I. Rabi was recently quoted by Fritz Stern, "knowing the libretto but learning the tune."[14] Rabi recalls that while traveling through Europe between 1925 and 1927, he encountered other young American physicists such as E. C. Kemble, E. U. Condon, H. P. Robertson. F. Wheeler Loomis, Robert Oppenheimer, W. V. Houston, Linus Pauling, Julius Stratton, J. C. Slater, and W. W. Watson. In her study of the American physics community, Katherine Sopka lists thirty-two Americans studying at European centers of quantum physics between 1926 and 1929.[15] Of these visitors, most of whom soon achieved major recognition, about 40 percent were supported by the new Guggenheim Fellowship Program, with the next most frequent support coming from Rockefeller-financed grants made by the International Education Board and the National Research Council. In short, in terms of the quality and number of its young scientists and the scale of institutional backing, the United States, without planning it, was getting ready, in a way no other country was doing, to become the recipient of the "brain drain," when the time for that would come.[16]

To the evidence given of the transatlantic building of mutual compatibility, competence, and colleagueship must also be added the important role of European physicists who traveled to the United States on lecture tours. During the twelve years following Einstein's first visit, many of the foremost physicists of Europe came to give seminars and lectures, including (in chronological order) Marie Curie, Francis W. Aston, Hendrik A. Lorentz, Charles G. Darwin, Arnold Sommerfeld, J. J. Thomson, Niels Bohr, Oskar Klein, Ernest Rutherford, Paul Ehrenfest, Arthur S. Eddington, Peter Debye, Max Born, Arthur Haas, Abram F. Joffe, Erwin Schrödinger, E. A. Milne, W. L. Bragg, Leon Brillouin, James Franck, H. A. Kramers, Hermann Weyl, Werner Heisenberg, P. A. M. Dirac, Enrico Fermi, Max von Laue, Otto Stern, Gregor Wentzel, Jakov Frenkel, R. H. Fowler, and Wolfgang Pauli.[17] Some of them visited more than once or served as guest professors for a term or a year.

Moreover, there was a relatively small but very significant influx of European scientific immigrants who settled in the United States before the upheaval of 1933, thereby further strengthening the foundation that was being laid. The list includes such distinguished names as W. F. G. Swann (1913), L. Silberstein (1920), P. Epstein (1921), A. L. Hughes (1923), H. Mueller and F. Zwicky (1925), K. F. Herzfeld (1926), S. A. Goudsmit and G. E. Uhlenbeck (1927), L. H. Thomas (1929), G. H. Dieke, Maria Goeppert, J. von Neumann, O. Oldenberg, and E. P. Wigner (1930), and R. A. Ladenburg, C. Lanczos, and A. Landé (1931).[18] Afterwards, the storm in Europe brought to these shores within the next five years such well-established or younger physicist-immigrants as O. Stern, H. Weyl, F. Bloch, G. Gamow, H. A. Bethe, J. Franck, V. Weisskopf, and E. Teller. They came to a country that was by no means unacquainted with or unprepared for their scientific interests or tastes.

In 1921, Einstein had remarked in passing on the custom of private philanthropy in the United States. He could not then have known how greatly it would aid the intellectual development of his future country of asylum, and its story still merits much detailed research. One example

was the role that the grants and policies of the Rockefeller Foundation
played in helping physics in the United States to come of age, particularly
with respect to the rise of quantum mechanics in the 1920s and to the
emergence and growth of nuclear physics in the 1930s. The Rockefeller-
financed agencies also helped greatly in the internationalization of physics. The foundation's aid to physicists involved at least seven factors:

1. Supported by Rockefeller funds, National Research Council (NRC)
Fellowships to study physics had been given to 190 U.S. citizens by
World War II. The list of young awardees contained a large proportion of
later world leaders in their profession.

2. Moreover, they studied at U.S. research centers that, in many cases,
had been transformed in the late 1920s through the (Rockefeller) General
Education Board's carefully placed gifts of about twenty million dollars
for the development of science teaching and research (e.g., at Caltech,
Princeton, Berkeley, Chicago, Harvard).

3. The International Education Board and the Rockefeller Fellowships
encouraged the circulation of physicists among laboratories and institutes
throughout the world precisely when the new quantum mechanics and
nuclear physics were coming into being. The fellowships could not have
been timed better.

4. Nor was physics the only concern of this philanthropy. After its
initiation in 1919, the NRC's Fellowships were awarded to over one thousand scientists in fields ranging from astronomy and anthropology to
psychology and zoology, and to over 350 medical scientists, at a total
cost to the Rockefeller Foundation of about five million dollars. The
scientists studied in nearly 150 universities and research institutions, with
some 20 percent going to foreign countries.

5. In contrast to the NRC Fellowships, which were awarded largely to
U.S. citizens, the Rockefeller Fellowships in the natural sciences, begin-
ning in 1924, were usually awarded to persons from outside the United
States. Of the nearly one thousand fellowships distributed before World
War II, 187 were in physics; the research was carried out in thirteen dif-
ferent countries, with one-third of the recipients doing their work in the
United States.[19]

6. In addition to the fellowships, monies of the Rockefeller Foundation also supported small grants in aid of individual scientific research.
Between 1929 and 1937, ninety-three such grants were made in physics.

7. One must add the foundation's gifts specifically for the pioneering

of large-scale scientific instruments, including the particle accelerators that opened a new era in both atomic physics and the application of physics to biology and medicine and, from 1928 on, the construction of the 200-inch telescope on Palomar Mountain, later named after its moving spirit, George Ellery Hale.

This brief sketch of some dull figures barely hints at the remarkable institutional achievement and the individual perspicacity of a few key officers in philanthropic foundations (e.g., Wickliffe Rose and Warren Weaver) who, quietly and with efficiency and economy, helped build the scientific potential of the nation between the wars.

The fruits of pragmatism

When those young physicists from Europe came to the United States on a visit, perhaps on a Rockefeller Fellowship in the 1920s, what did they find? The testimony of Werner Heisenberg is illuminating. During his visit at the University of Chicago, he gave a series of lectures on the principles of quantum mechanics and also spoke on many other American campuses, including Berkeley, MIT, Oberlin, and Ohio State University. The young man had just come from his new post at the University of Leipzig. There he had, in his first seminar on atomic theory, only a single student! His experience in the United States was quite different, as he noted in his autobiographical account, *Physics and beyond*.[20] In words that are very reminiscent of Einstein's account of his own first impression eight years earlier, Heisenberg wrote: "The new world cast its spell on me right from the start. The carefree attitude of the young, their straightforward warmth and hospitality, their gay optimism – all this made me feel as if a great weight had been lifted from my shoulders. Interest in the new atomic theory was keen."

Heisenberg added an account of an exchange with a physicist from Chicago, revealing how the American style of research at that time enabled young scientists here to forge ahead with remarkable speed on new territory:

> I told him of the strange feeling I had acquired during this
> lecture tour. While Europeans were generally averse and often
> overtly hostile to the abstract, nonrepresentational aspects of
> the new atomic theory, to the wave-corpuscle duality and
> purely statistical character of natural laws, most American

physicists seemed prepared to accept the novel approach without too many reservations. I asked Barton how he explained the difference and this is what he said: "You Europeans, and particularly you Germans, are inclined to treat such new ideas as matters of principle. We take a much simpler view. . . . Perhaps you make the mistake of treating the laws of nature as absolutes, and you are therefore surprised when they have to be changed. . . . I believe that once all absolutist claims are dropped, the difficulties will disappear by themselves."

"Then you are not at all surprised," I asked, "that an electron should appear as a particle on one occasion and as a wave on another? As far as you are concerned, the whole thing is merely an extension of the older physics, perhaps in unexpected form?"

"Oh, no, I am surprised; but, after all, I can feel that it happens in nature, and that's that."

Unlike Heisenberg's friends and opponents in Germany, nobody in the United States seemed to be caught up in those quasi-metaphysical debates on *Anschaulichkeit*. What he encountered here was a pragmatic attitude that brought with it a hospitality to new ideas and to those who could convincingly present them. (Another aspect of the same antimetaphysical approach to the new physics was the essentially American philosophy of P. W. Bridgman, as expressed, for example, in his widely circulating book, *The logic of modern physics*, published about a year before.)

Even during his brief visit, Heisenberg may well have discovered that in the United States the pragmatic ethos tended also to animate the decisions academic institutions have to make when young researchers enter a field of science that is as new to them as it is to their supposed mentors. Thus Edwin C. Kemble, the first scientist in the United States to write a doctoral dissertation on quantum theory (specifically on its applicability to the kinematic theory of gases and to infrared absorption spectra), readily obtained permission to do so from Bridgman, whose own work was of course in a completely different area, in experimental research on the properties of materials at high pressures. Those who wished to take the risk were allowed to set their own agenda. When the wave of forced immigrants began to arrive, just four years after Heisenberg's visit, the welcome that awaited them was based in large measure on the same intellectual and institutional factors that had made Heisenberg feel as if a great weight had been lifted from his shoulders.

To Millikan's Eldorado

We return now to Einstein. We left him in England in September 1933, pondering where to go. Ever since his first, stimulating visit to America a dozen years earlier, that country must have remained in the back of his mind as a possible future residence, if worse came to worst. That feeling would have been reinforced soon by the campaign of the far-seeing, energetic R. A. Millikan to bring Einstein to America as part of his grand strategy to make the new California Institute of Technology into a supreme research university. This, too, is a very American story. As Millikan recalled in his autobiography, he himself was persuaded in late 1920 or early 1921 by his "Pasadena friends," including George Ellery Hale and Arthur Fleming, to consider coming from the University of Chicago on a full-time basis to the newly organized Caltech. They "laid siege to me to persuade me to change my allegiance and accept full-time appointment in Caltech. . . . Dr. Hale was my most ardent wooer. He did not quite tell me that he would shoot himself if I did not yield to his suit, but I did actually have some misgivings about his health if I turned him down."[21]

By the late summer of 1921, when Millikan decided to make the transfer to Pasadena, there was in place a fine new laboratory of physics with an adequate budget, and the beginnings of an excellent staff. The next thing to do was to attract the promising young recipients of the newly established National Research Fellowships and those who were supported by the Rockefeller International Board. Another aim was to put Caltech on the map as a place that could attract distinguished foreign scientists.

Some of these ambitions show up in Millikan's own account of his first shopping spree, immediately after agreeing to join Caltech:

> In September of the year 1921 I went to Europe in response to
> an invitation to participate in the so-called Solvay Congress.
> . . . I used this visit to Europe to persuade Dr. H. A. Lorentz to
> spend the winters of 1921–1922 and 1922–23 at the Norman
> Bridge Laboratory [at Caltech]. I also brought back with me
> from Leiden, as a new member of our physics staff, Dr. P. S.

Epstein, an altogether outstanding theoretical physicist, and I
further went to Cambridge and arranged to have Charles
Darwin join us for the following year. Paul Ehrenfest, Arnold
Sommerfeld of Munich, and Albert Einstein later came on
similar temporary appointments, each, for at least two successive winters.[22]

As the archives of Millikan's and Einstein's papers show, Millikan's
attempts to bring Einstein at least for winter visits started in the early
1920s and continued with increasing energy and prospect of success.
Finally, Einstein agreed to come during the winter of 1930–31, after
receiving in Berlin a persuasive visit from Arthur Fleming, the chairman
of the Board of Trustees of Caltech. Fleming met Einstein probably at
the suggestion of R. C. Tolman, who was working at the Mt. Wilson
Observatory on just the kind of cosmological problems that interested
Einstein.

When Einstein set out for his first California journey on December 2,
1930, he left behind a country that was more and more obviously losing
its grip on its own destiny. By contrast, southern California must have
looked like a sunny and strange Eldorado, but even more remarkable
were the quality of the people and the work they were doing at Caltech,
the atmosphere of collaboration, and the excellent state of the research
equipment. The travel diary that Einstein kept on this trip gave a very
personal glimpse of his reactions:

> *January 2, 1931*: at Institute [with] Karman, Epstein, and
> colleagues. . . .
>
> *January 3*: Work at Institute. Doubt about correctness of
> Tolman's work on cosmological problem, but Tolman turned
> out to be right. . . .
>
> *January 7*: It is very interesting here. Last night with Milli-
> kan, who plays here the role of God. . . . Today astronomical
> colloquia, rotation of the sun, by St. John. Very sympathetic
> tone. I have found the probable cause of the variability of the
> sun's rotation in the circulatory movement on the [surface]. . . .
> Today I lectured about a thought experiment in the theoretical
> physics colloquium. Yesterday was a physics colloquium on the
> effect of the magnetic field during crystallization of the proper-
> ties of bismuth crystals.

For January 15, 1931, Millikan arranged a spectacular dinner at the new Athenaeum, with Einstein as the central attraction, seated next to Millikan and Michelson. The gathering also included the physicists and astronomers C. E. St. John, W. W. Campbell, W. S. Adams, R. C. Tolman, G. E. Hale, and E. P. Hubbell, as well as Mrs. Einstein, and two hundred members of the California Institute Associates. The latter group was from Millikan's point of view the central target, for he had founded it in 1924 as an organization that would pledge large sums of money to Caltech for a ten-year period – "men most able, interested, and active in promoting Southern California . . . [who would] put Caltech on their list of foremost Southern California assets."[23]

When Einstein's turn came to speak to the assembly, he could express his genuine friendship and pleasure at being in that company.[24] Addressing them as "*Liebe Freunde!*" and referring to his own general theory for relativity and gravitation, he said, "Without your work, this theory would today be scarcely more than an interesting speculation; it was your verifications which first set the theory on a real basis." He added an acknowledgment of each of the experimental contributions by Campbell, St. John, Adams, and Hubbell that had been an essential support for the acceptance of his work. While this meeting may not have been on quite the level of the brilliant colloquia Einstein was used to at Berlin, the company of scientists showed that America had come an immense distance scientifically since Ludwig Boltzmann's tongue-in-cheek account of his visit to California, *Die Reise eines deutschen Professors ins Eldorado*, just 25 years earlier.

"A bird of passage"

By mid-March 1931, Einstein was back in Berlin. Two months later, he left the troubled city again to give the Rhodes lectures at Oxford, where Christ Church College made him a senior member ("research student") for eight weeks per year for five years.[25] Here, too, Einstein found himself happy and in good company, with Frederick Lindemann playing the role of Millikan, although on a much smaller scale. The bonds here went back some years as well – to the first Solvay Conference of 1911, where Einstein and Lindemann had met. And, of course, England had had a special place in Einstein's heart ever since November 1919, when the announcement had been made at the Royal Society in London that observations by a British eclipse expedition under Arthur S. Eddington had

confirmed Einstein's prediction on the bending of star light near the edge
of the sun. As Einstein had confessed to Lindemann (in a letter written
on August 28, 1927), he was very aware that in England "my researches
have found more recognition than anywhere else in the world."

As we now approach the moment of Einstein's decision, his oscilla-
tion between the German- and English-speaking countries and the vast
differences in atmosphere that he found in them takes on a special poi-
gnancy. Not long after his Oxford Rhodes lectures, Einstein was invited
to go to Vienna for the first time in nine years and gave a lecture at the
Physical Institute of the university on October 14, 1931. Providing a view
of how things stood was a confidential report from the German Embassy
in Vienna to the Foreign Office in Berlin – one in a constant flow of reports by the nervous German Einstein-watchers.[26] There is a measure of
classic irony here, as the German Embassy, still intent on using Einstein
in its mission of *Kulturpropaganda*, noted how the good Viennese be-
haved, even though seven more years were to elapse before *they* would
throw themselves officially into the abyss of Nazism: "It is typical of the
manner in which in Vienna all things are dealt with from a party-
political point of view that the official Austrian authorities observed
special reserve with respect to Professor Einstein, because he is a Jew, and
considered oriented to the political left." Since neither the education
ministry nor the rector of the universities came to the lecture, and the
distinguished visitor was not received or even invited by any official state
authority (no doubt to Einstein's own great delight), the ambassador
described his frantic efforts to invite some officials at least to have break-
fast with him and Einstein. It was a moment for the stage.

Less than two months later, in December, Einstein was again on his
way westward, for his second visit to Pasadena. At home, the drum beat
was steadily getting more ominous. The National Socialists had become
the second largest group in the Reichstag. In July, the banks had col-
lapsed, and in October the National Socialists, the German National
Party, and the Stahlhelm had consolidated as the National Opposition
for the avowed purpose of forcing Chancellor Heinrich Brüning's resignation.

The travel diary in the Einstein Archives shows that, for Einstein, the
process of crystallizing his decision had begun in earnest:

> *December 6 [1931]:* Yesterday we left the Channel. It is be-
> coming definitely warmer, with rainy weather and considerable
> agitation of the sea. . . .

I have started to read Fridell's spirited *Kulturgeschichte*, Volume III, and Grünberg's Fairy Tales. In addition, I also read Born's Quantum Mechanics.

Today, I decided in essence to give up my position in Berlin. Well, then, a bird of passage for the rest of life. Seagulls accompany our ship, constantly in flight. They will come with us to the Azores. They are my new colleagues, although God knows they are happier than I. How dependent man is on external matters, compared with the mere animal . . . !

I am learning English, but it doesn't want to stay in my old brain.[27]

As this sketch has tried to indicate, by the beginning of the 1930s the process through which physics in the United States was coming of age had reached a satisfactory stage. There was now an adequate balance, such as had been achieved in chemistry a decade earlier, between experimental and theoretical work, adequate provision for undergraduate and graduate training, a strong professional society, a wide range of well-run research publications. Despite the worldwide economic depression, the intellectual, institutional, and financial base was sound. The laboratory facilities in several key places were remarkably good, and the opportunities for study and colleagueship on the national and international level excellent. Also, the interplay between academic and industrial ("pure" and "applied") science research that characterizes the modern United States was well launched. Even that uncertain indicator, the proportion of Nobel prizes awarded to U.S. nationals in physics, was impressive. The country was well on its way to fulfilling the brash prophecy Millikan had made in 1919: "In a few years we shall be in a new place as a scientific nation and shall see men coming from the ends of the earth to catch the inspiration of our leaders and to share in the results [of] our developments."[28]

Thus, even before the major influx of European physicists in the 1930s, the center of gravity of international activity had been shifting westward to the United States, preparing the country to be not only a physical refuge but also an intellectual attraction for the émigrés. And side by side with the professional preparation to receive and put to good use the influx of talent, there was, of course, the special, human quality of American life and spirit that Einstein had described after his first visit. As Victor Weisskopf said of the newly arrived European physicists in the

1930s, "Within the shortest time, one was in the midst of a society that was extremely appealing and interesting and active; and in fact, we felt much more as refugees in Europe than here."[29]

Indeed, the new arrivals, fleeing Hitler's world, came just in time to complete a transition that, at least for the physical sciences, had begun in the mid-1920s, the period Oppenheimer once called the "heroic time." It was therefore both symbolically and historically appropriate that the new, enlarged community of scientists soon was put to work on tasks that not only shaped modern science but also preserved modern life itself. For it is well to remember that these "graduates" of the transformation, and their students, were destined shortly to play a major role in the fight for survival of the civilized world against the war machine of the totalitarian Axis nations.[30]

In contrast to some of the other fields, the interaction of the refugees with the growing field of physics in the United States in the 1930s was a story of mutual benefit. Physics today in the United States is to a large degree the offspring of the happy union of the work of the new and older Americans. Einstein's move to the Western Hemisphere was not a cause but a symbol of a historic process.

PART II

On the history of
twentieth-century physical science

7

"Success sanctifies the means": Heisenberg, Oppenheimer, and the transition to modern physics

It is likely that no science had a more turbulent and fruitful three decades than did physics between the turn of the century and the end of the 1920s. Notions that were unquestioned at the beginning became barely remembered vestiges of a distant "classical" period. The very process of approaching scientific problems was dramatically altered. Even the names of some of the main actors – the Curies, Rutherford, Planck, Einstein, Minkowski, Schrödinger, Bohr, Born, and their generation – commonly stand for "revolutionary" innovations.

And yet an argument can be made that for all their hard-won advances, those physicists were at heart still closer to the classical tradition than to the conceptions most physical scientists today carry in their very bones. Even Bohr had to wean himself painfully from his correspondence-principle approach, with its base in mechanistic explanations; and his complementarity point of view did not turn its back on that tradition, but tried to incorporate it.

To find those who, in that transition period, dedicated themselves without visible ambivalence to the fashioning of a new physics, to find the real "radicals," one has to turn to the pupils of that older generation. Among these, none is a more appropriate candidate than Werner Heisenberg. Here was a young person who appeared to have no ties to old ideas that might impede or delay him. One probably has to agree with the point that "there is simply no ultimate *logical* connection between classical and quantum mechanics";[1] it was therefore perhaps unavoidable that a crucial intervention in the transition from one to the other came from one who seemed to have no *psychological* or *presuppositional* connection with the old physics, either.

Heisenberg's frank letters and interviews[2] greatly enrich our understanding of his many scientific and nonscientific publications, and so are important sources for identifying the personal responses of a modern scientist as he tried to deal with contrary demands and loyalties. He has given us a revealing picture of the context in which he discovered himself as a physicist, just as he was turning twenty, while a third-semester student at the University of Munich. Only four weeks after beginning Arnold Sommerfeld's seminar on "Theory of Spectral Lines on the Basis of the Bohr Atomic Model," the young man is given the honor of being allowed to try his hand at a problem. In fact, Sommerfeld himself had been defeated by it: the explanation of the experimental values of the anomalous Zeeman effect. In those days, before the discovery of electron spin and of spin–orbit coupling, it was not possible to associate the observed splitting of the doublet and triplet spectra in a weak magnetic field with the mechanisms of the otherwise successful Bohr-Sommerfeld quantized model of the atom. The separation between the components of the split spectral lines could be rendered by rational fractions (e.g., $\frac{2}{3}$, $\frac{4}{3}$). But how could these fractions be related to some atomic model governed by the familiar orbital mechanics, where whole quantum numbers characterized the initial and final states of the emitting atom?

In an interview in 1963, Heisenberg recalled this first test as a research scientist:

> Sommerfeld gave me the experimental values of the anomalous Zeeman effect. He had just published a minor paper which he told me was not important at all. He said, "Since we supposed from Bohr's work that every frequency is a difference between two energies, one should expect that in these funny numbers" – which one had in the anomalous Zeeman effect for the splitting – "that the denominator should be the product of two denominators belonging to the two states," . . . and he wanted me to find out what the initial and final states were, using the selection rules. . . .
>
> So after a very short time, I would say perhaps one or two weeks, I came back to Sommerfeld, and I had a complete level scheme. Then I came up with a statement which I almost didn't dare to say, and he was, of course, completely shocked. I said, "Well, the whole thing works only if one uses *half* quantum numbers." Because at that time nobody ever spoke about half quantum numbers; the quantum number was an entire number, you know, an integer. "Well," he said, "that

must be wrong. That is absolutely impossible; the only thing we know about the quantum theory is that we have integral numbers, and not half-numbers; that's impossible." . . . Since I was a complete dilettante and amateur, and didn't know anything, I thought, "Well, why not try half quantum numbers?" [3]

There is no indication that the week or ten days of work had cost the young man any deep anguish or soul searching. On the contrary; when Alfred Landé wrote Heisenberg not long afterward, to warn that this approach challenged the old quantum theory at its foundation, Heisenberg responded simply (October 26, 1921) that "one must give up much of the previous mechanics and physics if one wants to arrive at the Zeeman effect." Even the iconoclastic Wolfgang Pauli was worried; but he got the reply from Heisenberg (November 19, 1921), which announced a leitmotif: *der Erfolg heiligt die Mittel*, success sanctifies the means. When Heisenberg's paper – his first – appeared early in 1922, the reaction was the same as to many of his papers during the next few years: Its brilliance and daring challenged and perplexed his elders. [4]

Perhaps the readiness to give up the past came more easily to Heisenberg because, as he explained in the interview of 1963, his idiosyncratic, quite irregular education had brought him to quantum theory even before he had a good grounding in classical mechanics. But there is also some irony in the fact that this radical innovator was born in 1901 as the son of a scholar of Greek philology of the old German school. If we trace his personal trajectory we find that his development was affected by two very different forces: the vicissitudes of German political history and the breathtaking developments in physics. In both of these, Heisenberg saw himself as having to choose between the themes of tradition and innovation, or to put it in terms of their nearest thematic equivalents, continuity and discontinuity.

On the political side, the choice was predominantly for conservatism; he confessed once that he was attracted by the "principles of Prussian life – the subordination of individual ambition to the common cause, modesty in private life, honesty and incorruptibility, gallantry and punctuality." One of the earliest glimpses Heisenberg has given us of his youth in Munich shows him at the age of about 17. [5] Germany is in the grips of the spring 1919 revolution. The streets of Munich are battlefields. Far from engaging in youthful rebellion, young Werner and his friends do military service on behalf of the new Bavarian government's troops that are recapturing the city. Between guard-duty shifts, the youth retires to the roof of a nearby building, a theological college. He lies down in the wide gutter to soak up the warm sun and to catch up on his neglected studies, specifically on his Greek school edition of Plato's dialogues, as Kepler himself might well have advised.

And there he encounters, in the *Timaeus*, Plato's discussion of the theory of matter. Each small particle is said to be formed and determined by the mathematical properties of four regular solids. Heisenberg is fascinated by the ancient idea that matter in its chaotic diversity is explained by a few examples of mathematical form, and that thereby an orderliness is discerned behind all the seeming infinity of different behaviors of different materials. Going further, he finds himself speculating whether there can be found some similar sense of order to deal with the turbulent events of the day in the streets below. He asks himself, Does one have to discard the old order of traditional Europe and the "bourgeois virtues" taught by one's parents, or should the new order be built on the old? Strangely, more than most others, Heisenberg was to be confronted with such questions all his life, in physics as well as in politics.

"An intolerable contradiction"

By the time Heisenberg was twenty-one, the quality of his mind was unmistakable. In 1922, Niels Bohr had come to give some lectures at Göttingen, not long before going to pick up his Nobel Prize. Heisenberg was now there for a term. In the discussion period, this young student asked questions so trenchant that Bohr invited him to continue their talk on a long *Spaziergang*. Their three-hour-long discussion was the first in a long series of collaborations.

In essence, the problem that first brought Bohr and Heisenberg together was this. Bohr's classic paper of 1913 had served splendidly to show a connection to exist among a number of separate findings. For example, the atom, such as that of hydrogen, can be perturbed by high-speed collisions with other atoms, or by chemical processes, or by radiation; but it always returns to its original, "normal" or stable state. Another curious fact is that the light emitted by a hydrogen source has many, very specific frequencies – but always the same ones, the rest being somehow forbidden for that atom.

Bohr had dealt with the puzzle by proposing the well-known model of the atom as a nucleus surrounded by electrons. The electrons, Bohr held, would normally be in "discrete stationary states" during which there is no radiation. But when an orbiting electron descends between stationary states, radiation is emitted, and shows up as a line in the characteristic spectrum. While in the stationary states, the atom obeys classical laws of mechanics, those going back to Newton; while the atom emits such radiation, however, it chiefly exhibits quantum behavior, by laws first proposed by Planck in 1900. Thus Bohr's atom of 1913 was really a kind of mermaid – the improbable grafting together of disparate parts, rather than a new creation incorporating quantum theory at its core.

In its favor was that the model worked excellently to explain a number of phenomena, for example, the known lines of the hydrogen spectrum. But almost everyone, including Bohr himself, felt from the beginning that this mixed model was only the first step on a long road. Some were repelled by the way in which the discrete quantum rules were being imposed on the continuous laws of dynamics, in violation of classical physics. Otto Stern, after reading Bohr's paper soon after it appeared, had turned to a friend and said, "If that nonsense is correct which Bohr has just published, then I will give up being a physicist." Ernest Rutherford, writing to his protégé Bohr on March 20, 1913, had gently scolded him: "Your ideas as to the mode of origin of spectra in hydrogen are very ingenious and seem to work well; but the mixture of Planck's ideas [quantization] with the old mechanics makes it very difficult to form a physical idea of what is the basis of it."

It also became suspect that a number of ideas basic to Bohr's atom could be imagined or visualized, but not measured. Thus one could not, of course, actually see or test the presumed orbits of the electrons around the nucleus. Nor did the frequency of the assumed orbital motions turn up as anything observable, the frequency of the actually emitted light being connected only with the *differences* of energy in the electron's transition between stationary states.

Now during Bohr's Göttingen lecture in 1922, Heisenberg thought he detected that Bohr was inclining to question the suitability of his own atomic model. As Heisenberg recalled, it "was felt even by Bohr himself to be an intolerable contradiction, which he tried merely to patch over in desperation."[6] During their long walk, Heisenberg learned for the first time how "well-nigh hopeless these problems of atom dynamics then appeared."

Bohr knew already that the problem was one involving not only physics but also epistemology. Human language is simply inadequate to describe processes within the atom, to which our experience is connected in only very indirect ways. But since understanding and discussion depend on the available language, this deficiency made any solution difficult for the time being. Bohr confessed that originally he had not worked out his complex atomic models by classical mechanics; "they had come to him intuitively . . . as pictures," representing events within the atom. Thinking and talking about electron paths in the atom were easy, but the imagery had really been borrowed from macroscopic phenomena, such as watching actual electron tracks in cloud chambers, all on a scale billions of times larger than the atom itself. Similarly the properties of light allowed one to talk about it by analogy either with water waves or, in other experiments, in terms of energetic bullets (quanta). But how could light as such be both? Visualizability and model-based intuitability (*Anschaulichkeit*) of physical conceptions had always been a help in the past; by 1922 in this new realm it was beginning to appear to be a trap. Yet how could one do without it?

In 1921–22 there began a long journey through the wilderness for most of the physicists – even for Heisenberg, who has given a description of that period.[7] These were several years of

> continuous discussion, and we always saw that we got into
> trouble, because we got into contradictions and into difficulties. And we just could not resolve these difficulties by rational means. . . . So we actually reached a state of despair, even
> when we had the mathematical scheme – which was to every
> one of us to begin with a kind of miracle. . . . Out of this state
> of despair finally came this change of mind. All of a sudden
> [we said] well, we simply have to remember that our usual
> language does not work any more, that we are in the realm of
> physics where our words don't mean much.

Wolfgang Pauli declared in 1925, "Physics is decidedly confused at the moment; in any event it is much too difficult for me and I wish I . . . had never heard of it." But by 1928, the struggle had yielded the matrix form of quantum mechanics and the equivalent wave mechanics, and P. A. M. Dirac could write confidently: "The general theory of quantum mechanics is now almost complete. The underlying physical laws necessary for the mathematical theory of a large part of physics, and of all chemistry, at last are completely known." Indeed, although at first only the most adventurous spirits followed this road (and perhaps the majority of European physicists was not converted for several years more), physics had reached a new maturity, with quantum mechanics furnishing the tools for attacking a whole range of obstinate puzzles, from spectroscopy and magnetism to physical chemistry and nuclear physics.

A key event in this transition from despair to euphoria was Heisenberg's own work done in the spring of 1925.[8] He had asked Max Born, for whom he was now acting as assistant, for two weeks of leave. Heisenberg was having a bad case of hay fever and wanted to recover on the barren island of Heligoland.

In this lonely retreat, he worked out a formalistic approach to the understanding of atomic spectra, using a mathematics that more experienced scientists knew as matrix algebra. Heisenberg had totally eliminated the concept of electron orbits, or indeed of any "picture" of the atom. (When Born later saw it, he found this at first "disconcerting.") In its place, Heisenberg put a mathematical schema based on the laws of quantum physics, adjusted by introducing the data (e.g., the frequencies and intensities of the spectral lines) that long ago had been established by observation. In his revealing article "Quantenmechanik,"[9] he declared that the customary *Anschauung* – derived from ordinary space-time conceptions, which are in principle "continuous" – has to give way in the atomic realm to the apparent *Unanschaulichkeit* of "discontinuous" elements. He advised that the grasping of those elements, and of a "kind of reality" appropriate to them, was the real problem of atomic physics.

After the initial consternation among physicists about this approach, it turned out to give results that fit splendidly with the work of others, among them Bohr, Pauli, Dirac, and Schrödinger. Although a small number of physicists reject it to this day as nothing but a halfway house – and although any meaningful version of it has yet to penetrate beyond the walls of science faculties – the goal of a true quantum mechanics had been reached. Even the issue of visualizability, so important in German circles that it found its way into the very title of basic research papers,[10] seemed resolvable when a new kind of *Anschaulichkeit* was reached.[11]

The new quantum mechanics was the product of many hands in direct or indirect collaboration, of individuals and groups, of debates formal and informal. The week-by-week sequence of events, and the apportionment of specific credit, have been the subject of many published and unpublished reminiscences by most of the main actors, as well as of detailed treatises and occasional disputes among historians of science during the last two decades. It does not detract from the usefulness of such research into collegial relations and the operation of the "invisible college" if I focus instead on the individual, and ask here a question prompted by my interest in comparing aspects of the work of two brilliant physicists of about the same age, both trained by the same patron, active at about the same time in quantum mechanics, but with very different careers. The question of interest here is: In what consisted the novelty of Heisenberg's own contribution during the crucial period of 1925–27?

Heisenberg's claim to a key position in the transition to modern physics is hardly challengeable. Thus Max Jammer states flatly: "The development of modern quantum mechanics had its beginning in the early summer of 1925, when Werner Heisenberg . . . conceived the idea of representing physical quantities by sets of time-dependent complex numbers."[12] Such evaluations draw attention to a set of closely related methodological elements in Heisenberg's approach during those crucial years.

1. The theory for the description of quantum-physical events eliminates the immediately pictorial conceptions and interpretations (e.g., orbits, and later the electron itself) that had generally been associated with basic physical quantities. A typical Heisenberg dictum was "The program of quantum mechanics has to free itself first of all from these intuitive pictures. . . . The new theory ought above all to give up visualizability [*Anschaulichkeit*] totally."[13]

2. As signaled by the discrete nature of the observed spectral lines, the physics of the atomic realm is intrinsically discontinuous, requiring the elimination of notions based on continuous representations characteristic of macrophysical kinematics and mechanics. As Heisenberg wrote to Pauli, on November 23, 1926, "That the world is continuous I consider more than ever totally unacceptable," and Born and Jordan in 1925 referred to Heisenberg's work as a "true discontinuum theory."

3. The statistical interpretation as the way to think about nature is to be adhered to fully. Probability is a fundamental feature of phenomena in the submicroscopic regime and need not be imposed on quantum mechanics, as Heisenberg thought Born had done when introducing the probabilistic interpretation of the wave function in 1926. (Born had confessed to be following Einstein in interpreting probability to be a property of a sort of "phantom field.")[14]

Although other views (e.g., Bohr's complementarity, Einstein's realism) persisted as minority opinions, this set of three interpenetrating elements, developed in Heisenberg's work, came to define a position that was identified with the majority point of view on how to think about modern physics. The young man had shown the path by which to cross the Continental Divide.

"It is the theory which decides . . ."

Heisenberg has given us a number of discussions of Einstein and his work; none is more revealing than the account of an occasion in 1926 when, having attracted Einstein's attention, he was invited after a lecture in Berlin to walk home with him. In the flush of his new successes, Heisenberg reported he now believed one must build the atomic model only on the direct results of experimental observations. Moreover, he confessed that in this he thought of himself as following just the philosophy Einstein had used in fashioning the theory of relativity in 1905. But to his consternation, Einstein answered, "This may have been my philosophy, but it is nonsense all the same. It is never possible to introduce only observable quantities in a theory. It is the theory which decides what can be observed." [15] In experiments of any sophistication, we just cannot separate the empirical processes of observation from the mathematical and other theoretical constructs and concepts.

When Heisenberg told me this story, some years before he published it, he added: "Einstein was of course right. Indeed, I myself showed in the paper on the uncertainty relation, written soon afterwards, that the theory even decides what we *cannot* observe." It was a fair point. But Einstein's objection really touched also on another distinction. Unlike Einstein, Heisenberg had never been interested in firsthand, direct experimentation. He trusted that his theory would be safely built on experimental results obtained behind the walls of some other building. For the frequencies of spectral lines, this was (by the 1920s) safe enough. But with respect to the more sophisticated and ambiguous experiments at the frontier of physics and technology, it could be a dubious policy, as we shall see later.

In his book *Tradition in science*, Heisenberg retells the Einstein story and uses it to open the question whether the new physics, in whose growth he had played such a crucial role, had really overthrown "the traditional method in science." Is quantum mechanics, often referred to as a "revolution" in physics, really outside the tradition of work inaugurated in the time of Galileo? Is one right in seeing Heisenberg as the heroic conqueror of qualitatively new territory? Heisenberg did not think so. He said, "The fundamental method has always been the same"; and elsewhere; "I think science always has more or less the same structure. Science changes considerably in its philosophical aspect, but this I would not consider a crisis of science. It is not a revolution, it is evolution of science. . . . [Science actually proceeds by] a more gradual change, which afterwards could be called a revolution." [16]

The changes Heisenberg himself introduced involved nothing as fundamentally different from ongoing science as, for example, Goethe had demanded when he called for a return to a descriptive science, one based not on experiments that draw out new effects "artificially," but on the directly visible, natural phenomena themselves. Rather, Heisenberg held, the tradition of physics in the twentieth century has been continuous in three essential respects: in the *types of problems* scientists select, in the *methods* they employ, and most strongly in the *type of concepts* they use to deal with phenomena.

The tool kit of specific concepts changes in time, of course; thus, luminiferous ether had to be given up after 1905, and with it the dream of reducing electromagnetism to mechanics. Space and time turned out to be more dependent on each other than the Newtonians had thought. The absolutely "objective description" of a physical system turned out to be impossible. At each of these steps, tradition was eventually perceived to have been a hindrance to the progressive development because, as it is usually put, it "filled the minds of scientists with prejudices." But, said Heisenberg at the end of his career, this is a false view of the role of tradition in science, and "prejudice" is a derogatory label for a profound, more positive, and more functional mechanism in the scientific imagination. If we wish to study phenomena, we need a language. At the start of a deep investigation, the new words are not yet available, so we must use old, traditional (and therefore intuitable) ones, to think about the problem and to ask the initial questions in the first place. But these words are tied to the so-called prejudices that form a necessary part of the old language. It may be painful to realize that a natural, traditional conception like the orbit or path of an electron has no meaning when applied to the atom. But during the very act of coming to this seminal realization, the tradition against which the advance pushes is both a hindrance and a necessity.

Moreover, there is no escape from tradition in yet another sense: The opposite of a currently reigning tradition is another, perhaps equally ancient tradition. To illustrate that, we need only look at the topic that most preoccupied Heisenberg in his final years: the concept of the elementary particle in modern physics. He traces the modern theme of elementary entities as explanatory devices back 2500 years, to Democritus. But he notes that our elementary particles, such as the proton, are in fact not uncuttable as the Greek atom was thought to be; on the contrary, a proton has a finite size, and it can be divided or transformed. When energetic protons collide they can give rise to many pieces with sizes like those of the proton. Hence the proton is not truly elementary, but "consists of any number and kind of particles." Even the current candidate for the ultimate unit of matter, the quark, is really no better. In some theories it can be divided – into two quarks plus one antiquark, or similar combinations. Thus, in principle every particle needs for its full explanation all other particles.

Out of this vicious circle, as Heisenberg saw it at the end, there is only one way: to abandon the philosophy of Democritus and the concept of fundamental elementary particles. Instead, one should go back to precisely the antithesis of the Greek atomistic materialists, namely, the concepts derived from Plato, specifically his ideas of symmetry – thereby exchanging allegiance from one old thema to its old opposite, difficult though it now might be to do so: "We have returned to the age-old problem whether the Idea is more real than its material realization." It is not the embodiment of physical discontinuity, the material elementary particle, that beckons as fundamental explanatory bedrock, but the mathematical symmetry properties by which matter on the larger scale is construed.

This thought, Heisenberg confessed, is unfortunately very far from intuitable, and hardly intelligible to the mathematically untutored reader. He might have added that few physicists today are tempted to follow him in this direction. But this is where his pilgrimage had taken him, he who formerly had been the Democritus of our age, who had helped to build the modern atom and describe its essential discontinuities. In his final years he returned to Plato, as if pulled back to the *Timaeus* that had captivated him in his youth.

"We shall be in a new place as a scientific nation"

Let us return now to the crucial period in the 1920s when quantum mechanics was fashioned. I leave Heisenberg and Germany to trace a parallel view of work in progress on the same grand project, but in North America. Indeed, as recounted in Chapter 6, Heisenberg himself, on his visit to America in 1929, found that nobody seemed to be caught up in those fierce debates on *Anschaulichkeit* that had been raging among Heisenberg's colleagues. What he claims to have encountered here was a pragmatism that brought with it a hospitality to new ideas and to those who could convincingly present them.

Heisenberg may well have noted another, more institutional difference: In the 1920s, when some of the younger scientists in the United States pounced on the new quantum physics, they evidently did not have to worry too much about the incredulity or displeasure of their older mentors with respect to those strange new ideas. Since the older generation in the United States was trained not only in classical physics but in addition predominantly in experimental physics, young scientists who wanted to do their Ph.D. thesis on theoretical quantum physics were pretty much on their own. But emotional and financial support was made available to them, in a way that would have been more difficult in a more hierarchical system. The tendency was to put one's money on the younger people, as R. A. Millikan was fond of expressing it, whereas in Europe it was more likely that the young candidate or *Dozent* worked on a problem put to him by the *Ordinarius*.

Something had happened during the decade of the 1920s – the period Robert Oppenheimer later called the "heroic time" – to which few Europeans had paid attention. In convenient shorthand terminology, the period of catching up with Europe has been called "the coming of age" of physics in the United States. Although the issue of national differences is usually overplayed, in this case the existence of an identifiably American effort during those years has been documented by the work of many scholars.[17]

It all happened in a remarkably short time. To be sure, in earlier years, the work of a few outstanding contributors in the United States, mainly experimentalists, had achieved world renown, and that of some others, such as the theorist J. W. Gibbs, had been unjustly neglected. On the whole, however, America at the beginning of the 1920s seemed, with respect to theoretical physics, an "underdeveloped country," far from ready to play a major role on the world stage of physics. The atmosphere has been caught in a few lines by John H. Van Vleck:

> The American Physics Society was a comparatively small
> organization, with only 1,400 members in 1921 . . . and only a
> small number of the communications were theoretical. Very
> few physicists in this country were trying to understand the
> current developments in quantum theory . . .

The problem I worked on was trying to explain the binding energy of the helium atom by a model of crossed orbits which [Professor E. C.] Kemble proposed independently of the great Danish physicist, Niels Bohr, who suggested it a little later. In those days the calculations of the orbits were made by means of classical mechanics, similar to what an astronomer uses in a three-body problem. The Physics Department at Harvard did not have any computing equipment of any sort, and to get the use of a small hand-cranked Monroe desk calculator, I had to go to the Business School. I felt very blue when the results of my calculation did not agree with experiment.[18]

But by the mid-1920s this picture had begun to change rather dramatically. "Although we did not start the orgy of quantum mechanics, our young theorists joined it promptly."[19] It was as if they had been waiting in the wings: Carl Eckart, Robert S. Mulliken, Robert Oppenheimer, Linus Pauling, John Slater, Van Vleck himself, and a rapidly increasing number of others came forward with widely noted contributions, on the very stage where Europeans such as Heisenberg and Pauli, working at well-established centers, had so recently been near despair.

By the end of the decade there was in the United States in physics – as had begun to be achieved in chemistry a decade earlier – an adequate balance between experimental and theoretical work, adequate provision for training at all levels, a much-strengthened professional society, and a spectrum of well-run research publications. The interplay between academic and industrial science, between "pure" and "applied" research, was well launched. R. A. Millikan had made a brash prophecy in 1919: "In a few years we shall be in a new place as a scientific nation, and shall see men coming from the ends of the earth to catch the inspiration of our leaders and to share in the results [of] our developments."[20] It would have been quite sufficient to predict that after a long period of intellectual and institutional subservience to European contributions and styles, American physicists, after absorbing the necessary research attitudes, would be joining the Europeans in the front ranks in terms of the quantity and quality of their contributions.

"Climbing a mountain in a tunnel"

These points, and more, can be illuminated by a sketch of another individual career, that of Robert Oppenheimer.[21] Born in New York in 1904, thus three years younger than Heisenberg, young Oppenheimer, the son of a well-to-do businessman and of a trained artist, received a fine education in scientific and humanistic studies (he later called himself "a properly educated esthete"). In school, writing stories and poetry, learning everything easily, he developed a baroque and exaggerated style that shows up in many of his early letters, and which may even have penetrated his early work in physics. He later confessed that he "probably still had a fascination with formalism and complication," before he fully realized the need for "simplicity and clarity."

It is significant that by his own account Oppenheimer's aim on entering Harvard College was to study chemistry or mineralogy, "with an idea of becoming a mining engineer." He had come to science first of all by becoming fascinated, as a child of five or seven, when on a visit to his grandfather in Germany he was given a "perfectly conventional tiny collection of minerals." In his family one did things in style; so as a high school graduation gift the boy was staked to a prospecting trip to Czechoslovakia – indeed to the mining center in Joachimsthal, famous for its pitchblende from which uranium salt had long been extracted for the manufacture of glass. It was from here that tons of pitchblende residue had been obtained by Pierre and Marie Curie, who then discovered in it the mysterious substance they called "radium." The same area was to be the main supply of uranium, many years later, for the Nazi attempt to make an atomic bomb. Indeed, the German embargo on the export of uranium ore from Joachimsthal was one of the telltale signs noted by Einstein in his letter of August 2, 1939 to Roosevelt, warning that the Germans were on their way.

Oppenheimer visited that mine at a time almost precisely halfway between the Curies' labors and Einstein's letter – one of the mysterious symmetries in his life trajectory that normally are associated only with staged plays. At that excursion, he contracted a nearly fatal case of dysentery, which exacerbated a predisposition to bouts of colitis, melancholia, and periods of deep depression. Again, a strange irony: Having been forced to spend a year at home to recuperate and thereby miss a year before entering college, young Robert was sent on a refreshing trip to New Mexico and the West, with Herbert W. Smith, his admired teacher at the High School of the Ethical Culture Society, who continued to play the role of mentor, serving as companion. The boy fell in love with the wild Southwest – and incidentally came upon a beautiful mesa on which was located the Los Alamos Ranch School. Indeed, the main events in Smith and Weiner's book are balanced between Robert's boyhood visit and, twenty years later, the return trip when Oppenheimer took General Groves to that spot one November afternoon, to propose it as the site for the secret laboratory.

Although the center of gravity of Oppenheimer's college life was clearly in chemistry, mathematics, and physics, he also stretched in other directions: philosophy, literature, continuing his study of languages. To Smith he writes: "I labor, and write innumerable theses, notes, poems, stories, and junk. . . . I make stenches in three different labs, listen to Allard gossip about Racine, serve tea and talk learnedly to a few lost souls, go off for the weekend to distill the low grade energy into laughter and exhaustion, read Greek, commit faux pas, search my desk for letters, and wish I were dead." At an age when Heisenberg walked in Göttingen with the likes of Niels Bohr, the various fragments of Oppenheimer's soul were still waiting for a magnetic field that would line them up.

That magnet came in his encounter with P. W. Bridgman, the experimentalist who later won the Nobel Prize for his researches on high-pressure physics and who was then also developing his operationalist philosophy of science. Looking back, Oppenheimer recalled that as far as science goes, Bridgman's course presented "the great point of my time" at college; for Bridgman "didn't articulate a philosophic point of view, but he lived it, both in the way he worked in the laboratory which, as you know, was very special, and in the way he talked. He was a man to whom one wanted to be an apprentice."

There is, however, important asymmetry. While young Heisenberg shocked Arnold Sommerfeld with his confident innovations in theoretical physics, Oppenheimer was heading for a terrible disappointment with the type of physics he had planned as his life's work. Bridgman later reminisced with some amusement that Oppenheimer had come to work on an experimental problem in his laboratory and at the beginning "didn't know one end of the soldering iron from the other." The delicate suspensions of galvanometers had to be replaced constantly when this apprentice was using them. But he stuck with it and did a quite respectable job of measuring the pressure coefficient of the electric conductivity of alloys to about 15,000 atmospheres – a kind of apotheosis of his initial interest in mineralogy and crystal physics.

Yet for a would-be physicist in the early 1920s, to be not very good in the laboratory, a test Heisenberg never had to face, raised in an American university some doubts about his ultimate promise. In Bridgman's letter of recommendation, intended to help Oppenheimer to go on to Ernest

Rutherford's laboratory at Cambridge, he had to confess Oppenheimer's "weakness is on the experimental side. His type of mind is analytical, rather than physical, and he is not at home in the manipulations of the laboratory." He was not accepted for work under Rutherford and had to settle for a place with the septuagenarian J. J. Thomson. Oppenheimer himself, in a letter of about that time, had discovered that "my genre, whatever it is, is not experimental science." That realization may help to explain, in some part, the depression and identity confusion that overtook Robert when, after continued poor success in laboratory matters, now at the Cavendish, he realized that he still had not made the central discovery that every young person must make: to "find and obey the demon who holds the fibers of one's very life," to quote the splendid words of Max Weber at the end of his essay "Science as a Vocation."

Like many creative persons in their youth, Oppenheimer faced a terrible problem: He felt himself to have the ability to look for central questions, but neither saw them yet with clarity nor had the ability to survive psychologically, without deep anguish, the disappointment of remaining still outside the center. A few years later, in one of his beautiful letters to his brother Frank, he hints at the internal resolution that, to a large degree, came to him upon passing over that threshold: "I know very well, surely, that physics has a beauty which no other science can match, a rigor, an austerity and depth." Its study induces, and demands, a kind of mental discipline that at its best helps one to "achieve serenity, and a certain small but precious measure of freedom from the accidents of incarnation, and charity, and that detachment which preserved the world which it renounces. . . . We come a little to see the world without the gross distortion of personal desire."

Smith had warned an American friend in Cambridge that Robert, on leaving his earlier, pampered life at college, might find "frigid England hellish, socially and climatically," and that if allowed to be overawed, "he'd merely cease to think his own life worth living." That evidently diagnosed part of the problem, but not the fundamental part. Rather, it was first Robert's bad luck, and later his rescue, that this labile, vulnerable, and brilliant young person passed through a personal crisis in the mid-1920s, at the very time when physics itself was passing through a period of despair. The difficulty and fast pace of the science at that time of great excitement clearly exacerbated what can be called at its best a serious problem of temperament; and at the same time, it put to a test what one of his friends of that time called Robert's "ability to bring himself up, to figure out what his trouble was and to deal with it."

Even at the best of times, becoming a competent scientist is a hard test – as Oppenheimer himself once counseled, it is "like climbing a mountain in a tunnel: you wouldn't know whether you were coming out above the valley or whether you were ever coming out at all." But as it happened, for Robert the turbulences of the history of physics and of his own psychohistory had come together in a most frustrating way, and it culminated during a Christmas holiday trip to Paris in an episode that showed the severity of the psychological crisis. The editors of the letters say that, apparently without warning,

> Robert suddenly leapt upon [Francis] Ferguson [a friend and fellow student from Harvard days] with the clear intention of strangling him. . . . The uncharacteristic display of violence, combined with Robert's despair over his inept performance in the laboratory, and confidences about unsatisfactory sexual ventures, convinced Francis that his friend was seriously troubled.

Subsequently, there were other melodramatic episodes, periodic visits to a Cambridge psychiatrist, and evidence of continued depression. That spring, Robert told his friend John Edsall that "he had *dementia praecox* and that his psychiatrist had dismissed him because in a case like this further analysis would do more harm than good." To another friend he said at the time he was leaving his psychiatrist because the fellow was too stupid to follow him, and that he knew more about his own troubles than the doctor did.

A word needs to be said about the use in those days of the phrase *dementia praecox*. The same symptoms might now be called borderline schizophrenia, a term that lacks the stigma of deterioration, debility, and irreversibility. Through the work of Erik Erikson and others on the identity formation of young persons, we are now more ready to deal with such episodes. For cases in which, in the past, young people were judged to suffer from a "chronic malignant disturbance," we now suspect, in Eriksonian terms, a somewhat delayed developmental crisis. The young person is engaged in two tasks. One is the working through of a "negative identity" – the sum of the fragments which the individual has to submerge in himself as undesirable or irreconcilable; and the other is the development of a positive identity to preserve the promise of a traditionally assured wholeness. Erikson speaks specifically about the rage that can be aroused by threatened identity loss and can explode in arbitrary destructiveness, as well as the more traditional explanation of dangerous or radical alternation between loving and hateful tendencies during transference.

Some day there may exist enough material and skill to attempt a psychobiographical study of Oppenheimer. But it is clear already from his letters that, once he had made the decision not to go back into a laboratory but to throw himself into the grand theoretical problems then attracting some of the best European physicists, Oppenheimer had essentially embarked on a course of self-therapy. Those of us who, as teachers and mentors of young would-be scientists, watch them over a period of a few crucial years not infrequently see some version of this process of self-identification taking place before our very eyes. For Oppenheimer, joining in the great intellectual drama of transforming physics – "coming into physics" – helped resolve his dilemmas, once he achieved some success. Some psychological damage remained, however, not least a vulnerability that ran through his personality like a geological fault, to be revealed at the next earthquake. Thus, when Oppenheimer was put on secret trial by the AEC Hearing Board in 1954, the supposedly private record made public by the AEC shows clearly how ineffective he was under fire, how he acquiesced in his own destruction. He later wearily explained: "I had very little sense of self."

But the letters of the young Oppenheimer, and reminiscences recorded later by both Oppenheimer and his friends of that period, show that the years 1926–27, spent at the University of Göttingen, produced that inner alignment that formed the strong armature of his mature self. He met like-minded, challenging young people such as Courant, Heisenberg, Wentzel, and Pauli – a support group with which, as he later said elegantly, "I began to have some conversations." The formal and informal education proceeded at an incredible pace. Thus, a few months after arriving he could write (perhaps with much satisfaction) a letter to his former professor, Bridgman, explaining why Bridgman's rather classical theory of metallic conduction – on the basis of which Oppenheimer had embarked on his ill-fated laboratory work – had to be changed to take the new quantum mechanics into account. At the age of twenty-three, he had essentially achieved his goal of "making myself for a career."

Oppenheimer's mentor at Göttingen was, as it happened, Max Born – no easy taskmaster; Enrico Fermi had been so intimidated by him that he had toyed with the idea of giving up physics, before being rescued by finding sympathetic tutelage under Paul Ehrenfest in Holland. But Born had encountered Robert on a visit to the Cavendish Laboratory, and they discovered that they both were interested in the same problem, the light spectrum radiated from excited molecules. Within a few months, Oppenheimer had finished his Ph.D. thesis, in an article on the quantum theory of continuous spectra, received at the *Zeitschrift für Physik* on Christmas Eve, 1926. Before a year was over, he had written four more articles, one of which, with Max Born, was a quantum-theoretical treatment of the behavior of molecules that remains in use, six decades later ("The Born-Oppenheimer Approximation").[22]

Appropriately enough, having passed through his ordeal, Oppenheimer came back to the United States as a kind of young guru of the new physics. He chose for his main appointment a university in which there was no one working in the new quantum mechanics, where he could begin a career of leadership of a major school of theoretical physics. But it is most significant that he also cultivated genuine working relationships with experimental physicists, the happy harvest of his earlier struggles in laboratories, and also a characteristic of the possibilities in American universities, which rarely separated the theoretical and experimental divisions in the manner that was still rather usual at the time in European instruction and research.

Differences in scientific style

The personal trajectories of Heisenberg and Oppenheimer, when set forth in more detail than would be appropriate here, show remarkable symmetries, intersections, and divergences. Both came to intellectual maturity about the same period, anguishing about the same problems of physics. They were touched by many of the same people (thus a chance meeting with Niels Bohr, while Oppenheimer was still struggling at the Cavendish Laboratory, was fateful to Oppenheimer too; he said later: "At that point I forgot about beryllium and films and decided to try to learn the trade of being a theoretical physicist"). And it is rather uncanny that each ended up being in charge of the project that aimed to produce a nuclear weapon for his respective country during the 1940s.

Oppenheimer's range of interests was to be of great importance to his effectiveness as leader of the Los Alamos project; at that point, the very fate of the civilized world may have depended on it. For, precisely as Einstein and his colleagues had feared in August 1939, the Germans did start on a project to build a nuclear weapon; they began considerably before the antifascist alliance, and had the advantage of excellent and experienced nuclear scientists such as Otto Hahn, in whose laboratory nuclear fission had been first discovered.

The outcomes of those projects were, of course, of historic importance, and the subsequent fate of the world would have been very different if the Americans had failed and the Germans succeeded, or both succeeded, or both failed. Consequently, these wartime projects have been much analyzed and many reasons have been found for accounting for the actual results, from the lack of sufficient support at high political levels in Germany, to the Germans' failure to engage in the large industrial operation necessary for the actual production. Such factors undoubtedly played some role. But one element in the eventual outcome of these "applied science" projects, not negligible but not much remarked on, may be found in the difference of scientific styles characterizing the heads of the two projects, and the sociology of the scientific communities within which they worked.

In barest outline, the point in question is this. In Heisenberg's project, the top responsibility and many of the essential ideas were those of a theoretician, who, in the tradition of his country's university system, had been able to keep a distance from the experimental side of physics. The evidence points strongly to the possibility that the design of the early uranium pile on the German side, strongly influenced by Heisenberg's own ideas, was quite impractical. Also, at least two crucial experimental measurements – of the diffusion of neutrons in the graphite lattice of the pile, and the number of neutrons freed during fission – seem to have been botched at an early stage, and not effectively challenged for a long time or at all, possibly because the hierarchical system, transferred from the university to the weapons project, made such challenges more difficult there.

On the other side, one of Oppenheimer's first acts on assuming leadership – even a condition he made on accepting the assignment – was that an experimental physicist should be provided to assist him. In addition, of course, Oppenheimer's access to, and knowledgeable interaction with, other experimentalists was not a problem for him at all. The interlacing of the theoretical and experimental aspects was complete under Oppenheimer's influence and natural for all who worked with him. In a letter to John H. Manley (the experimental physicist who assisted him, often by mail until the actual focusing of the project, in the spring of 1943, at the Los Alamos site), Oppenheimer writes, "We are up to our ears in every kind of work," especially in a careful determination of the fission-neutron yields obtained from uranium bombarded with neutrons.[23]

Before it was all over, Oppenheimer had to dig deep into his own reservoir of knowledge, both of theory and of experiment, had to stretch and supplement his command of physics ranging from theoretical nuclear physics to metallurgy and ballistics, and, by assignment and example, had to do the same for many others. Those who worked with him then seem to agree that his taste and skills, across this spectrum, provided the essential glue that held the whole enterprise together despite great strains. Without implying a theory of causation, one has no great difficulties in seeing some of the roots of later skills and interests in the documented early period.

In certain other ways, Oppenheimer and Heisenberg were rather similar, not least in their unwillingness to oppose military authority, and in not asking too many questions or allowing dissent on weapons-use policy in wartime. The final irony was that after the war, Oppenheimer became the victim of a group in the military who wanted him to be even more subservient to their projects, whereas Heisenberg and his colleagues succeeded in making most people believe they really never seriously tried to develop nuclear weapons.

I have sketched the outlines of what a "dense account" would be like that focuses on two persons caught up in the outburst of creativity in the physics of the mid-twenties. If we had a fully constructed dense account, how would it be best used toward the goal of studying the mechanisms that propel the creative work in the sciences? In closing, a brief personal remark on methodology seems appropriate here. Following the assembly of a dense account, the second stage is to select for attention a determinate, small portion of it, a portion that on the basis of that account suggests itself as a promising keyhole to the laboratory of the mind under study.

The third stage is to dive into that segment and, typically, select one or more particular documentary artifacts, be it a publication, a laboratory notebook page, a letter, a sketch, a photograph, a piece of apparatus, or nowadays a computer printout or voice record – an artifact prepared during the nascent moment or phase so chosen. A document of this sort might be one of Heisenberg's early letters written while he formulated his approach to quantum mechanics,[24] or Oppenheimer's first paper of lasting scientific quality.[25] Depending on one's preference, the more conventional dimensions of scholarship lead one to establish (a) the historic state of science at the time of the production of the document, (b) the time trajectory of the state of public scientific knowledge leading up to (and, if possible, beyond) the time chosen, (c) the sociological, (d) the cultural developments outside science and the ideological or political events that may have influenced the work of the scientist, and (e) the epistemological or logical structure of the work.

Recently, we have seen also more serious attempts to deal with the personal aspects of the scientific activity, and with the psychobiographical development of the scientist under study. The discussion of Oppenheimer's self-therapy of an evident identity disturbance in the mid-1920s, through his progressive attainment of mastery in quantum physics, may well deserve a place in the full description of the forces at work in the production of one of his early papers.[26] Similarly, young Heisenberg's readiness to make radical departures is made more complex if we also hear the ticking of the time bomb of his early Platonic enchantment.

Last but not least, another tool for the analysis of a scientific work is what I have termed thematic analysis. Here I would point only to the previously noted example, in Heisenberg's work, to establish a physics that is thematically fully built on discontinuity, thereby freeing physics from what he called, in a letter to Pauli (January 1925), the "swindle" of allowing physics to work with a mixture of quantum rules and classical physics, as Bohr and Sommerfeld still tolerated.

When these various dimensions of modern analysis, originating from different directions, are brought to bear on a specific case in the history of science, one can hope to see emerging from the dense account – akin to the ethnographer's "thick description" – an entity with an orderly structure of its own. At the very least, the striking diversity we have seen in the responses of a Heisenberg and an Oppenheimer when both are caught in the same system of tensions indicates the riches that still wait to be discovered in the continuing study of that turbulent period in the making of modern physics.

8

Do scientists need a philosophy?

When George Sarton began his work that led to the founding of the modern study of history of science, technology, and medicine, he was concerned with questions of nineteenth-century mathematics. But searching for the predecessors of innovators was like being on quicksand. Soon he found himself at the beginning of scientific thought in Greece. From there he worked his way systematically forward, reaching the medieval period in some of his last works. By contrast, a growing fraction of today's scholars in the profession Sarton started has chosen to establish its research on the ground of recent and contemporary science, turning the historically trained eye on living cases – the rise of solid-state physics or molecular biology, modern industrial research teams, or the elementary particle physics groups at CERN.

There they are struck by two kinds of discrepancy. First – at least among physical scientists, to which group I shall largely confine myself – the immense forward thrust today is neither enlightened nor diverted by epistemological debates of the kind that engaged so much energy and attention in the past, through the first half of this century (e.g., on the fundamentality of discontinuity, indeterminacy, causality, wave–particle duality, the continuity of scale, etc.). Only a very few scientists now write on ideas that were once at the center of brilliant debates among the élite. It is as if there were now no major puzzles left on the level of the electron paramagnetic resonance (EPR) paradox or the Bohr–Einstein debates.

The second surprise is the marked differences between the popular perception of science, which thinks of it as in a state of periodic revolution, and the contrary opinion of those who supposedly are the revolutionaries. Even among scientists one sometimes hears remarks on the "revolutionary" nature of past achievements that have become part of the corpus, such as relativity theory. But by and large they disavow the revolutionary model in favor of the evolutionary one when they speak of their own work rather than of textbook science. Thus Steven Weinberg writes on the history of physics since 1930: "The essential element of progress has been the realization, again and again, that a revolution is unnecessary."[1] In this self-evaluation, today's scientists only follow in the footsteps of their predecessors, from Copernicus down to our century.

These two kinds of discrepancy are in fact related, and can be resolved together. For a convenient starting point we can turn again to Einstein's remark, now apparently out of tune with current thought even in the most active laboratories, that "Epistemology without contact with science becomes an empty scheme. Science without epistemology is – in so far as it is thinkable at all – primitive and muddled."[2] This notion reflected an old tradition among scientists at the forefront; one thinks here of the Newton–Leibniz debates, or of the mutual effects of *Naturphilosophie* and nineteenth-century science (sometimes with grotesque results). It was not usual for a great scientist to come to the deep puzzles of his field with no interest in or naked ignorance of philosophy. Einstein recalled with pleasure the profound impression which Ernst Mach's writings made on him as a young student, and he listed some of the other authors he and his young friends studied together for self-education: Plato, Spinoza, Hume, J. S. Mill, Ampère, Kirchhoff, Helmholtz, Hertz, Poincaré, and Karl Pearson (as well as Sophocles and Racine). J. T. Merz, writing about this period, commented that the German man of science was a philosopher; but it was not much different in other countries. Thus the American physicist-philosopher P. W. Bridgman recalled that in his last year of secondary school at the turn of the century, he read Mach, Pearson, Clifford, and Stallo. At about the same time, young Niels Bohr was taking an important course in philosophy from Høffding, and was deeply affected by Kierkegaard.

The historical record is clear: right through the first few decades of this century, a good fraction of the young scientists who came to prominence later prepared itself (whether consciously or not) for the interaction of scientific and philosophical questions, and perhaps for eventual candidacy to the charismatic chain of scientist-philosophers and, in Wilhelmian Germany's terminology, as *Kulturträger* – carrier of responsibility for the nation's cultural life that extended far beyond technical scientific work. We may call this the "classical" situation. The result could be found in the scientific journals of the first few decades of this century, for example, in the debates on the theory of scientific knowledge between Planck and Mach in 1909 to 1911, published in the *Physikalische Zeitschrift*; or in Minkowski's essay on space and time, in his explicit use of imagery taken from Plato; or in the difficulties the Curies had in bending their positivism enough to consider Rutherford's ideas on transmutation; or in the debate surrounding Jean Perrin's insistence on molecular reality; or in Heisenberg's struggle with vestiges of Kantian demands of *Anschauung* and *Anschaulichkeit* in atomic physics (even in the titles of some of his scientific papers in the 1920s);[3] not to speak of the well-known epistemological discussions between Heisenberg, Bohr, Born, Schrödinger, Einstein, and de Broglie.

In different degrees these men saw themselves as both scientists and culture carriers, with the duty, or the psychological need, to fashion a coherent world picture. The most ambitious expression of that hope was the projected thirty-six-volume *Encyclopedia of unified science*, planned in the 1920s by Otto Neurath, Einstein, Philipp Frank, Hans Hahn, and Rudolf Carnap. At least in the main Western countries before about 1945, one would have been surprised if serious, aspiring physicists had not been exposed to, and intellectually civilized by, some of the "tribal books" by scientist-philosophers of the kind young Einstein read, or by others such as those named above, or by Duhem, Schlick, Russell, Eddington, and Jeans.

This classical preparation is now dead. The "tribal books" are no longer read – with a very few exceptions the whole genre has disappeared, giving way to occasional autobiographies in the style of James Watson's *Double helix*, or textbook expositions of straight scientific content, or, in the Marxist countries, party texts. When Sheldon Glashow was asked recently what he and his fellow students had read outside science in their formative years, he named science fiction, Immanuel Velikovski, and L. Ron Hubbard. It is not an untypical list. Moreover, when it comes to more recent and current works in the philosophy of science, the editor's summary in *Springs of scientific creativity: Essays on founders of modern science* seems close to the mark when he reports that working scientists on the whole now regard those products "as a debilitating befuddlement, and shun the vagueness and generality of much that has been said and written."[4]

These uncomfortable facts raise an important paradox. For despite the decay of explicit allegiance to the scientific-philosophic tradition, science in our day is without doubt as powerful and interesting as it has ever been, both as a product and as a process. The intellectual constructs are more far-reaching in their control over the phenomena, and the techniques more sophisticated than had been thought possible. A relatively few fundamental conceptions and metaphors provide the armatures that hold up complex structures in widely different specialisms.

One would therefore expect that the commonalities among the sciences, their shared philosophical underpinnings, would be more evident in scientific activity than ever. But this is precisely what we do not find. Therefore, two questions arise: How did this come about? And how can science be done so well without the conscious contact with epistemology that characterized the classical mode?

The first question can only be touched on here. It would be folly to think that the deep questions which explicitly preoccupied scientists for centuries are of no further interest because they have been answered. Rather, they seem to be temporarily suspended. We are at present in one of those periods of historically unusual euphoria about the prospects for continued rapid advance in the physical (and biological) sciences. The road seems fairly clear and relatively unimpeded. Under such conditions it is not surprising that there is no great incentive for most scientists to become introspective. But sooner or later science has always run up against some severe, apparently insurmountable conceptual obstacles. That is when the euphoria turns – for a time – to despair. The word *despair* is just what you find in the descriptions Planck, Einstein, Bohr, and Heisenberg have given of their own state of mind at crucial times.

This is bound to happen again. I expect that in that valley of despond, some of the best scientists will turn again to self-conscious epistemologizing, and they will say, as they have in the past, that scientists do need a philosophy. For the time being, however, such preoccupations are in a state of hibernation. In their place, new extra-scientific sources of stimulation and strength have appeared during the past few decades, new sociological externalities that are considered more nourishing than a continuation of the self-conscious introspection of the previous decades. These new factors include the large increase in the number of scientists, of funding sources, of administrative structures that represent long-range commitments. These in turn brought the better support of bright students, the greater strength of professional societies and their journals, the greater freedom for scientists in the West to travel and become part of the international community.

The ever-closer collaboration of science, technology, and engineering has also had a transforming effect. This is obviously the case for the

experimentalists, whose apparatus design, data reduction skills, computerization, and organization in big teams were all markedly shaped by
what they learned during their engineering-directed service in the laboratories of the Second World War and its Cold War aftermath, and by the
ready-made equipment available from industry. A striking example is the
direct link between military technology and the confirmation in 1959
of Gell-Mann and Nishijna's prediction of the cascade zero in particle
physics.[5] The huge seventy-two-inch liquid-hydrogen bubble chamber
used for the test was constructed and operated under Louis Alvarez's
excellent team of structural cryogenic and accelerator engineers, putting
to use skills developed in military laboratories from the Manhattan Project to the Eniwetok hydrogen-bomb test (in fact, the compressors used in
the seventy-two-inch chamber had initially been made and used during
that test in 1952). But not only the experimentalists were affected by
those links. Thus Julian Schwinger,[6] who worked on electromagnetic
problems of microwave and wave guides in the early 1940s (as, in fact,
did Sin-itiro Tomonaga on the opposite side), later applied the methodological lessons of this work to the effective-range description of nuclear
forces, leading to the concept of renormalization. Willis Lamb similarly
traced his discovery of the "Lamb shift" to his wartime work on mag-
netrons.

Such externalities can account only in part for the recent somno-
lescence of the old philosophizing impulse. Another factor in the flight
from what is now considered "debilitating befuddlement" may well be
the perception by the large majority of scientists, right or wrong, that the
messages of more recent philosophers, who themselves were not active
scientists, are essentially impotent in use, and therefore may be safely
neglected. The possibility of this harsh judgment may be illustrated by
the testimony of two well-placed observers. The first is Hilary Putnam,
himself a philosopher of science at Harvard University. In his essay
"Philosophers and human understanding,"[7] he argues that in the end the
main schools of scientific philosophy, despite their early promise and
their hold on the imagination of major scientists in what I have called the
"classical" period, turned out to be failures. The early positivists, building on Frege and Mach, inspired the vain hope that the scientific method,
including its inductive core, might turn out to be an algorithm, a mechanical proof procedure that permits "rational reconstruction." Today,
Putnam reports, it is widely agreed that this is impossible, for there is
always a need for judgment of "reasonableness." (Ironically, this liberalization has allowed anarchism to put in its strident claim here and there.)

But scientific understanding is human understanding after all, differing only insofar as research scientists may reasonably hope to be able to converge on the same conceptions. The implicit model for scientific progress is not solving a puzzle once and for all, but the evolving emergence of tolerable solutions from which better problems will become available for future work.

The verification principle of the logical positivists, that nothing is rationally verifiable unless it is critically verifiable, did indeed serve well to free scientific thinking from the fetters of explicit metaphysical doctrine in the nineteenth century. But Einstein had sensed the limits of this service earlier than most. As he wrote to his friend Michele Besso (May 13, 1917), "I do not inveigh against Mach's little horse; but you know what I think about it. It cannot give birth to anything living, it can only exterminate harmful vermin."[8] Summing up the results within philosophy six decades later, Putnam's judgment is more sweeping: that the work of the logical positivists and the recent post-positivists has been refuted, and is self-refuting to boot. That is of course not to say that rational argumentation and justification are impossible. But the attempt to consider them certifiable by appeal to public norms is an absolutistic delusion. To this, the student of case histories of modern physics adds a specific point, that if the claims of falsificationism had merit in earlier science, when the chain of hypotheses was shorter, it hardly corresponds to successful scientific practice today. On the contrary, it is striking how often a theory is delayed or abandoned because of initially credible experiments that turn out to have been flawed. In this manner, Schwinger's close approach to a successful electro-weak theory was blocked in the 1950s, and the acceptance of weak currents was long delayed.

In sum, Putnam's judgment amounts to a declaration that by its own criteria much of recent philosophy of science has been a degenerating research program. Coming from the other direction, the voice of a practicing scientist reinforces that of the practicing philosopher. In his essay in the same collection, "Rationality and science," Henry Harris, the Regius Professor of Medicine at Oxford, provides an urbane analysis that amounts to a withering indictment parallel to Putnam's. He finds, for example, that the works of the most senior among current philosophers of science "not only undermine the classical picture of scientific method . . . [but] also eventually undermine his own position, so that nothing in the way of a coherent logical structure for science remains." The methodological rule specifying which hypothesis to prefer (i.e., the one with greater empirical content) is, "on the face of it, an attractive notion, but is of no use at all to the practising scientist" – in part because the "empirical content" can't be known in advance, but unfolds as the hypothesis is explored. And the conception that the work of scientists replaces one hypothesis by another in unending series, whether commensurate or not, neglects that they "provide facts," facts that don't have to be changed; blood does circulate.

Moreover, a discussion of what induction from oft-repeated observation can tell about the future is also off the point, for a scientist does not blindly repeat his own or another's experiments; he deliberately introduces changes to get more information. To make the point, Harris says simply, "He kicks the problem around," and he adds the initially shocking thought that what appears as philosophical naiveté (as in the intuitive acceptance of rational realism) can in fact be a fruitful mode of operation at the bench. Indeed, if scientists in the past listened to philosophers, the converse should be true in our day: the claims of a "logic" of scientific discovery do not stand up to a "detailed study of what scientists actually do." No list a priori makes a hypothesis intrinsically more probable to be right than its rivals. The best strategy is, again, an evolutionary one, in which the value of hypotheses is found a posteriori.

Professor Harris, a philosophically sophisticated scientist, ends with the judgment of laboratory common sense: "Rationality helps, but it is not a prescription for making discoveries." The "rational scientist," therefore, is a "thorough-going empiricist who never troubles his head about the logic of what he is doing." He makes mistakes, but also gets things right. He "has no reservations about the ability of scientific procedure to verify and to falsify scientific propositions,"[9] and he submits his publications in the hope that others will build on what is valuable in them.

What emerge from such accounts are some characteristics of the post-Second World War styles of the scientific imagination – styles productive of superb results even if unappealing to some of us who were brought up in the earlier setting. (In this, as in other ways, there may be more than an analogy with the contemporary scene in the pictorial arts, music, and literature.) If the older role models were the philosophically introspective Poincarés or Bohrs, the newer ones are the apparently philosophy-immune successors of the brash young experimentalist Rutherford – who had no hesitation in proposing the near-alchemical conception of transmutation, typical of his constant stream of hypotheses and simple meta-phorical explanations – or the totally agnostic Enrico Fermi, of whose immensely successful theoretical work Bohr was reported to have said in dismay that it was attained in "too elementary" a way, too "cheaply." Einstein's constantly repeated epistemological credo was, after all, also a plea for liberalization from school philosophies, for the useful role of "free invention." And Bridgman's simple dictum that the scientific method is "doing one's damnedest, no holds barred," signaled a self-confidence, a skepticism of methodology, and an impatience with earlier authority that have become earmarks of modern scientific practice.

The success of this style seems to have entered even into the selection process by which young collaborators are chosen. Maxine Singer, chief of the Laboratory of Biochemistry, National Cancer Institute of the Na-tional Institute of Health in the United States, recently spelled out the desired characteristics of young scientists brought into the laboratory. A high value is placed on the "degree to which they challenge their senior colleagues" in scholarly discourse. She sees the necessity of preserving such "troublesome inclinations for their motivating force" as may be found in evidence of ambition, aggressiveness, even belligerence. She believes it is up to the social organization of science, in the words of Jacob Bronowski, to "transmute these brutish energies into disciplined inquiry by the community as a whole."

Eric Ashby, fifteen years ago, made a similar point. To prepare a per-son to participate in research, we must first make him familiar with "the orthodoxy," but then instill in him the principle of "constructive dissent: it is this discipline of dissent that has rescued knowledge . . . from re-maining authoritarian and static." Dissent, shading into unbuttoned ir-reverence: one recognizes the traits of a young Francis Crick or Richard Feynman.

Self-propelled by an unepistemological confidence, encouraged by the ability of his mentor to suspend disbelief in the face of "disconfirming evidence" for long periods, the scientist now has early and continuing psychological support for risking hypotheses that, in my view, would have had much difficulty in the first half of this century passing through the filter of presuppositions. The unkempt style of today shows up in the very terminology proposed for new scientific concepts, forcing the *Physi-cal Review* editors on occasion to reject risqué neologisms – a problem that, say, Max Planck, as editor of the *Annalen der Physik* could not even have imagined. Moreover, the older *sine qua non* of sound work such as the sturdiness of hypotheses connecting the observed events to the underlying theory, or the quick repeatability of new phenomena, has become technically almost impossible in many advances – in research that involves large teams, when complex and expensive apparatus exists in the given form only in one place, when the search centers on a very evanescent signal in the noise. An exemplar of this syndrome is the exhibition of a single (or "gold-plated") event in the bubble chamber, as in the proof of the existence of the omega-minus particle or the demonstration of the K^+ decay process.[10] We have come a long way from the reluctance of J. C. Street, who in 1937 had only a single cloud-chamber picture of the muon, and so hesitated to claim the discovery of the new particles.

But if unepistemological confidence now were all, what prevents the process of innovation from degenerating into mere fantasy? If that were the main strategy for scientific research, it would act in most cases as a centrifugal tendency, leading soon to a crossing beyond the frontiers of good science. What prevents physics from becoming a cousin to astrology? Why not anarchy? Some sort of sound epistemology must be at work after all, even if it is subterranean or not fully conscious. Therefore we should look for offsetting, centripetal tendencies. And this is just what we shall find. But we have not yet exhausted the characteristics of the modern style. The almost improvisational heuristic (in Whewell's sense of the word heuristic: serving to discover) sometimes seems like pole-vaulting over obstacles. Thus the twenty-nine-year-old Sheldon Glashow, in his paper on "Partial symmetries of weak interactions" on which his 1979 Nobel Prize would be based, announced simply: "The mass of the charged intermediaries must be greater than [zero], but the photon mass is zero – surely this is the principal stumbling block in any pursuit of the analogy between hypothetical vector [bosons] and photons. *It is a stumbling block we must overlook.*"[11]

The new physicist is unimpressed by those among his forebears who, having read Mach or Pierre Duhem, had trained themselves assiduously to avoid theories containing empirical unmeasurables; he is equally neglectful of more recent philosophies of science that earnestly invoke demarcation criteria and use them to claim that elementary particle physics is a "degenerating research program." None of these dampens speculative proposals as long as there is felt to be "good reason" – not in the sense in which a "good hypothesis" is one that has already been corroborated, or even in Carnap's sense, where "good reason" is equivalent to high probability. Rather, "good reason" is part of an expression of the risk-taking, "what-if?" improvisational heuristic that allows proposals to be made without regret even when they have highly implausible aspects, or when tests are not likely to be possible in the foreseeable future. Thus C. N. Yang and R. Mills pioneered gauge field theory in a paper of 1954, despite the prediction inherent in their theory that there should exist charged but massless particles. And Glashow wrote in the *Physical Review Letters*:

> We present a series of hypotheses and speculations leading inescapably to the conclusion that $SU(5)$ is the gauge group of the world – and that all elementary particle forces (strong, weak, and electromagnetic) are different manifestations of the same fundamental interaction involving a single coupling strength. . . . *Our hypotheses may be wrong and our speculations idle, but the uniqueness and simplicity of our scheme are reasons enough to be taken seriously.*[12]

"Reason enough" is a simple short-circuit across acrimonious debates on warrants for rationality. The closest philosophical predecessors of the reason-enough style are David Hume and Charles S. Peirce. Hume distinguished between what is rational and what is reasonable, and allowed criteria of "reasonableness" even when the basis of the original judgment was ultimately only intuitive. Peirce, the nineteenth-century American mathematician-philosopher and originator of pragmatism, thought of creative work in terms near to Galileo's *il lume naturale* of reason, and akin to Kepler's readiness to let his presuppositions interact with the empirical material before him. Peirce's is not the logic of discovery from books, such as Descartes's or Bacon's, but a logic-in-use. His process of abductive inference is powered by the unashamed proposal of forward-looking hypotheses that are scrutinized, and made corrigible by experience and disciplined thought – not only by its originator, but by the community of scientists, engaged in the self-correcting process of public discourse. The warrant for any scientific innovation lies in the future, in the outcome of further inquiry. He wrote: "The best than can be done is to supply an hypothesis, not devoid of all likelihood, in the general line of growth of scientific ideas and capable of being verified or refuted by future observers." To this future-and-community orientation, Peirce adds an action orientation. The concepts used in scientific discourse were to be what later came to be called operational: "The meaning of a concept . . . lies in the manner in which it could conceivably modify purposive action, and in this alone."[13]

This attitude, verging on the outrageous when first proposed, has since become an element of the implicit lore and tool kit of most scientists. Few among them would be able to cite a source either for this operational credo or for the discussion among philosophers of its highly problematical nature. They are now far more interested in the communal support-system in which they speculate and experiment. They see the scientist's chief duty to be not the production of a flawlessly carved block, one more in the construction of the final Temple of Science. Rather, it is more like participating in a building project that has no central planning authority, where no proposal is guaranteed to last very long before being modified or overtaken, and where one's best contribution may be one that furnishes a plausible base and useful material for the next stage of development.

This methodology-in-action-and-for-the-future is well described in a metaphor proposed by Putnam. He modifies Otto Neurath's picture of science as the enterprise of constructing a boat while the boat floats on the open ocean:

> My image is not of a single boat but of a *fleet* of boats. The people in each boat are trying to reconstruct their own boat without modifying it so much at any one time that the boat sinks, as in the Neurath image. In addition, people are passing supplies and tools from one boat to another and shouting advice and encouragement (or discouragement) to each other. Finally, people sometimes decide they do not like the boat they are in and move to a different boat altogether. And sometimes a boat sinks or is abandoned. It is all a bit chaotic; but since it is a fleet, no one is ever totally out of signalling distance from all the other boats. We are not trapped in individual solipsistic hells (or need not be) but invited to engage in a truly human dialogue, one which combines collectivity with individual responsibility.[14]

This lowering of explicit epistemological barriers has resulted in sociological effects peculiar to contemporary science. A larger number of practitioners feels invited to participate; more students, even undergraduates, can take part in research on frontier problems, and appear as co-authors of publications. The size of teams is growing steadily, with the maximum now around 150 and heading to even higher levels. Similarly, the number of competing and mutually reinforcing teams is growing, as more daring problems are being attempted and the collaborative benefits of problem-solving are proving themselves in practice. The absence of idiosyncratic epistemological commitments also has the advantage of easing international collaboration, as the differences between national styles have been disappearing. Correspondingly, persisting differences between schools of thought within a given nation are now rare. The walls between disciplines have also become more permeable. On this last point, the useful intrusion of technology and engineering into physics has been mentioned; a similar finding extends to the other sciences and to mathematics.

As we noted, the centrifugal tendencies, if left to themselves, should be tearing science apart. They cannot do so because they are only one part of the total armamentarium. As soon as we also include the unacknowledged and often subterranean modes of scientific thought, we find a centripetal tendency at work also, and Einstein's dictum on the necessary links between science and epistemology turns out to be correct after all. In brief, the free leap made during the process of innovation is still bounded by an adherence, unselfconscious but strong, to long-established and enduring conceptions. The scientist does need, and in fact does use, a philosophy of science, whether or not it is held consciously and openly. Moreover, controversies between scientists are at bottom still about differences concerning which of these old conceptions to give one's full allegiance to. These attachments are what makes contemporary science a recognizable offspring of earlier science. We can safely predict that they will also connect the future state of the field, despite all apparent changes, firmly with the present.

The enduring elements to which I refer are somewhat like the old melodies to which each generation writes its new words. They are the *thematic concepts* (such as evolution, devolution, or steady state); *methodological themata* (e.g., the practice of expressing regularities in terms of constancies or of extrema; or forming rules of impotency); and *thematic hypotheses* (such as the postulation of the discreteness of electric charge, or the wrong hypothesis of continuity for light energy, widely held for years after contrary evidence was at hand). I have discussed at some length in case studies[15] how such thematic materials can guide individual decisions – whether to success or not – either during the nascent phase of scientific work, or during the controversies between rivals. Thus, from the very beginning of modern science, a presuppositional allegiance to the plenum, or to either atomism or the continuum as a ground of explanation of phenomena, has shaped the way scientists have used the other main components of their discourse, namely the empirically available content and the analytic devices of logic and mathematics. Such a presupposition can explain how Max Planck, at an early stage, could predict confidently that the assumption, by different schools, of finite atoms and of continuous matter respectively would "lead to a battle between these hypotheses in which one of them will perish," and he added his bet that despite the "great success of the atomic theory [it] will ultimately have to be abandoned in favour of the assumption of continuous matter."

In this, Planck agreed with Einstein, who put the hypothesis of atoms and quanta to superb use but nevertheless thought that the basic explanations will ultimately have to come from the continuum. Among the other themata which a study of Einstein's theory construction reveals are these (as noted in Chapter 2): primacy of formal rather than materialistic explanation; unity or unification, and cosmological scale (applicability of laws throughout the total realm of experience); logical parsimony and necessity; symmetry; simplicity; causality; completeness; and invariance. His attachment to these themata explains in specific cases why Einstein would obstinately continue his work in a given direction even when tests against experience were difficult, or unavailable, or apparently disconfirmatory.

Although themata are rarely verbalized and hence cannot be found in the index of textbooks, an analysis of contemporary physicists' writings will yield most of the thematic concepts that were active in Einstein's days, plus a few other well-established ones such as the methodological thema of using metaphors, or of establishing conceptual hierarchies. There are now also a few differences from Einstein's list, such as the new presupposition in favor of fundamental probabilism, the antithema to classical causality. The stability of the scientific enterprise despite the profound changes during the past three centuries is largely due to the longevity of most reigning themata, as well as of the choices given by thema–antithema couples; the relatively small number of them; and the remarkable rareness of the need to introduce a novel thematic concept (complementarity and chirality being the last major new entries in physics in this century).

It will suffice here to mention only what may be the most ancient and persisting of these thematic conceptions, acting as a motivating and organizing presupposition to this day. It is of course the attempt since Thales – the "Ionian Fallacy" – to unify the whole scientific world picture under one set of laws that will account for the totality of experience accessible to the senses. One aspect of this commitment is the hope, ever new in detail but the same in essence, to achieve a unification of all the forces of nature. Oersted was committed to finding it before doing the experiments that revealed the link between electricity and magnetism; Faraday called it a "dream" he hoped to realize for all forces, including gravity; Einstein devoted more time to this dream than to anything else; Julian Schwinger called it the "grand illusion"; and in its current version it is in full swing today in the attempts to fashion versions of a Yang-Mills gauge field theory able to account for every particle, every force, through a single principle.

It was precisely in the pursuit of the unification of electromagnetic effects and phenomena associated with weak interactions that Glashow in the early 1960s said he would suspend his disbelief in the face of apparent paradox, and would "overlook" that stumbling block. It was in the service of the conception "that all elementary particle forces . . . are different manifestations of the same fundamental interaction," together with the thematic belief in the uniqueness and simplicity of the scheme, that Glashow and Georgi wrote in 1974 there was "reason enough" to take the scheme seriously, years before any tests became feasible, and despite their confession of being forced to build "outrageous ideas" into the theory.

That illustrates the chief point made under this heading: the apparently unepistemological style today is still in the service of an ancient quest, transmitted from generation to generation: the pursuit of a few basic themata – by their very nature unverifiable and unfalsifiable presuppositions – that help to guide the search for order, though always disciplined by eventual accountability to sharable experience. The modern philosophers' apparatus of strict demarcation criteria, of the logic of justification, of the supposed incommensurability of successive stages of science, has not been able to deal with the persistent thematic side of the scientific imagination. Yet it is the latter, the old internally directed or center-seeking part of the process, which serves as the counterpart to the new, externally directed or center-fleeing element in it. Together they stabilize each other, while leaving the necessary elbow-room for the imaginative act.

We return finally to the issue with which we began, whether science is in a state of constant or frequent revolution, as popularly perceived, or is not, as reported by the scientists at the frontiers. The eye and ear of the outsider tend to miss evidences of the continuity of the scientists's allegiance to a few well-established, persisting themata even through drastic changes of analytic or phenomenic detail, a continuity that reminds and assures the individual scientist of his connection to his historic forebears. Despite all superficial differences, an elementary particle theorist might be (and in fact has been) overheard to say: "We might now stand in a position analogous to that of Oersted, Ampère, and Faraday"; he can locate himself on the trajectory. In just this way, Einstein constantly protested[16] that the relativity theory was only a "modification" of the existing theory of space and time, "not differing radically" from the development initiated by Galileo, Newton, and Maxwell.

To regard one's own work as truly revolutionary would require the discovery by the scientist concerned that the whole set of thematic presuppositions on which he and his contemporaries have been relying turns out to be in need of replacement by the corresponding antithemata. That would indeed make the new incommensurable with the old. But such a wholesale change is quite unlikely to happen. (The one scientist who came close in this century – as we saw in Chapter 7 – was the young Heisenberg.) The main thrust has been and undoubtedly will continue to be the continuation or slow evolution of the few core ideas. That is not to say that scientists all hold the same set of thematic beliefs, or are equally well served by them, or cannot differ drastically on some deep issues, or cannot discover when a thematic choice is not functional and must be abandoned. But the individual spectra of thematic commitments active at any time in the scientific community show considerable overlap, and hence ground for substantial agreement. Therefore even a far-reaching change such as that from Maxwell's work to Einstein's requires for the individual or the community no conversion, "Gestalt switch," or similar dramatic discontinuity of all beliefs, but merely the eventual accommodation of a few components out of the otherwise largely invariant set of current themata.

This analysis of current dimensions of scientific innovation might prompt the question: Where has the energy of explicit philosophizing among scientists gone? Perhaps that is a pseudo-problem. But if there exists a natural tendency to such conscious introspective activity, it may be that it has been merely transposed from concerns of the individual to the problems of the scientific community at large (a shift parallel to that concerning the value of hypotheses). Thus the anxious individual inquiry into the warrant for rationality has been replaced by discussions among some scientists of questions coming from another branch of philosophy, namely ethics. (In a sense this returns scientists to the concerns of Socrates, and to the idea, in seventeenth-century discussions, of the parallelism of scientific and spiritual progress.) The professional societies of scientists (the American Physical Society, the American Chemical Society, etc.) have become notably involved in questions of ethics and human values, such as the access to science of previously disadvantaged groups; the rights of scientists to object to unethical practices; the human rights of colleagues in totalitarian systems; the desperate need for arms control, as well as for a sharing of scientific resources with Third World countries.

To a degree unimaginable a few decades ago, scientists are discovering that there is a morality which the enterprise of science demands of itself – even if such concerns are as yet expressed by only a small fraction of the total community. Indeed, with about one-third of the world's scientists and engineers working directly or indirectly on military matters while the arms race proceeds unchecked, this transfer of attention from epistemological to ethical problems may be too little and too late. At this ominous junction of science and history, as we watch the growing reign of the irrational in world affairs, the debates of former times to give precision to scientific rationality seem curiously antiquated. Perhaps this redirection of philosophical concerns signals a growing awareness that the process of scientific innovation is not in danger – but that humanity is.

9

Science, technology, and the fourth discontinuity

The title I have chosen borrows, I should confess at once, from an article by Bruce Mazlish.[1] Mazlish reminded his readers that Sigmund Freud, in his lectures at the University of Vienna between 1915 and 1917, identified the elimination of three conceptual discontinuities that marked the development of modern Western thought and, in each case, caused turmoil and anguish. The first was Copernicus's view that the earth, and therefore man, was not the center of the universe, but only – as Freud put it – "a tiny speck in a world-system of a magnitude hardly conceivable." The next was Darwin's, who "robbed man of his peculiar privilege of having been specially created, and relegated him to a descent from the animal world." And the third shock, Freud explained, came from his own work, which showed each one of us that "he is not even master in his own house, but that he must remain content with the various scraps of information about what is going on unconsciously in his own mind."

Not everyone will agree with the identification and numbering of these major transitions. But it is true that the work of the charismatic scientists, from Copernicus to Einstein, has amounted to breaking down barriers that had previously been taken for granted: the comfort of fundamental differences thought to exist between terrestrial and celestial phenomena, between man and other life forms, the conscious and the unconscious, the child and the adult, or space and time, energy and matter. In each case, a culture shock resulted from the discovery that such barriers did not exist, that the discontinuity gave way to a continuum.

The adjustment to these recognitions was painful, and indeed is not complete to this day in any of these cases among the general public. It is not merely that each of those barrier smashers offended man's innate "common sense" and narcissism; even more ominous and frightening was the prospect of new freedoms that came with the elimination of the discontinuities – and, for many, the fear of exercising these freedoms.

The fourth discontinuity Mazlish considers is that between man and machine. It is his thesis that man has begun to realize that he is continuous with the tools and machines he constructs, that he is not only learning how to explain the workings of one in terms of the other, but is in fact forming a closer and closer physical, symbiotic relationship with the machine. At the same time, however, "man's pride, and his refusal to acknowledge this continuity, is the substratum upon which the distrust of technology and industrial society has been reared."[2]

To illustrate his point, Mazlish discusses the nightmare of the servant-machine rising against its master – the myth of Frankenstein, or more accurately, the myth of Dr. Frankenstein and his unnamed monster. If you have not recently looked into Mary Shelley's novel, you may not recall why the monster turned to murder. The living thing Dr. Frankenstein had assembled was in fact human and virtuous, but of such horrible appearance that its creator, and others whom it tried to befriend, fled in panic. The wrath of the monster, in the first instance, was thus caused by the abandonment of his creator's responsibility for his own work.

"I was benevolent and good; misery had made me a fiend. Make me happy, and I shall again be virtuous." With this plea the monster persuades his maker to go back into the laboratory and put together a second monster, a helpmeet with whom he would leave the world of men and retire happily into the wilderness of the Americas. But when Dr. Frankenstein is about to breathe life into the female creature he begins to have scruples. He reasons that at best this horrible pair will settle down to have children, "and a race of devils would be propagated upon the earth which might make the very existence of this species of men a condition precarious and full of terror. Had I right, for my own benefit, to inflict this curse upon everlasting generations?" And with this, even while the male monster is secretly watching him, Dr. Frankenstein destroys her. Now, of course, vengeance knows no bounds.

To contemporary sensibilities, the story contains some high ironies. Today, eager biologists and their corporate sponsors argue before the Supreme Court that they should be allowed to patent new life forms that are the products of both nature and man, such as genetically engineered bacteria. The legal question, fascinating as it is, is only one ramification of current, spectacular research in bioengineering. Such research is now heavily circumscribed by requirements for ethical and environmental impact statements – precisely the type of consideration that the unfortunate Dr. Frankenstein entered into so belatedly.

Indeed, this brings me to the point where I differ with my friend Bruce Mazlish concerning the locations of the fourth discontinuity. I think we crossed the man–tool boundary he is concerned with at a much earlier stage of human development, and the rabbis of medieval legend even anticipated the construction of a golem. Instead, the discontinuity being eliminated now is the difference between three previously separate, fundamental imperatives, those that animate progress, respectively, in science, technology, and society. To translate it into modern terminology, the novelty of Dr. Frankenstein was not that he made a recombinant man-machine monster, but that he became aware of the necessity for timely environmental and ethical impact statements of R & D. (Not even the good Lord himself, while engineering the transformation of clay into Adam, and eventually into the improved model, Mark II, or Eve, appears to have thought of doing that.)

No one would maintain, of course, that in the past science, technology, and social advance were completely separate from one another in every case. That first great invention of mankind, agriculture, serves as the oldest counterexample. But by and large, the barriers between them were thought to be reasonably clear. Even in the period leading to the industrial age of today, they were considered at most semipermeable, with science and technology acting unilaterally on the social process. Thus, when Francis Bacon announced that "knowledge and human power are synonymous," he had in mind that what we would now call research and development could be used together to subdue nature in the service of man.

But more and more, since the end of World War II, the realization has been spreading that the fates of science, technology, and society are linked in a much more complex and multilateral way. As one indicator, it has been recently estimated that nearly half the bills before the U.S. Congress have a substantial science/technology component. It is now also obvious that the establishment of the continuum of interactions between science, technology, and society has begun to shape each of these three elements at least as much as it does its own dynamics. And it is surely not an accident that some of the best early writings on this subject came from the father of the electronics-communication revolution, Norbert Wiener himself.

Two years after the publication of his great work, *Cybernetics* (1948),

Wiener brought out *The human uses of human beings: Cybernetics and society* (1950, now being reissued), which was designed, he explained, to bring out the "ethical and sociological implications of my previous work." The main point of the second work was, Wiener stressed, "a protest against the inhuman use of human beings," by which he meant above all the failure to bring the person – the citizen, consumer, worker – into a cybernetic relationship with the social organizations on which he depends. With this he opposed all unidirectional mechanisms in social institutions, arguing that uncontrolled resonances, or even fascistic totalitarianism, could be avoided only by the conscious design of institutions in which feedback is a primary rather than merely a cosmetic or perfunctory function.

Chiefly in reaction to the weapons race in the early Cold War years, Wiener was perhaps one of the first to grasp fully the malignant possibilities to which science and technology can be diverted. It was, however, not merely the spectacle of war-minded executives, legislators, and others that brought him to this realization. In his third book, *God and Golem, Inc.: A comment on certain points where cybernetics impinges on religion* (1964), he probed below the level of public policy to study the root motivations, which he identified as three drives: knowledge, to power, and to "worship." The old relationships between these three – or for that matter, the historic paucity of interactions – had become completely changed by the advances of science and technology. Deep ethical problems were now coming to the fore, such as the meaning of *purpose* in man and machine when we deal with machines that "learn," that "reproduce themselves," and that become part of and coordinated with living persons. Wiener warned that wise action for dealing with the problems of our time is doomed so long as different factions continue to give absolute primacy to only one of the three components – treating knowledge "in terms of omniscience," power "in terms of omnipotence," and worship "only in terms of one Godhead." Such sharp separation distorts reality.

Francis Bacon, often considered the godfather of modern Western industrial society, launched the enterprise precisely with such distortions. In his *New Atlantis* the narrator tries to find the secret of the utopia that beckoned so seductively. At long last he is received by the ruler of the House of Salomon – the chief, so to speak, of the research and development laboratory forming the very heart of the New Atlantis – and, in a confidential audience, he is told: "God bless thee, my son; I will give thee

the greatest jewel I have. For I will impart unto thee, for the love of God
and men, a relation of the true state of Salomon's house. . . . The end of
our foundation is the knowledge of causes, and secret motions of things;
and the enlarging of the bounds of human empire, to the affecting of all
things possible."

All things possible! Marlowe's *Faustus* had exclaimed: "Lines, circles,
signs, letters and characters – / Aye, these are those that Faustus most
desires, O what a world of profit and delight / Of power, of honor, of
omnipotence / Is promised to the studious artisan." To this day, we see
all around us the Promethean drive to *omnipotence through technology*
and to *omniscience through science*. The effecting of all things possible
and the knowledge of all causes are the respective primary imperatives of
technology and of science. But the motivating imperative of society continues to be the very different one of its physical and spiritual survival. It
is now far less obvious than it was in Francis Bacon's world how to bring
the three imperatives into harmony, and how to bring all three together
to bear on problems where they superpose.

In graphic terms, one can represent the relationship between them,
and their changes, by drawing three circles, each representing one of the
elements: science, technology, and society. Initially, the three circles en-
croached on each other rather little; the whole point of Galileo's tragic
flight was to insist on the autonomy of science. In principle, the three
elements could stand apart, like three clover leaves, and any bargaining
between could be correspondingly simple. But in time, each of the three
circles grew in size and also moved to increase the area of overlap with
the other two. Unlike Galileo's telescope and Faraday's electromagnet,
the microcomputer is located squarely in the direct overlap of all three
circles – just where the inherent contradictions of the three old, incommensurable imperatives are most active and just where the new, media-
ting institutions are least mature.

To be sure one can argue (and historians love to do just that) to what
degree either science or technology ever had been or could be "pure."
There are classic debates between Marxians, idealists, and all the others
caught in the middle, on that point. But the reality of today is captured
more appropriately by the remark of Sir Peter Medawar in his new book,
Advice to a young scientist: "The direction of scientific endeavor is determined by political decisions as . . . acts of judgment that lie outside
science itself." And there is no doubt that the most advanced technologies
of our day, those represented by large-scale integrated circuits and by

genetic engineering, not only are the products of powerful scientific and technological as well as social forces, but are also the focus of deep concern as each imperative contends for supremacy over the two others.

As a result, the ancient optimistic view that technological advance takes place in the service of the quality of life has been under increasing challenge. In the words of Bruce Hannay, vice-president for research and patents at Bell Laboratories, speaking to the American Academy of Arts and Sciences, there is "a steady decline in the general acceptance of the desirability of technological change." Undoubtedly, technological advance has decreased the costs and increased the availability of energy, food, information, and basic medicine to large masses of people. Therefore the quality of life should in principle be recognized to have been advanced substantially. Yet there has been a parallel development of pathologies, which have caused the gifts of high technology, from computers to nuclear reactors, to become in many quarters the very symbol of technological changes undermining societal objectives, the gift that is a Trojan horse. The increasing demands of a high-technology-oriented arms buildup is reorienting – and indeed militarizing – both scientific research and our social priorities.

Under such shadows, what meaning can one assign to the phrase "quality of life"? For an operational definition of the components of the phrase, I would be tempted to go back to Franklin D. Roosevelt's definition of the "four freedoms" (in his message to Congress, January 6, 1941): freedom of speech and expression, freedom of worship, freedom from want, and freedom from fear. It was a mark of Roosevelt's insight to couple the last two "freedoms from." To the extent that science and technology pursue their own imperatives, they may indeed give us – as by-products, so to speak – freedom from want (in the sense of freedom from basic material needs). But except for the removal of superstition through greater scientific literacy, science and technology by themselves cannot increase the freedom from fear. On the contrary, the very instincts for survival and self-preservation, which animate social action, will demand assurances that the imperatives of science and technology are made, if not subservient to, at least to harmonize with, that of society – and the more urgently precisely as the powers of science and technology increase. I do not think that there is a widespread fear of scientists and engineers as such – but there is a widely shared perception that they, like Dr. Frankenstein, are still apt to make their impact considerations as an afterthought, and perhaps too late.

The emergence of "Combined-Mode Research"

Yet at the very same time antithetical forces are at work, too. There are some indications of a very hopeful sort that at this very time a new relationship is emerging between science, technology, and society that may eventually help to rearrange the forces in an era dominated by the recognition of the fourth discontinuity. Although each of the three cloverleaves will undoubtedly retain an area of its own relative autonomy – and, in the case of basic research science, can do no less if science is to thrive – a model of interaction is emerging in the area of overlap, where the discontinuities have been disappearing. To this, I shall now turn, in the hope that if I exhibit the general case, you will find it easier to draw corresponding consequences for the subject of your specific expertise and concerns.

To summarize my point before illustrating it: when, as a historian of science, one studies the "center of gravity" of the choice of basic research problems on the part of good scientists, one can discern a marked shift in emphasis. At the time of Kepler, Galileo, and Newton, the researcher seems to have asked himself chiefly what God may have had in mind when creating the physical world. By the time of Maxwell the burning question had become what Faraday might have had in mind with his obscure ponderings about the field and how one might improve on them. This is still, and to a degree will remain, the type of puzzle that excites the basic researcher. But alongside, an alternative and complementary motivation for certain research scientists is making its appearance. The stimulus comes now not only from considering one's Creator or one's peers but, more and more frequently, *from perceiving an area of basic scientific ignorance that seems to lie at the heart of a social problem.*

Work motivated in this manner positions itself squarely in the area of overlap between science, technology, and society, without giving up its claim to being indeed basic research. Such an investigation might be termed "Combined-Mode Research," since it can be considered a combination of the discipline-oriented and problem-oriented modes. Note that it is not to be confused with such programs as Research Applied to National Needs (RANN) and similar ones, which encourage the application of *existing* basic knowledge to the meeting of supposed national needs; that has its place, and I am not arguing here against it. But I am, instead, speaking of the opposite of RANN and similar programs, namely, of *basic* research, located intentionally in uncharted areas on the

map of basic science but motivated by a credible perception that the findings will have a fair probability – perhaps in a decade or more – of being
brought to bear upon a persistent national or international problem.

With this we have, of course, reached disputed territory. For the better
part of three centuries the consensus among basic research scientists has
been that "truth sets its own agenda." Any intrusion of the consideration of utility that might eventually accrue from basic research has been
thought to be incompatible with the agenda of the true scientist. From
that point of view, omniscience first, omnipotence later. Did not the
seventeenth-century giants teach us that reductionism is the way to success in the natural sciences and that applications to the seamless and
endless complexities of societal problems are best left to serendipity or to
later generations?

And in any case, have scientists not been remarkably blind when it
came to occasional attempts to forecast practical applications of their
work? Thus it is said Kelvin could see no use for the new Hertzian radio
waves except possibly communication with lightships, and Rutherford
stoutly refused to the end to see any significant practical applications of
his exploration to the atom. And last but not least, have we not learned
from the Lysenko episode, and the constant drumfire of attacks on basic
science by legislators in the West, that our first job is to fight for the
preservation of "pure" scientific research? The anguished assessment of
the chemist Francis W. Clarke in 1891 still largely holds true today:
"Every true investigator in the domain of pure science is met with mo-
notonously recurrent questions as to the practical purport of his studies;
and rarely can he find an answer expressible in terms of commerce. If
utility is not immediately in sight, he is pitied as a dreamer, or blamed as
a spendthrift."

Certainly, the science of plate tectonics did not arise out of an effort to
predict earthquakes, or genetics out of a desire to create a better harvest
in the vegetable garden. On the contrary, it happened the other way
around. Indeed, the history of science and technology is full of case
studies that could be collected under some heading such as "how basic
research reaps unexpected rewards."[3] It follows that anyone inclined to
mix considerations of utility with the choice of basic research problems
may be risking both the granting agency's money and his career as a
scientist.

But now there are signs that this has been too simplistic a dogma when applied across the board. While it still is, will be, and must be true for the majority of basic research scientists, at least a small fraction of the research programs can be and, in fact, are now centered in what I have called earlier Combined-Mode Research, in research where the imperatives of science and of society overlap instead of claiming mutual exclusivity. And while, from the viewpoint of social utility, basic research in the "pure" mode could be called "Project Serendipity," research in the combined mode might be called "Project Foresight" (by asymmetry with the ill-fated "Project Hindsight," which attempted to sketch the influence of the past on the present; I am concerned here rather with the influence of the present on the future).

It is appropriate to interject that this recognition is not a *normative* proposal on my part. I am not speaking as a science planner or missionary but as a historian of science who is describing what he sees happening in science today. I am also not assessing the long-range future of these developments. It might well turn out that Combined-Mode Research, which by its very definition is difficult, requires more patience than our society now has for waiting for the promised payoff in social benefits.[4]

But there simply is no doubt that, under our very eyes, a mutation has been taking place, and programs are growing up specifically designed to seek fundamental new knowledge and scientific principles, in the absence of which current national or international needs are difficult to ameliorate or even to understand properly. It is, after all, not too hard to imagine plausible research areas that can hold the key to well-known societal dysfunctions. Even the "purest" scientists are likely to agree that much remains to be done in the field of cognitive psychology, the biophysics and biochemistry involved in the process of conception, the neurophysiology of the senses such as hearing and sight, or molecular transport across cell membranes, to name a few. As a result of such basic work we could plausibly expect in time to have a better grasp on such complex societal tasks as childhood education, family planning, improving the quality of life for the handicapped, and the design of food plants that can use inexpensive (brackish) water. Other basic research examples that come readily to mind might include the physical chemistry of the stratosphere; that part of the theory of solid states that makes the efficient working of photovoltaic cells still a puzzle, bacterial nitrogen fixation and the search for symbionts that might work with plants other than legumes, the mathematics of risk calculation for complex structures, the physiological processes governing the aging cell, research on learning and on career decisions of young people, the sociology underlying the anxieties some segments of the population have about mathematics and computers, and the anthropology of ancient tribal behavior that seems to be at the base of genocide, racism, and war in our time.

Any specific list of examples of this sort is open to challenges. But it is not difficult to imagine a consultative mechanism designed to identify the research areas that could benefit from such cultivation. There is no doubt that institutional innovations in this direction are sorely needed; unless the fundamental decision of siting such research is made with the full participation of a wide spectrum of experienced and trusted research scientists, the effort would degenerate quickly.[5] The last thing science, or society, needs is some political command center for the approval or disapproval of basic research.

A first list of "Combined-Mode" plans

I turn now to a roundup of specific evidences of the rise of support for research in the combined mode, for research driven by (or targeted as the result of) perceived national need. In citing examples of the past few years I do not wish to imply that this movement has no history whatever – just as the very recent reversals of this policy do not imply it has no long-run future. Particularly in the biomedical area there have been clear and continuing instances, from Pasteur to the founding and successes of the NIH institutes. But Combined-Mode Research in the biomedical field has in a sense been "easier" to start and to support (partly because of the immediate self-interest of the patrons – an example of the process of science and technology affecting the setting of social priorities – and partly because much of what has passed for basic research in that field is really closer to mission-oriented applied research on systems whose fundamental complexities have hardly been charted). The real test is outside the biomedical field.

A suitable point of departure for our accounting is provided by an address delivered in early 1978 by Frank Press, then director of the Office of Science and Technology Policy (OSTP), and the science and technology adviser to the president. Dr. Press described the science-policy planning that went into the budget for the federal funding of basic research for the fiscal year of 1979. In addition to the Office of Management and Budget (OMB), the heads of NASA and NSF, leaders in science and engineering from universities, industry, and the government, the process also involved consultation with the members of the Cabinet:

> During the course of our interactions on research with the departments and agencies, the President queries the Cabinet members on what they thought some of the important research questions of national interest were. Here are a few examples cited by the Cabinet officers:
>
>> Can simple chemical reactions be discovered that will generate visible radiation? How does the material pervading the universe collect to form complex organic molecules, stars, and galaxies? What are the physical processes that govern climate? . . . What are the factors – social, economic, political, and cultural – which govern population growth? . . . How do cracks originate and propagate in materials? How do cells change during growth and development? What are the mechanisms responsible for sensory signal processing, neural membrane phenomena, and distinct chemical operations of nerve junctions? . . . What predisposing factors govern cellular differentiation and function in plant and animal?[6]

These are of course questions of basic research for the "purest" Ph.D. theses at the best academic departments; and yet they are also precisely targeted in the areas of perceived national need.

The same intention surfaced also in the reorganization of the NSF's applied research programs in 1978, when, as part of the new Directorate for Applied Science and Research Applications, a division of Integrated Basic Research (IBR) was formed. Unlike the older, applied-research activity, which aims to encourage and accelerate the application of existing basic scientific knowledge to a wide range of potential users, the division of IBR was formed to provide "support for basic research that has a high relevance to major problems" in selected topic areas in the basic research directorates. The operational meaning of these intentions became clearer in the Tenth Annual Report of the National Science Board of the NSF entitled *Basic research in the mission agencies: Agency perspectives on the conduct and support of basic research*, released by the president on August 2, 1978. In his covering letter to the Congress, President Carter specified that he had "encouraged the agencies to identify current or potential problems facing the Federal Government, in which basic or long-term research could help these agencies. . . . [The report also should be helpful] with setting priorities for future federally-supported research and development, and in making our spending in this area more effective."

In the memorandum to science writers and editors, on the same day, the NSF itself started its description of the report with the straightforward and familiar sentence, "Basic research is useful." But this is no longer the vague, old promise, as in Vannevar Bush's *Science: The endless frontier* of some two dozen years earlier, that disease, ignorance, and unemployment would somehow be conquered if basic science were supported on a large scale – without any conscious attempt to link the input and the output. Rather, the new linkage is proposed to come about in a way which Vannevar Bush never considered in pursuing his main purpose, which was to provide what he called "special protection and specially-assured support" for pure research, so as to avoid what he saw to be the perverse law governing research, namely, that "applied research invariably drives out pure." The NSF and OSTP initiatives of 1978, instead of concentrating on either pure or applied research in relative isolation, start with the novel and rather daring attempt to gather the perceptions of the various science-related mission agencies concerning their "priorities and gaps in their research agendas."[7] A list was assembled of the "problem areas that appear to merit national attention and that require basic research (if for no other reason than to complete our understanding of the problem)." The authors of the report were evidently aware of the difficulty of deciding on the relative importance of research programs even within a delimited area of science and of the increasing difficulty of such priority setting as the time span for planning increases. But as a public attempt at the identification of priorities and gaps, the document is of considerable interest and undoubtedly will be long studied by historians of science and technology when they consider the development of science policy in this country.

Also, as one might expect, the 16 agencies, ranging from the Department of Agriculture to the NSF, and from Department of Housing and Urban Development to the Veteran's Administration, are by no means equally adept in responding to the National Science Board's questions such as: "What promising or vital areas of research, not now supported but involving basic research, warrant increased emphasis and support by your agency?" But there emerge entirely plausible proposals nevertheless. Thus the Department of Agriculture provides a lengthy list of priorities, from agricultural research (starting with nitrogen fixation, photosynthesis, and genetic engineering for plants) to social science research (ending with "impact assessment"), all of them "areas of science in which a basic research approach is required," not only for agricultural and forest technology, but also, thoughtfully, for "the quality of life in rural communities and homes." The Department of Commerce's list includes, typically, atomic and molecular science (chemical reaction rates, ozone layer dynamics, very high temperature plasmas) and ends with a call for "over-all resources for the broad spectrum of basic research, free from competition from short-term applied projects."

There are enough passages of this sort in the total report to make the "purest" basic scientist feel right at home. It is, of course, in the nature of the exercise that such a feeling would not be generated on every page. Thus the list of priorities of the NIH is unexceptional (genetics, immunology, virology, cell biology, neuroscience) and, for the main "gap" area, neurobiology ("the ultimate challenge to medical research, representing the very pinnacle of our understanding of the human organism"). On the other hand, the same understanding of "basic research" does not seem to bolster such entries as that of the Army ("improvement of helicopter performance") or of the Maritime Administration ("propeller design").[8]

These initiatives of early 1978 were by no means the last. They were strengthened in the discussions, concerned with the preparation of the budget for fiscal year 1980.[9] This tendency (to be further discussed in Chapter 11) gathered further momentum with the release by the National Science Foundation of the two-volume report, *Five-year outlook: Problems, opportunities, and constraints in science and technology.*[10] As the news release of the NSF accompanying the publication recorded, the report "discusses national issues that we will face during the next five years from a scientific and technical perspective. This is the first time such a long-range outlook on science and technology has been prepared."

The study was mandated by the National Science and Technology Policy, Organization, and Priorities Act of 1976. It represents a major effort to identify and describe "in depth problems of national concern that are most likely to need special attention through the mid-1980s and later, and to which science and technology can contribute in the coming years." The heart of the report is a book-length monograph prepared by the National Academy of Sciences, to which are added statements from twenty-one U.S. agencies, papers by individual specialists, and a synthesis presenting selected problems of U.S. society and the opportunities for science and technology to solve them. These were prepared by the NSF staff, under such headings as energy, materials, transportation, demography, space, agriculture, health, electronic revolution, and the hazards of toxic substances in the environment.

While the Vannevar Bush report, 35 years earlier, had concentrated on the necessity to support basic research regardless of how it eventually might fulfill the distant promise of helping to "create a fuller and more fruitful employment and a fuller and more fruitful life," the *Five-year outlook* incorporated rather explicitly both the "pure mode" and the combined mode of basic research. Thus, in the covering statement of the then-director of the National Science Foundation, Richard C. Atkinson referred to the many examples in the volumes illustrating "the contribution of long-term research to the solution of national problems," but of course also calls for the continuation of support for research where the primary goal is "a better understanding of nature" regardless of whether any link can be discerned with "specific societal problems." He was signaling a preservation of a necessary balance between the old and the new expectations from basic research, a signal made the more necessary as the Congressional act that required the periodic preparation of a *Five-year outlook*, by its very specification of a relatively short period of preview, stressed only one of the two modes.[11]

Reading these volumes, one noted that the invisible hand, which has long been thought to be sufficient for guiding the process from the basic research laboratory to the specific application, was gradually becoming more visible. The language is quite frank: "Research should focus on the following long-range opportunities" (biological processes that develop food plants with less fertilizer and less fresh water, environmentally safe methods for controlling animal and plant pests and diseases, better understanding of acid rain); "we must emphasize at once research that will provide the scientific knowledge" for the development and commercialization of advanced energy technologies, such as nuclear fusion and direct solar conversion; "we must expand" the science on which risk assessments are based.[12]

Elsewhere there are long and detailed lists of fundamental research that must be encouraged, lists that show considerable overlap with those prepared two years earlier for the Tenth Annual Report of the NSB. Two "problems" highlighted are "how to use the capabilities of computer and communication sciences and technology to serve a wide range of commercial, public, and personal needs; [and] how to resolve the social, ethical, and regulatory issues that are emerging as a result of the electronic revolution."[13]

Volume 2 includes two separately commissioned, thoughtful essays on the impact of new communication technologies, particularly on privacy. Here, too, we encounter the conflicting imperatives of technology and society in the area of overlap. One obvious signal comes from measurements of the public attitudes in the United States about the overall state of privacy and how it is affected by the increasing use of computers by governmental and commercial agencies with which the public has to deal. The findings gathered by the survey commissioned by Sentry Insurance, designed and carried out by Alan Westin and the Louis Harris organization in 1978, are certainly sobering. Over 80 percent of those surveyed disapproved of the wide access that police (and others) have to personal bank account information without a court order, and – surely not unrelated – nearly two-thirds agreed with the statement that "If privacy is to be preserved, the use of computers must be sharply restricted in the future."

Since 1974, there has been a striking shift in the response of the general population to the proposition that "Americans begin surrendering their personal privacy the day they open their first charge account, take out a loan, buy something on the installment plan, or apply for a credit card." In 1974 slightly less than half agreed. By 1978, the fraction had risen to slightly over three-quarters.[14] If only for reasons of self-interest, the computer industry now, as we have entered the era of the fourth discontinuity, might be expected to concentrate some of its research and development talent on this area of abuse and fear. History will judge the prophets of the silicone chip by the degree to which they are able to provide intelligent machines that not only make life more interesting and fruitful but also enhance personal freedom.

I have concentrated on examples taken from public documents published in the United States. Analogous and in some ways even more telling examples could be drawn from the international literature. For example, in Sweden, where such groups as trade unions have begun to take an energetic interest in science policy issues, there has been a shift to nudge science planning in accordance with social goals defined along sectoral lines.[15]

A report rather similar to these discussions for the United States was published by the National Council of Science and Technology (CONACYT) of Mexico in 1978.[16] In laying out the rationale for assigning high priorities to some research areas and not to others, CONACYT canvassed scientific representatives from the public, private, and academic sectors, including its various ministries. On this basis, areas of encouragement for basic research were identified (in addition to fields demanding encouragement for applied research and the development of technology).

The space available has not allowed me to triangulate to the same point from yet other directions. The fact that the era of the fourth discontinuity demands an additional mandate for the pursuit of basic research as a contribution to the fulfillment of human needs has produced a number of related developments: one is a widening of the purview of the professional societies and corporate activities, seen most simply in the large increase of discussion of social concerns in the annual reports of the presidents of professional scientific societies.[17] Another is the rise of educational programs on science, technology, and society.[18] These are generally carried on in the spirit of giving young people who will be scientists or managers a double competence that was perhaps best indicated in Einstein's words when he addressed the students of the California Institute of Technology in 1931:

> It is not enough that you should understand about applied science in order that your work may increase man's blessings. Concern for man himself and his fate must always form the chief interest of all technical endeavors . . . in order that the creations of our minds shall be a blessing and not a curse to mankind. Never forget this in the midst of your diagrams and equations.

All the evidence seems to me to point to the fact that, in our time, a historic transition can occur in the direction of basic research policy. In time's own laboratory, a new amalgam has been forming that will challenge the inherited notions of every scientist, engineer, and social planner. Undoubtedly, we shall witness battles to preserve those autonomies that are and always will be essential. Undoubtedly, there will also be over-enthusiastic projects that cannot deliver on their promises. But if the movement allows at least a portion of the total research in science to be done in the Combined Mode, the spectrum of research may well be greatly extended, its links to technology and society become more fruitful and certain, and its mandate reinforced. As scientists, engineers, industrialists, or educators, we should welcome this possibility; for any professional activity has a just claim to more authority when, and only when, it is widely seen to honor both truth and the public interest.

PART III

Science, education, and the public interest

10

The two maps

Much of my work has had its origin in the notion that science should treasure its history, that historical scholarship should treasure science, and that the full understanding of each is deficient without the other. If I were asked to indicate the chief motivation behind this once unusual, not to say perverse, idea, I would have to do so in the form of a stark statement that can be supported, on this occasion, only sketchily. It is this: At a time when passionate unreason around the globe challenges the fate of Western culture itself, the sciences and the history of their development remain perhaps the best testimony to the potential of mankind's effective reasoning. Therefore, if we do not trouble ourselves to understand and proudly claim our own history, we shall not have done full justice to our responsibility as scientists and as teachers.

The Oersted invisibility

A good way to illustrate this point is to glance at the work of Hans Christian Oersted (1777–1851), who is little known even among physicists, although an argument can be made that he was a modern kind of scientist. He announced his great discovery, that a magnet needle can be deflected by what was then called a galvanic current, on July 21, 1820, in a broadside which he had privately printed and distributed to the foremost scientists of Europe.[1] The publication led to a veritable explosion of scientific work, for example, the great discoveries in electromagnetism by Ampère and Faraday. Technical applications followed quickly also, starting with an electric telegraph.

Oersted, then 43 years of age, was professor of physics at Copenhagen. An autodidact, he had first studied pharmacy and languages and had also

This chapter constitutes the response on receiving the Oersted Medal at the joint American Physical Society–American Association of Physics Teachers meeting, Chicago, 22 January 1980.

contributed to chemistry and other sciences. He was an early evolutionist in biology, a man with wide interests in science and outside. Among his works are some on the relation of science to poetry and to religion, in which he shows himself to be a gentle but persuasive rationalist.

Oersted saw his chief task to be the discovery of the unities in nature. Far from having stumbled accidentally on the fact that a magnetic field surrounds currents, as the popular myth still has it, he had sought for years for the effect which practically everyone else, during the two decades of widespread experimentations with voltaic cells from 1800 on, had missed. At least eight years before his discovery of 1820, he had declared his faith that light and heat and chemical affinity, as well as electricity and magnetism, are all "different forms of one primordial power," and he had announced that attempts must be launched "to see if electricity has any action on the magnet." Oersted had been deeply impressed by the philosophical works of Immanuel Kant in which he found the argument that all physical experiences are due to one force (*Grundkraft*). He also accepted many views of his friend Friedrich W. J. Schelling, a leading exponent of *Naturphilosophie*, who provided an enthusiastic program to find the unity of all natural phenomena, and specifically the unity of physical forces, in such proposals as: "For a long time it has been said that magnetic, electrical, chemical, and finally even organic phenomena would be interwoven in one great association. . . . This great association, which a scientific physics must set forth, extends over the whole of nature."

R. C. Stauffer, in whose paper the quotation from Schelling is given, also cites the response of Mme de Staël, on concluding her discussion with him on *Naturphilosophie*, that "systems which aspire to the explanation of the universe cannot be analyzed at all clearly by any discourse: words are not appropriate to ideas of this kind, and the result is that, in order to make them serve, one spreads over all things the darkness which preceded creation, but not the light which followed." Be that as it may, the lack of Cartesian clarity did not impede the influence of *Naturphilosophie* on Oersted (or for that matter, on Ampère, Faraday, Julius Robert Mayer, and others).

The thematic presupposition of the unity of forces led Oersted to look for the connection between electricity and magnetism through a convincing experiment. His success identifies him as the modern initiator of the grand unification program – that great source of motivating energies of modern physics, with a direct genetic influence on Faraday,

Maxwell, Einstein, and on to the recent Nobel Prize winners. In contemporary physics meetings, Oersted, at least in spirit, would surely have felt very much at home.

Moreover, Oersted's apparatus for demonstrating the unity of nature's phenomena was eminently sensible, at least in terms of his presuppositions. He sent progressively larger electric current through high-resistance platinum wires. First the wire got hot, then the wire began to give off light, and then he saw the effect on the magnet needle. Some day I hope to look at Oersted's laboratory books in Denmark to see if he also looked for a gravitational effect as Faraday did later; it would have been a likely extension in the thinking of a unifier.

In his later years, Oersted dedicated himself to many social causes, such as the freedom of the press, and to science education. One of his biographers notes that when Oersted died in 1851, the students arranged for a torchlight cortege in a huge funeral procession in which 20,000 people are said to have participated.

Now there is no doubt that Oersted's advance in physics changed history in two ways. It opened up physics itself to a succession of unifying theories and discoveries without which the modern state of our science would be unthinkable. And his key discovery, embodied in every electric motor, also triggered the engineering advances that have produced the modern technological landscape. One would expect such a person to be visible in works of history. Yet, despite his role in the initiation of vast changes in science, engineering, and through it our very society, Oersted is virtually absent from both science books and history texts, not least those which young persons typically encounter in school. If one looks into historical encyclopedias for what happened in 1820, one finds a plethora of events of a quite different sort. For example, Karl Ploetz's compendium of chief dates of history notes for 1820 that "Austrian troops reestablish order in Italy"; and the big *Encyclopaedia of world history*, edited by William Langer, records for 1820 that King George III, having earlier been declared insane, dies.

Let me dwell a little on this phenomenon, which we may call the *Oersted Invisibility*, for we are getting to a significant truth about the influences that shape the intellectual formation of young people throughout the world. William Langer's widely used *Encyclopaedia* was not meant to be a synthetic work and does not represent the more sophisticated work of scholars in history today. But the two main concerns that animate the book are still alive in the teaching of history as most young people encounter it in their formative years: the chronological presentation of historically important "facts" and the "periodization" of the sequence of facts into labeled categories. As to the first, Langer says in his introduction that it is his function to provide "a handbook of historical fact" – as it turns out, chiefly political, military, and diplomatic history. He explains that for him to have gone also into other achievements such as science would have taken him too "far afield." In my copy (second edition, 1948) there *is* a section entitled "Scientific thought and progress," precisely 2¼ pages out of the total 1270 – or less than 0.2 percent (which happens to be better than the proportion of fundamental science support in our GNP). As an attempt to put some order into the chaos of facts, periodization of history into segments entitled the "Age of   ," the "House of   ," the "Revolutionary period," and so forth, provides a seemingly well-bounded shape to the various fragments of time. But it encourages only occasionally a comment that parts the dark curtain behind which historians are debating. (Was Thucydides right in considering the war of 431–421 B.C. and the war of 414–404 B.C. to be in reality one period, that of the Peloponnesian War? How far back do the causes of the revolutionary outbreaks of A.D. 1848 go?)

The net effect, at least on most young minds, must be that the purpose of history appears to be the provision of a well-labeled place for every miscreant if only his factual mischief was on a big enough scale, and to give some space to every ruler, whether effective and beneficient or not. In such books, genealogies of the mighty abound, from the Ch'in Dynasty and the succession of the Merovingian kings to the sequence of czars and presidents. One is reminded of Voltaire's complaint that "for the last 1400 years the only persons in Gaul apparently have been kings, ministers, and generals."

Scientists and their works are virtually taboo. If you look for Newton in my copy of Langer, you will find him mentioned once, on page 431, under the heading "Third Parliament of William III": "Isaac Newton, master of the mint." The four-word description is not what you or I might have chosen; but at least he is mentioned, which is more than can be said for Galileo or Kepler or even Copernicus. (Others have done worse: Arnold Toynbee's list of "creative individuals," from Xenophon to Lenin and Hindenburg, included not a single scientist.)

On the other hand, Langer's book does give a detailed survey of, say, the vicissitudes and successions of the Ottoman emperors, from Bayazid II, remembered by Langer for being "the least significant of the first ten sultans," through the exploits of Selim I ("The Grim") and Selim II ("The Sot," described as an "indolent ruler, much given to drink"), and so on through the last of them, Mohammed IV, who is described as a boy of ten, followed by a period of anarchy. It might almost sound amusing, until one asks what the costs were for the unmentioned mass of humanity doomed to be born, to live, and to die in the dark back alleys of history.

Every holy war gets its place in Langer's and so many books like it. There is a tiny admixture of humane figures, a Marcus Aurelius, a Jefferson, or a Ghandi; but they are lost in the succession of genocidal maniacs, from Emperor Tiglath III of Babylon, who innovated the idea of consolidating his conquests by "deporting entire populations," to Josef Stalin. As John Locke asked: "What were those conquerors but the great butchers of mankind?" Moreover, from one development to the next, any rational conclusion or extrapolation is seemingly hopeless. No wonder that even many of today's historians, in the words of J. H. Plumb, "have taken refuge in the meaninglessness of history."[2] Thus the Ploetz compendium – which was the model for Langer, and, as it happened, also the book I used in high school in Vienna – started on a promising note, from a rational point of view, with the entry for July 19, 4241 B.C. as the precise date the good Dr. Ploetz proposed for the introduction of the calendar in Egypt. But it, too, rapidly went downhill through the whole sorry list of massacres and delusions, and ended with an entirely unprophetic but appropriate entry, for September 27, 1934, which ran as follows: "Declaration by England, France, and Italy, guaranteeing the integrity and independence of Austria." I can well believe the famous story about another boy, some 15 years earlier but in a school not far from mine. "You are a clever chap," the history teacher said to Viki Weisskopf, "but you don't know any dates!" "Oh, I do," he replied. "I know all the dates; I just don't remember what happened on them."

The bifurcated maps

I have of course been a little hard on William Langer, a distinguished diplomatic historian, who later became a valued colleague of mine, and who, perhaps partly because of my teasing, did put a good deal more science into the last edition he edited. But the main point I wish to make still stands: wherever their schooling may take place, young people encounter, through the historically oriented set of courses as usually taught, a view of the accomplishments and destiny of mankind that almost celebrates the role of passionate unreason. And if the student is one of the relatively few who also takes substantial science courses, he or she will encounter from that direction a very different picture of mankind's interests and attainments. Indeed, the opposition between these stories is so great that it must seem to many students that there are really two different species involved.

Let me put it in terms of an image from my own schoolroom in Vienna, an image surely not qualitatively different from the one you carry in your mind, or which, *mutatis mutandis*, is being found even today in your town's school. The curriculum at our Gymnasium emphasized history, literature, and ancient languages, a triad that merged into one message. Latin and Caesar's *Wars*; Greek and the *Iliad*; medieval German and the *Nibelungenlied*; the chief tragedies of German theater, the Bible, the Edda sagas, and the painstaking probing of the ever-unfinished sequence of historic battles. It was a powerful and blood-stirring brew – but with only occasional traces of the rational processes.

On the other hand, in our science classes we encountered an entirely different universe. Here was the finished and apparently unchanging product of distant and largely anonymous personages, unchallengeable monuments to their inexorable rationality – but with only occasional traces of historic development. Just as the historian neglects science – Richard Hofstadter said of the historian "he may not disparage science, but he despairs of it" – the scientist is silent about history – the record, a scientist put it to me once, of the errors of the forgotten dead.

It was a cultural schizophrenia which I could not formulate, but which I also could not dismiss. I can capture it best by recalling two very different maps that were hanging in front of my class, to stare at and wonder about, year after year. You probably saw them, too. On the left side was always a geopolitical map of Europe and Asia. (America and most of Africa were evidently on some other planet.) In some storage closet there must have been a whole set of such maps because the left map was regularly changed, a new one for each period, and with each change we students could see the violent, spasmodic, unpredictable pulsation of shapes and colors in the wake of the thrilling story of conquests. On the right side was a very different map – the Periodic Table of the elements: the very embodiment of empirical, testable, reliable, and ordered sets of truths. That map was never changed, although there was a rumor in the benches that some of the blank spaces were being filled in, and even that a previous student of that very Gymnasium, a man named Wolfgang

Pauli, had shown that the different features of the table were the consequence of some underlying great idea, rather than some accident on which nature had settled.

To the mind of the child exposed to these two maps, the utter differences between them and thus between the subjects they represented were so profound as to seem unbridgeable. On the left side, the forces shaping history were the four horsemen of the Apocalypse. On the right side the forces shaping science were, to use modern terms, the four forces of physics. To the young mind, it seemed like a division that demanded some sort of decision. We know from the autobiographical notes of scientists that this dichotomy can help to focus a career choice. In Einstein's famous remarks, it is the choice, in early youth, between a world "dominated by wishes, hopes, and primitive feelings," and on the other hand, "this huge world which exists independently of us human beings," the contemplation of which "beckoned like a liberation." It is only much later in life that it dawns on such a student at what great costs this separation was being nurtured, that these two kinds of destiny are in fact intertwined, that these two developments stem from two potentials within the same person.

In the meantime, however, students in our classes were left to wonder what the moral point of this bifurcated education would be. Concerning our personal destinies there was in fact a sharp division of opinion between our parents, on one side, and our history books as well as those few teachers who had any interest in discussing such matters, on the other. The parent's theory was that the purpose of the curriculum was chiefly to prepare for the school's final examination, without which one could not go on to university. However, our physical education teacher spoke explicitly for the view that presented a different scenario for us: We would be there to fight one day, to revenge the loss of our territories in World War I, and of our honor in the treaties of Versailles, Trianon, and St. Germain. We were being readied to change that map on the left once more.[3]

The great historians we studied seemed, from that second point of view, useful preparations. For example, the father of historical writing, Herodotus of Athens in the Age of Pericles, declares the aim of his great book on the history of Greece to preserve, as he puts it, "the great and wonderful actions of the Greeks and the barbarians [from] losing their due meed of glory." What he means is set forth right at the start. A certain Gyges, upon being made King of Sardis, "made inroads on Miletus and Smyrna. . . . Afterwards, however, although he reigned for 38 years, he did not perform a single [further] noble exploit. I shall therefore make no further mention of him, but pass on to his son and successor . . . Ardys. This Ardys took Priene, and he made war upon Miletus." Our class quickly got the idea that here was an altogether more glorious type. Even better was his son Alyattes who, Herodotus says, had inherited from his father that war with Miletus, and who "performed other actions very worthy of note." Herodotus tells us one of these: In warfare, "he cut down and utterly destroyed all the trees and all the corn throughout the land and then returned to his own dominion. . . . The reason that he did not demolish also the buildings was that the inhabitants might be tempted to use them as homesteads from which to go forth to sew and till their land; and so each time that Alyattes invaded the country he might find something to plunder." Indeed, worthy of note.

Turning to a more recent work, we found that Georg Wilhelm Friedrich Hegel, in his great *Philosophy of history*, struggled with the meaning of history and concluded that "the final cause of the world at large, we allege to be the consciousness of its own freedom on the part of spirit, and *ipso facto* the reality of that freedom." I must confess that this formula, when presented in history class, was not easy to unpuzzle; but it had a nice ring to it. History as the evolution of freedom seemed an appealing idea. But just at that point in our studies, this train of thought was deprived of a good deal of its credibility when one Friday evening in March, the portion of the map of Europe showing Austria turned suddenly brown, and our history teacher, like many teachers in the other subjects, turned up in Nazi regalia on the following Monday.

Hegel also had said, quite correctly, that the most profoundly shaping experience on the mind of the West was Greece of the Homeric Period. The reading of the *Iliad* was meant to be our most permanent memory; I must confess I liked it best of all our classes. Just for this reason let me use it further to crystallize my point. You recall the early scene on the beach before Troy. Agamemnon, leader of the Greeks who are laying siege, has lost his own mistress, robs Achilles of his, and thereby launches 1000 pages of dactylic hexameter. Agamemnon explains himself in a way which Homeric Greeks accepted: "Not I was the cause of this act, but Zeus." It is the habit and prerogative of the Gods to put *atē* (temptation, infatuation, a clouding of consciousness) into a person's understanding. "So what could I do? The Gods will always have their way." Achilles, the outraged victim of Agamemnon's action, takes the same view. Without Zeus, he agrees, Agamemnon "would never have persisted in rousing the *thumos* (passionate, violent response) in my own chest."

In fact, throughout the *Iliad*, temptations and infatuations, *atē*, and passionate and violent response, *thumos*, are the chief forces motivating human action. As E. R. Dodds explains in *The Greeks and the irrational*, *atē* is, in fact, a partial, temporary insanity, "and, like all insanity, it is ascribed not to physiological or psychological causes, but to the external, 'daemonic' agency." Only very rarely does one encounter in the *Iliad* a glimpse of rationality – dispassionate reason sensitive to long-range consequences – as a motivating force. This role falls to Athena, the goddess of good counsel, who occasionally intervenes in the brawl and carnage (even at the risk of conflict of interest, since on the side she is also a deity of war). There is the famous scene in Book I when Athena, visible only to Achilles, catches him by the hair and warns him not to strike Agamemnon. A moment of hesitation overtakes Achilles – that hesitation which Simone Weil identifies as the saving margin of civilization. Achilles puts back his sword. For a moment, there is sanity. But on the whole, the glorious poem takes place against the background of a dark, archaic world dominated by will and force, dreams and oracles, blood lust and atonement, portents and magical healing, orgiastic cults, superstitious terror, and the obbligato of reckless massacres – the world governed by the irrational self, the *thumos*, the strong force that surfaces today as tribalism, racism, and the longing for combat.

After the Homeric Age – rather precipitously, as history goes – the landscape does change. It is the period identified as the Greek Enlightenment, during which there was, in a phrase Dodds quotes approvingly, a "progressive replacement of mythological by rational thinking among the Greeks." The rise of this first enlightenment, in the sixth century B.C., coincides with the first stirrings of Western science as we understand it.

There were bound to be enthusiastic mistakes; for example, Protagoras, an early Sophist, has gone down in history as perhaps the first optimist who thought that virtue could be taught, that history could be cured, that intellectual critique alone could rid us of "barbarian silliness" and lift us to a new level of human life. But at least his intention was right. Science, and rational thought which produced science, are beachheads in the soul that otherwise would be largely given over to *atē* and *thumos*. The existence of such beachheads allows one to hope for a change in the balance of potentials in the individual and, therefore, in the balance of forces that have raged over the geopolitical map since prehistoric times.

I submit that this fact defines, today more than ever, an essential part of the task of all who claim to be teachers. To neglect it is to invite peril. The subsequent history of Greece itself reinforces this point. Around 430 B.C., at the end of the reign of Pericles and the beginning of the Peloponnesian War which finished with Athens's surrender and Sparta's triumph, that first Age of Enlightenment gave way. Teaching astronomy, or expressing disbelief in the supernatural, now could have grave consequences. There followed some thirty years of trials for heresy, with victims such as Anaxagoras, Socrates, Protagoras, perhaps also Euripides, and of course an unknown number of less prominent ones. In the long series of conflicts between reason and passion, Athena, the Weak Force, had lost.

One of the causes for this turn of events appears to be that, from the late period of Plato onward, intellectuals situated themselves not *in* but *beside* society. Dodds writes: "As the intellectuals withdrew further into a world of their own, the popular mind was left increasingly defenseless . . . ; and, left without guidance, a growing number relapsed with a sigh of relief into the pleasures and comforts of the primitive." Greece saw again a great rise in cults, in magical healing practices, in astrology, and other familiar symptoms. It ushered in the long decline or, to give its proper name, the "Return of the Irrational." It was as if the bicameral mind had become aware of its rational strength – and been frightened by the possibility of freedom from the death dance of history, and freedom from the external gods to whom one could spin off the responsibility for the excesses of *atē* and *thumos*. (I am quite aware that a group of philosophers, from Nietzsche to Spengler, from Husserl to our day, puts the blame, on the contrary, on science itself, on what they call "excesses" of the rational. Their arguments are saved from dismissal as utter absurdity by the unhappy fact that science, too, has frequently lent itself to be a weapon in the service of our Dionysian and antihuman drives.)

The decline of Greece shows parallels with our present predicament. We are at the tail end of the second experiment with rationalism, the fruit of the scientific revolution of the seventeenth century and the era of enlightenment that followed. That schoolroom I described, and all the others across the globe, may really have functioned as a trap designed to keep us from escaping a destiny that is archaic except for its modern, much larger scale. The most persuasive evidence that the human mind has the power to progress, individually and cumulatively, from ignorance and confusion to sensible, testable, sharable world conceptions – the kind of triumph of man's rational potential of which his science is eloquent testimony – was all but sabotaged by the method of presentation, itself an institutionalization of our reluctance to honor the imperatives of sound reason. Indeed, the very facts of science which we had to memorize seemed the work of some deity that plants them in its passive victim, even as Zeus planted the *thumos* in Agamemnon. Never once was the liberating idea presented to us that the findings and very methods of science are the results of an historical process by which mere humans seek sense and expose nonsense, and that the potential for this process is in each of us.

Helping Athena

All that was long ago. Many have worked hard on improving the effectiveness of educators. But the admonition of Max Weber, in his magisterial "Science as a vocation," has become no less urgent. Rational thought, he reminded his audience, has the moral function of leading to self-clarification, of helping the individual "to give himself an account of the ultimate meaning of his own conduct," and so to be able better to make decisive choices. Without it, life "knows only of an unceasing struggle of these various gods with one another . . . , the ultimately possible attitudes toward life [remain] irreconcilable, and hence their struggle can never be brought to a final conclusion." Today, as we watch the reign of the irrational in world affairs, I find it hard to believe that we have succeeded in making a qualitative change for the better in the products coming from our classrooms. Moreover, in our century the would-be conquerors who are writing themselves into the pages of the new history books have learned how to hire and use for their bloody work students coming from our science classes. It is an ominous conjunction of science and history, making it that much more difficult – and essential – to hold on to the vision that the trajectory of history can bring us to a time when, at long last, the goddess Athena in our very soul wins out consistently over the dark passions.

This ancient aim (which Werner Jaeger, in his great work *Paideia: the ideal of Greek culture*, identified as the chief hope of education in antiquity) should now be a special concern of those who, through their own life's work, have learned how one distinguishes between fact and delusion, between the demands of eternal law and of internal longing; of those who care most to find out how things work and cohere. The immense authority of, say, an Oersted came of course from the painstaking and repeatable demonstration of a beautiful and useful discovery. The history of science can show that might comes from being right, rather than, as in the rest of history, more often than not the converse. Bringing science and history together in that kind of conjunction – in scholarly research and in the classroom, for scientists and for nonscientists – is one effective way to enlarge the beachhead of reason.

There are others, there have to be others; but it has been my motivating (and perhaps now no longer quite so perverse) view that this way helps focus a young mind exactly on the point where the confrontation should take place, between the habits responsible for the kaleidoscopic sequence of follies on one side, and, on the other, the kind of passionate yet sane thinking that shaped the development and testing of, say, the table of the elements or of elementary particles.

I have no illusion that more chemistry, more physics, more mathematics will, by themselves and soon, produce wise leaders and wise followers. Protagoras *was* too simple. There is no quick cure for the barbarian silliness within us which shows up so grotesquely in the acts of the present-day Agamemnons, generalissimos, premiers, shahs, and ayatollahs. We must also not be misunderstood to be defending some inevitable benignancy, purity, and progressiveness of science. Paradise will not come upon invoking the name and deeds of Mendeleyev or Wolfgang Pauli. And yet, enough of us must act nevertheless as if something of this sort can happen eventually; for otherwise it will not change. As we, and the future teachers who pass through our hands, face those young students, there is the opportunity to assert and demonstrate the rational powers of Athena as a complementary and balancing element in the productive life of the human spirit.

The risk of failure is high in all educational efforts, and there are always other, more immediately rewarding things one might do instead. But it is a risk worth taking. For those young students in the schoolroom, wondering about the forces that grip the world map and soon caught up themselves in its convulsions – those were you and I; those are now our children; and those should not have to be our children's children, forever.

11

From the endless frontier to the
ideology of limits

The public discussion in the United States about constraints on scientific research seems to have come upon us with startling suddenness. Ironically, it surfaced at the very moment when scientists have the right to think that the fundamental advances being made are better than ever, and when scientists are loudly asked to help solve the vast problems in such areas as energy and the environment. Scientists find themselves rather bewildered. For decades, they have proudly accepted P. W. Bridgman's well-known operational definition, "The scientific method is doing one's damndest, no holds barred." Now, they are asked to add the phrase, " – except as laid down in guidelines issued in Washington and by the local town fathers." The old image of science as the "endless frontier," on which a whole generation has been brought up, seems to be giving way in some quarters to the notion of science as the suspected frontier.

Many observers date these changes from the first public expressions of concern by biologists about the possible side effects of doing research on recombinant-DNA molecules. If the debate is of such recent origin, can it last? In a country where even the most sensational and preoccupying activities can suddenly disappear in silence, are we really dealing here with a serious challenge, rather than merely a highly visible but short-lived excitement?

The answer seems to be a clear "yes." In this chapter I shall sketch reasons for believing that we have only begun to struggle with such problems. For whether one likes it or not, the disputes concerning the wisdom or danger of placing "limits on scientific inquiry" may have been inevitable and were perhaps overdue. Depending on the specific cases that clamor for attention, the intensity of the discussions may wax or wane; but they have a certain preordained character, and in maturing form will remain with us for a long time to come.

There are more than half a dozen chief factors determining these events today, each of which can be expected to continue to exert its force in the future. I shall now take up each one in turn.

The new visibility

The simplest of the reasons for the expected endurance of the issue has to do with the attainment by the scientific establishment of a kind of critical size. The increase in the scale of scientific and technological activity in the industrialized countries has made for visibility, and with that – as should be the case, at least in a democracy – has come a reexamination of the mechanisms of accountability. The annual cost of basic research in the United States [in 1982] was over $10 billion (two-thirds of it supplied by the federal government) and a total of about $100 billion is being spent on the national research and development effort (including defense and space projects, which, for better or worse, the public identifies closely with science). Clearly, any enterprise that employs on the order of three million scientists and engineers, and commands 20 percent of the relatively controllable portion of the federal budget, must be subject to mechanisms of accountability with respect to its performance and justification, in terms that taxpayers or their representatives can appreciate.

This rising level of activity, and hence of the power of science and engineering to change our world, acts in two quite opposite ways on public perception. Along with splendid new discoveries and some welcome gadgetry, there is also a higher level of risk, compared with the level of risk in the past, in periods of slower technological change. Patent dangers, outright abuses, and significant mistakes may constitute only a small proportion of ongoing work; but when the rate of change of scientific knowledge and of technological advance is rapid enough, that small proportion does add up, in absolute terms, to form a visible set of cases. The public alarm system seems to be set at absolute levels, not by relative measures.

Moreover, as the rate of change is increased, it has the unsettling effect of decreasing the time interval during which the change being imposed on society can be monitored, evaluated, or even absorbed intellectually by the public – just when this public has learned to insist more and more on having some role in consequential decisions. To illustrate: in 1976 the lay citizens' group appointed by the City Manager of Cambridge, Massachusetts, to advise the City Council on guidelines for the use of recombinant-DNA molecule technology declared on page one of its generally thoughtful report:

> The social and ethical implications of genetic research must receive the broadest possible dialogue in our society. That dialogue should address the issue *whether all knowledge is worth pursuing.* [Emphasis added.] It should examine whether any particular route to knowledge threatens to transgress upon our precious human liberties. It should raise the issue of technology assessment in relation to long-range hazards to our natural and social ecology. Finally, a national dialogue is needed to determine how such policy decisions are resolved in the framework of participatory democracy.

One may try to dismiss this as a pastiche of clichés. Back in the minds of such persons is the fact that the implicit promise of science has been a rosy world free of disease, with new industries and new jobs waiting for us, a world giving us secure peace, and perhaps even the joy of understanding what those scientists are discovering in the noble quest. Little of that has come to pass. On the contrary, the very solutions proposed, such as nuclear reactors to supplement our shrinking energy supplies, now loom for many as threats to health and peace. As will be stressed in more detail further on, there is a widespread disillusionment with the explicit or implicit promises of technological solutions to social problems. And even those who were not attracted to the activist image of science and technology, who preferred the older idea of the scientists as lonely thinkers and the engineers as inspired tinkers – occasionally very useful people but essentially harmless – have had to readjust to the much greater and confusing variety of the current roles that scientists and engineers play, as employees of universities, industries, consulting firms, government laboratories and "think-tanks," or as advisors to state government, the Congress, or the President. If science ever was a charismatic profession dominated by abstract spirits, those days are gone forever.

Short of suffering crippling decreases in financial support, work in science and engineering will not be less visible and less watched in the future. It is, however, reasonable to hope that the fraction of abuses, mistakes, surprises, and other alarming problems will drop as the professionals involved become more and more sensitized to the possibility of such problems. That may be one good lesson drawn from the list of horrors of recent years, including the side effects of pharmaceuticals and food additives, the dubious mathematics of risk calculations, and alas, the estimate that at least one-quarter of the world's scientists and engineers is engaged fairly directly in exploiting the "pure" advances for weapons-related research. The celebration of the basic intellectual triumphs of science has had hard competition.

The old credo

If the newly visible risks are a part of the public's perception about science that is not easily changed, the current self-perception of scientists may be only slightly more flexible. While today there is more diversity of views in what is lightly called the scientific community than has been the case since the 1930s, the largest proportion of scientific professionals is watching these debates with considerable apprehension. Many scientists believe that the integrity of their activity must be strenuously protected against what they perceive to be potentially serious threats from an unchecked, or perhaps ill-informed, limits-to-inquiry movement. Whether they originate from fellow scientists or other citizens, calls for any explicit limits go against long-standing traditions of academic and research freedoms as still understood by most scientists. They contradict the predominant philosophical base of science as an infinitely open system in which it is not to be feared that one will run up against questions that are, in principle, unanswerable, not to speak of questions that are unaskable. They are also contrary to the psychobiographic drive of most of the younger scientists and the reward system that has shaped the older ones. They clash violently with the world view of scientists brought up on the notion that science and optimism are virtually synonymous, that somehow the findings of science will be for the good of mankind.

One recalls Jacques Monod's remark in *Chance and necessity* that the ethic of knowledge is the commitment to the scientific exploration of nature. The limits movement conflicts basically with such commitments; moreover, it comes, ironically, just at a time when the results of scientific and technological work, both in terms of quantity and in terms of quality, contain more examples of ingenuity and beauty by any standard than ever before.

Some of the most successful and visible scientists, when calling for as unrestrained an autonomy as is compatible with safety, tend to regard the very discussion of proposals for "limits" as dangerous and self-fulfilling. The period of accommodation, on either side, therefore, is likely to be a long one, the more so when scientists find themselves confronted with extreme proposals, or with banners such as the opening to a recent draft statement from the World Council of Churches: "However robust, the faith of yesterday in the power of science and technology is today misplaced and therefore misleading. In fact, the reductionist, triumphalist, manipulative approach of science and technology has itself been in large part responsible for our predicament."

In fact, however, scientists or engineers today are on the whole far more ready to take care that the ethos and practice of science include protective limits and constraints than is popularly recognized. A number of internal and external constraints have been found functional and even essential (except that the red tape associated with certain regulations is threatening to get out of hand). Thus, to cite some controls internal to the practice of science, it is taken for granted that quantitative results, if obtainable, are preferred over qualitative ones; that operational definitions rather than metaphysical ones be used; that important experiments be repeated; that an attempt at relating theory and practice be made; that the work be ultimately published; that credit for priority, collaboration, and support be given; and so forth. As every modern researcher knows, getting a share of scarce resources – for example, time, funding, and manpower to do an experiment involving one of the large machines – or getting the approval of the local committee charged with the regulation of hazardous biological agents, can lead to epic fights.

Among external mechanisms for discipline and accountability, scientists consider it natural that peer review be undertaken on funding priorities and on the soundness of procedure; that the peer review process itself be open to inspection and evaluation; that informed consent be obtained from experimental subjects; and that research involving toxic or radioactive substances, or human beings used as "animals of necessity," be placed under careful regulation; and that environmental and other impact statements be provided for large engineering constructions.

One can discern three levels of external restraints or controls on research, to which scientists tend to react with increasing alarm. Those arising from the hazardous nature of research are long-established and respected on all sides. Constraints that arise from decisions to invest in one area of research rather than another are more controversial, in good part because this has to be ultimately a political decision (made in the United States, in effect, more by the Congress and the Office of Management and Budget than by the scientists concerned). Third, controls may be imposed by nonscientists because they have qualms about scientists by themselves being able to estimate properly the hazards of a given research and/or the ethical problems involved in the procedure or its possible findings. The last is evidently the area of greatest dispute.

Until recently, most of the limits on inquiry were self-imposed by the scientists, and invisible to the public. Even the self-denying ordinances on doing harm, on publication of results, or on the actual pursuit of research – from Hippocrates to Leo Szilard to the Asilomar conferees – were of a very different kind from those now being discussed or proposed in forums far from the laboratory bench. Indeed, even in the laboratory most of the constraints listed above are today still largely passive and invisible rather than being the subject of conscious examination or having an explicit place in the teaching of young scientists. This is only one more example of the well-known resistance among scientists to self-consciousness in the study of actual scientific practice; but in this instance, it also slows the chances for convergence between the interest of scientists and their monitors.

The endless frontier revisited

Having pointed out some difficulties which the "old credo" of science has in today's world, one must hasten to acknowledge again its immense power and usefulness. It has helped nurture a strong science/technology community in the United States, and elsewhere. It has helped us to fashion two activities of great strength. First, as the meetings and publications of the major scientific societies amply show, the quantity and quality of basic research in many fields is higher than at any other time in history. The work of these scientists, largely but not exclusively done in the universities, is usually motivated and measured by the standards of "pure" or "basic" science rather than public need. It is the product of a largely autonomous, self-governing system, not directed by the calculus of risk and benefits. If there are other affected interests, most of those are placed at a distance, and the scientists are insulated from them. The hope for social utility as a by-product of one's discipline-oriented research may be in the background. But the ruling motto is that "truth must set its own agenda."

Moreover, some of us, myself included, continue to insist whenever we can that the very business of developing a rational and functional model of the universe through basic research is *itself* a major social goal for any civilized society, and fulfills a public need on that account. We point to the triumphs in molecular biology or biophysics, in cosmology or elementary particle "zoology," and in many other fields, and we say that whatever changes may be made, the quality of the product must not be put in danger. Of course, there are costs: severe specialization (although it is not an unmitigated evil: at the very least it helps to define the peer group and hence the peer review system, and in any case it is often the only way to do anything useful at all); reductionism (but it helps to select manageable problems); and the vague discomfort that the public, which is paying for all of this work, no longer knows what we are doing, since the conceptions are increasingly sophisticated and our patience with educating the public is not notably high.

To turn to the second mode of current excellence, there is another portion of the scientific/technological community that knows how to apply the basic scientific findings to a multitude of "public needs," needs that are articulated by a variety of constituencies, from the householder to the Joint Chiefs of Staff (and occasionally, merely by advertising agencies): the list includes vaccines, nuclear reactors, 10^4 industrial chemicals per year, computers, moonshots, new food plants, photovoltaic cells, insecticides, the cruise missile, not to speak of frozen dinners and shaving creme coming out of cans at the push of a button. The accent is more on problem-oriented development than on "truth"-oriented research, although there is often an interaction between truth and utility, as in materials sciences and medical research.

The great problem that has surfaced in this second mode is of course that such work, done on behalf of expected customers in a competitive market, provides at most only first-order solutions, into which are built sometimes awesome, second- and third-order problems. The larger the scale of the technology involved, the more likely are these second- and third-order problems to surface eventually. The reason for these by-products is partly that the complexity of the situation is almost by definition very great (hence, they are called "unforeseen" new problems), partly that the training of even the applied scientists is generally not geared to such sober and, on the whole, boring business as technology self-assessment; and, above all, that the client – the state, the industry – is usually in a hurry, is reluctant to pay for more than first-order solutions, and until recently has made little attempt to get advice on hidden flaws until they have become obvious, and therefore less manageable.

This dichotomy of styles of research and development, the separation between them, and the unfavorable by-products, are the results of a long development of science and technology in the Western world. But at least as far as the United States is concerned the most formative part of that history has been the most recent period – specifically, the conscious choices made by the politically most powerful segments of the scientific community when the outlines of the present system of science support were set up at the end of World War II. That system was the result of a battle between the ultimately victorious forces, led by Vannevar Bush, the head of the Office of Scientific Research and Development, on the one hand, and the forces identified with the New Deal Senator Harley Kilgore on the other.

The issue between them was really an old one: What should be the main thrust of science and hence the main justification for its support in our democracy? As Daniel Kevles has noted in his book *The physicists*, since its start in the United States the debate has been polarized along the same axis, one that may be quite roughly characterized as knowledge for its own sake versus social usefulness.

To be sure, one must not only look at the extremes of these alternatives, for the progress of both "pure" and technological application has become more and more strongly intertwined. The pre–World War II debates between Hogben, Crowther, and Bernal on the one hand, who argued for the primacy of the function of science in the promotion of human welfare, and, on the other hand, their opponents such as Polanyi, who favored the pursuit of science dissociated from any conscious connection with practical benefits, created so sharp a dichotomy that the arguments had to be ultimately inconclusive. Today, we know well that advances in the most recondite and speculative branches, such as cosmology, can depend on data obtained with space- and computer-age technology, while the most mundane electronic gadgets trace their lineage to the early papers on pure quantum physics by Planck, Einstein, and Bohr.

Nevertheless, it matters greatly where the center of gravity is placed in a large national institution for the support of science, and how large a spectrum is encouraged. Kilgore's declared aim was to set up a national research and development foundation which would assure that at least a major part of federally supported science research be linked to a social purpose, and that progress in science be planned to some extent by such an agency. Vannevar Bush and his colleagues, however, were opposed to this model, and were against such notions of Kilgore as a major revision of patent rights flowing from government-sponsored research, or geo-graphical criteria for the distribution of funds, or the inclusion of support for the social sciences in a super agency for research. They also had to worry about attacks from the other wing. For at that end of the political spectrum, the opinions about Bush's own proposals were, as Bush noted in his autobiography, *Pieces of the action*, that "we were inviting federal control of the colleges and universities, and of industry for that matter, that this was an entering wedge for some form of socialistic state, that the independence which has made this country vigorous was endangered." [1] Bush himself had similar fears about Kilgore's ideas; as Bush's associate Homer W. Smith put it, the scientist needs "the intellectual and physical freedom to work on whatever he damn-well pleases."

One can understand why, particularly at that time, the working scientists may have felt keenly that way. World War II had just ended, and the scientists were concerned that the armed services were going to keep their hands on the direction of research. Even the *New York Times*, under the heading "Science and the Bomb," declared editorially on the day after the bomb devastated Hiroshima (in its August 7, 1945, edition) that the scientists had better shape up and learn a lesson from the events:

> University professors, who are opposed to organizing, plan-ning and directing research after the manner of industrial laboratories because in their opinion fundamental research is based on "curiosity" and because great scientific minds must be left to themselves, have something to think about. A most important piece of research was conducted on behalf of the Army by precisely the means adopted in industrial labora-tories. And the result? An invention is given to the world in three years which it would have taken perhaps half a century to develop if we had to rely on prima donna research scientists to work alone. The internal logical necessity of atomic physics and the war led to the bomb. A problem was stated. It was solved by teamwork, by planning, by competent direction, and not by a mere desire to satisfy curiosity.

Such omens, and the memory of scientists having to buckle under in totalitarian countries, were a source of considerable concern to many American scientists. While Kilgore wanted the best science required for the national needs of the United States, Bush thought it safest to ask for "the best science, period." In a free society, he hoped, it would automatically serve the national need. The model for that to happen was at hand, and Bush confessed freely that he used it to advantage in lobbying for his planned agency:

> There were some on Capitol Hill who felt that the real need of the post-war effort would be support of inventors and gadgeteers, and to whom science meant just that. When talking matters over with some of these, it was well to avoid the word "fundamental" and to use "basic" instead. For it was easy to make clear that the work of scientists for two genera-tions, work that had been regarded by many as interesting but hardly of real impact on a practical existence, had been basic to the production of a bomb that had ended a war.[2]

Other institutions – industry, the Office of Naval Research, the planned Atomic Energy Commission – could be counted on to support research "along the lines of their special interests," but this would not be true for "basic research, fundamental research."[3] This gap was to be filled by the National Science Foundation. Without a dedicated agency of this kind, Bush held, "a perverse law governing research" would assert itself: "Under the pressure for immediate results, and unless deliberate policies are set up to guard against this, *applied research invariably drives out pure.* This moral is clear: it is pure research which deserves and requires special protection and specially assured support."[4]

Thus, the public credibility of Bush and his fellow scientists in their drive to invent institutions and obtain previously undreamt of financial support for basic research came at the time not from the persuasiveness of their own basic research performance – that work had been laid aside during the war years – but rather from the superb job that had been done in applied science and development during the war. And although the irony is clear now, it was then quite natural that the top scientists, turned wartime engineers, whose work had started the nuclear reactor wastes flowing into the fragile tanks at Hanford, Washington, and elsewhere, now wanted to hurry back to "best science." They did not even stay to ask what important scientific puzzles would have to be solved before the nuclear wastes could be safely disposed of. In fact, a full seventeen years had to elapse from the first design, around 1940, of nuclear reactors by some of the best scientists of the time (Fermi, Szilard, and their col-leagues) until the publication of the first major study of high-level waste management.[5]

Ironically, too, Vannevar Bush obtained his mandate to write what became the ground plan for the National Science Foundation from Franklin D. Roosevelt, who was personally most attracted to the *applied* side of the plan. In his original letter to Bush, Roosevelt asked for a way to put to use "the information, the techniques, and the research experience" developed during the war in order to wage a "war of science against disease," to "aid research activities by public and private organizations," and to devise ways of "discovering and developing scientific talent in American youth."[6]

In his response, the seminal report *Science, the endless frontier*, Bush followed that rhetoric closely. He gave low visibility in print to the pursuit of pure science for its own sake, relying heavily instead on the tenuous promise that somehow science will be of social use. It is, to this day, a splendid document. It captures the well-deserved, utopian hopes at the end of a terrible war, and in many details was visionary, necessary, and right. But it does strike us now forcefully that in the report no mechanism was specified or encouraged for the benign spinoff-effects of basic science. It was not revealed how the central alchemy would occur that would yield the promised long-range benefits. And of course no mention was then made of the other, darker side of the coin, the possibility of negative consequences, not to speak of the need for impact statements and the like. On the contrary, the very first paragraph of Bush's report starts with this clarion call:

> Scientific progress is essential.
>
> Progress in the war against disease depends upon a flow of new scientific knowledge. New products, new industries, and more jobs require continuous addition to knowledge of the laws of nature, and the application of that knowledge to practical purposes. Similarly, our defense against aggression demands new knowledge so that we can develop new and improved weapons. This essential, new knowledge can be obtained only through basic scientific research.
>
> Science can be effective in the national welfare only as a member of a team, whether the conditions be peace or war. But without scientific progress no amount of achievement in other directions can insure our health, prosperity, and security as a nation in the modern world.[7]

In itself, this program and this promise, and even this language, are not all that new. For some centuries, the announcements of great projects of science have had a remarkably similar ring. For example, the preamble of the Act establishing the American Academy of Arts and Sciences in 1780 begins with the presupposition of beneficence:

> As the Arts and Sciences [effectively, a range of subjects from cosmology to medicine and horticulture] are the foundation and support of agriculture, manufactures, and commerce; as they are necessary to the wealth, peace, independence, and happiness of a people; as they essentially promote the honor and dignity of the government which patronizes them; and as they are most effectively cultivated and diffused through a State by the forming and incorporating of men of genius and learning into public societies for these beneficial purposes. Be it therefore enacted. . . .

Two hundred years later, we find the same belief expressed in the slogans of our highest officers, for example: "Basic research is the foundation upon which many of our Nation's technological achievements have been built," and "Basic research is useful."[8]

The promise of beneficence emerging from such documents has been routine, and (at least until lately) almost universally accepted as plausible. In Vannevar Bush's plan, the lack of specificity – how disease, ignorance, and unemployment would be conquered if basic science is supported on a large scale – also had a number of good reasons. Not the least may be that the very image of science as the "endless frontier" avoids the need for operational details of this sort. In his book *The American character* (1956), D. W. Brogan made the astute observation that American and English thought differ in the connotations of the idea of the "frontier":

> The frontier in English speech is a defined barrier between two organized states; in American it is a vague, broad, fluctuating region on one side of which is a stable, settled, comparatively old society, and on the other, empty land, a few savages, unknown opportunities, unknown risks. American history has been a matter of eliminating that debatable area between the empty land and the settled land, between the desert and the town. This elimination has now been completed, but it is too early, yet, for the centuries-old habits to have changed and much too early for the attitude of mind bred by this incessant social process to have lost its power.

The implication of the beneficence, unforeseeable in detail but inevitable, of the effects trickling down from basic science to useful application continues to be held before the public to this day. An effect of this sort of course exists. But it is really still little understood, is difficult even to reconstruct historically, and is not assured of long-term institutional support as industry fitfully increases and decreases its support of basic research. In good part for these reasons, few scientists would want the agencies that are now committed to nourishing basic research to change that commitment in any major way. Yet the public has by no means forgotten about the implied IOUs; and as the combined bill for both elements of the dichotomy, "pure science" and "useful science," has greatly increased over the years, and as the bridge from one to the other has not become clearer, people from Senator Proxmire on to the right and left have become more critical. In addition, as Don K. Price correctly remarked, "The danger of political constraints on science now comes not so much because politicians disapprove of the methods of science, but because they take seriously what some scientists tell them about the way in which scientific discovery leads to practical benefits. Once they believe that, it is inevitable that they should try to control practical outcomes by anticipating the effects of research, and manipulating it in one way or another."[9] It may therefore be the final irony that the implied but difficult-to-deliver promises in the Bush report, on the basis of which the dichotomy was achieved and the pure-science ideal triumphed, became a factor in the current assertion of external constraints on science.

The precariousness of the new bargain

Even as the old institutional arrangements for debate and agreement now seem inadequate, so does the old bargain between science and society itself. The generally accepted model of the linkage, in typical cases, involved a barter of an odd sort. That is, as in the Bush report, society could expect in the long run to get certain though not well-defined material benefits from the work of researchers in science and technology, and those individuals in turn expected to benefit by receiving some moderate financial support and considerable administrative freedom from society. In that barter the two sides were in most instances distant and rather unengaged. Each gave up to the other things that it could easily afford to yield, or that it did not particularly treasure. Each side had no great incentive to understand the other, and could be satisfied with the gains it was eventually receiving for itself. The system worked well, and was moderately well understood.

But the model of interaction that has been developing lately is rather different, and very much in need of study. (As Harvey Brooks writes,[10] we do not even know how science and development are connected with social policy: which drives which?) The new model involves (again in barest outline) the giving up by each side of substantial, treasured items, and the exchange requires much more mutual involvement and understanding. Each side now has to barter away some of its autonomy, and each side expects to obtain substantial, monitored benefits and assurances. Much of the argument on where the limits of scientific inquiry should be is a by-product of this encounter.

The mood I see developing in this encounter corresponds to the impression I gained at the end of a series of meetings of a faculty seminar at MIT. If there was a consensus, it was, in the words of one of the participants, that just as we are wary of the "slippery slope" in biomedical ethics, so we must resist slipping inadvertently into increased controls over fundamental science, since such controls can easily lead to abuses of their own.

The image shared by most scholars there was that we are now dealing with *two* slippery slopes, joined at the top by a razor's edge upon which we are precariously balanced. There will thus be hard bargaining on such questions as these: *Who* will control what, and at what level? (At the level of the institution, of the individual researcher, of funding?) How can "public participation" be arranged without clashing with the very meaning of science as a consensual activity among trained specialists? (In the old bargain, the inherent contradiction between science and democracy did not have to be faced. Now, it does.) When today's largely unwritten code of ethics, maintained by peer pressure, is turned in for a set of congressionally developed guidelines, how can enough room be left in them for personal choices that will need to be made as conditions and knowledge change? (This is analogous to obtaining "variances.") Or will such choices be smothered by blanket regulations, formulated by bureaucratic entities that have considerable inertia? In that ominous event, how can the freedom and momentum be preserved upon which any imaginative group must be able to draw to do its best?

Need for new institutional forms

The fact that the questions posed in the last paragraph have today no persuasive answers indicates that there is a real paucity of institutional forms for dealing with them. As one reads proceedings of conferences or congressional hearings in which representatives of the scientific establishment and of the public struggle with the new problems of accountability, one is impressed by how often these old institutional forms for mutual negotiation really have become inadequate for the complex task that is required. Instead of the model of an interdisciplinary investigating team in which the members assist one another in a common task from which all would benefit, one finds all too often the model of unilateral pronouncements inherited from the classroom, or of adversary proceedings as learned in the courtroom.

Institutional forms are slow to grow, and until enough credible ones have been designed and tested, the public may well continue to push for ad hoc solutions that may not at all appeal to the scientists. It is a very hopeful sign of their new maturity that the scientists, too, are in fact hard at work fashioning new institutional devices and experiments. A brief list would include the role scientists have played in the Asilomar conference, the various citizens advisory panels, federal funding of scientists to work with public interest groups or with projects on the public understanding of science, the revision of codes of ethics by some professional societies (to make such codes something more meaningful than the traditional, rather self-serving documents), the federally funded courses concerned with ethical and value impacts of science and technology, the sections within professional scientific societies such as the American Physical Society (its Panel on Public Affairs, its Forum on Physics and Society, its summer study groups on nuclear waste management and other problems, its congressional fellowships, etc.). A glance at the table of contents of the journal *Science* and of the meetings of its parent organization, the American Association for the Advancement of Science, will show how profound the changes are which have taken place over the past dozen years in the attention of that large group. And of course there are many other signs – the emergence of the Society for the Social Study of Science, the increase in modern case studies in the field of the history and sociology of science, and academic programs with titles such as Science, Technology, and Society.

To illustrate the novelty and, in the face of controversy, daring of such activities in which scientists themselves are participating, I will single out only one example. It concerns the discussion within research institutions on the question of impact statements. In seeking funding or the acceptance of proper research procedures, scientists and engineers have for some time now accepted their liability for submitting impact statements dealing with environmental impact, financial responsibility, manpower policy, affirmative action, and the like. Nobody has liked such bureaucratic impositions, and they can become detrimental to the execution of the work when they are too burdensome; but at least they are now being routinely attended to. Lately, there have been some discussions that sooner or later ethical impact or values impact statements may be added in certain areas of inquiry, and that therefore scientists and engineers should begin to become familiar with such conceptions before they are imposed unilaterally from without.

To a degree, this was the message of the Bauman amendment in 1976 by which the Congress would have reserved for itself the right to re-examine each grant made by the National Science Foundation after peer review. Depending on the effect or impact this research might be thought to have, in the opinion of any Congressman, an embargo could have been put on a particular grant after it was made. As part of a similarly motivated action, in the authorization bill for appropriations for the National Science Foundation issued in March 1977 the Congress directed the NSF to provide for the protection of students in try-out classes that use pre-college educational materials. The aim was to make sure that "interested school officials" could make certain that steps had been taken "to insure that students will not be placed 'at risk' with respect to their psychological, mental, and emotional well-being" while serving as human subjects in these educational developments.

The idea of asking researchers for a statement of the prospective impact of their research has been circulating quietly. Thus in March 1974, a faculty committee was set up at Massachusetts Institute of Technology to see how faculty members could make a more thorough and self-conscious study of "the impact of MIT research insofar as that research has influence on matters such as the physical environment, the economy, national security, and other important social concerns." One function of the committee was to develop methods "to assist individual investigators in the preparation of impact statements." In making its report to the faculty, the committee – one containing such widely respected members as Frank Press and John Deutch – said "An honest effort to estimate the plausible consequences of his own research is as properly to be expected from a researcher as how he is expected to estimate the cost of carrying out his work, and is in the long run perhaps more significant."

When it came to the vote, the MIT faculty felt that the time was not ripe for this experiment. However, it was made clear in the discussion that the issue should not be dropped, and that a more specific proposal should be developed for future action. I expect we shall hear more about this and similar ideas, aiming to fashion new institutions for dealing with the problems of limits that arise when a scientific pursuit or engineering project has the potential of encroaching on other widely held social values – and to do so before the problem of setting limits has entirely escaped from the hands of the scientists themselves.

Ideology of limits

A conflict of potentially large consequence has begun to be felt in the developed countries of the West between the old "ideology of progress" and a new "ideology of limits" that goes much beyond limits to scientific inquiry. One of the lessons of the sensible decisions not to build the SST in the United States and not to go to a plutonium economy is that a nation's leadership now *can* invoke an almost unprecedented self-denial of technology, a turning away from the "can do means must do" imperative. To a certain degree, an embargo on some aspect of science can be characterized as being a call for a stop to the pursuit of knowledge of *how to* (rather than knowledge of *why*), and is therefore closely similar to the self-denial of these new technologies.

This development fits in with the general new awareness of the existence or necessity for *limits* (a word that may yet come to characterize a main lesson of the 1970s) – limits to natural resources, in particular, energy supplies; limits to the elasticity of the environment to respond beneficently to the encroachment of the man-made world; limits to food supplies; limits to population; limits to the exercise of power (including presidential power); and, within science itself, limits to growth of its manpower and institutions.

The pervasiveness and reality of this new climate of opinion are being more widely studied. Thus a major report made for the Joint Economic Committee of the Congress, entitled *Limits of growth*, volume 7 of *U.S. economic growth from 1976 to 1986: Prospects, problems, and patterns*, specified the existence of a number of popular new "emphases and trends" that are expected to influence growth during the next decade. These include the continued rise of risk aversion, concern with health and environmental protection, and antitechnological and anti-industrial attitudes. It is clear that for a small but vocal part of the population of the

West, the image of this century has been changing from the era of frontiers waiting to be exploited, to the era of the globe as a crowded lifeboat.

New conception of progress

The last and perhaps most decisive component in the current shaping of events has to do with the rise of a new conception of progress in science itself. Here too a substantial change in Zeitgeist has been taking place, and it surely reinforces the effectiveness of the limits-to-inquiry movement.

To be sure, the old notions of the purity and objectivity of science and of its claims to truth in some absolute sense have been under attack for a long time. But this has been largely a philosophical debate, with little effect on most research scientists whom such epistemological – perhaps to some degree even quasi-theological – developments reach with a long time delay and with few credentials. Thus scientists see themselves engaged in an enterprise, in which they consider progress and cumulation to be perfectly well identifiable, and in which the chief marks of progress are still taken to be the classical ones: greater inclusiveness of separate subject matter, and greater parsimony or restrictiveness of separate fundamental terms.

Until not too long ago, the popular conception of science as inexorably "progressing" had been a component of thought even in the historiography of science. George Sarton went so far as to assert in 1936:

> *Definition*: Science is systematized positive knowledge or what has been taken as such at different ages and in different places.
>
> *Theorem*: The acquisition and systematization of positive knowledge are the only human activities which are truly cumulative and progressive.
>
> *Corollary*: The history of science is the only history which can illustrate the progress of mankind. In fact, progress has no definite and unquestionable meaning in fields other than the field of science.[11]

But just in the last few years these progressivist assumptions have come under a variety of attacks from scholars in the history and philosophy of science. Among those, I select two who agree with each other on little, except on the rejection of the old notion of progress. Each has a wide following (and their published opinions, excerpted below, only indicate the flavor of the writings by others whose versions lack the acknowledged subtleties of the more fully developed arguments).

The first school, headed by the historian of science T. S. Kuhn, regards the notion of scientific progress as a self-fulfilling definition, therefore essentially a tautology. He asks, "Viewed from within any single community . . . the result of successful creative work *is* progress. How could it possibly be anything else?" The upshot is, we are told, that we may "have to relinquish the notion" that the large changes which science occasionally undergoes could carry scientists and those who learn from them "closer and closer to the truth." An implicit consequence is not far away: when it is widely believed that scientific progress is defined only by whatever the scientists are doing anyway, then limiting some specific work in the presumed interest of other, more urgent human values will seem a far less intolerable intervention.

The other school, headed until his recent death by the philosopher of science Imre Lakatos, quite explicitly aims to set up, as Lakatos put it, "universal criteria" for distinguishing "progressive" research programs from those he regarded as "degenerating." He wrote that he did so partly in order to help editors of scientific journals to "refuse to publish" such papers, and research foundations to refuse support for such work. (He warned, "Contemporary elementary particle physics and environmentalist theories of intelligence might turn out not to meet these criteria. In such cases, philosophy of science attempts to overrule the apologetic efforts of degenerating programs." [12]) To help the layman to decide what scientific project may or may not be pursued or supported, this school announces finally that it will "lay down statute law of rational appraisal which can direct a lay jury in passing judgment."

This is not to imply that public officials who want to stop scientific research or engineering projects within their cities are motivated by reading books in the history or philosophy of science. Rather, the theories of progress in those books seem to me to be a part of a current of thought in which working scientists who still have the earlier, progressivist notions now appear, to some, as dangerous meddlers. We have in fact entered a period where old and new theories of progress in science are vigorously competing – in the mind of the public, among those engaged in the study of nature, and among scholars who study science as an activity.

This current, like the other forces I have cited, is not amenable to sudden change. The momentum of the debate is large, and the divisions on fundamental issues are deep. It is therefore not reasonable to expect quick solutions. A better hope – as has been attempted here – is to gather a greater variety of interested parties to see what we may learn from one another, and thereby improve the chances for wise action and accommodation. For the issue is still the same one which Watson Davis, with a different aim, identified when he said over forty years ago: "The most important problem before the scientific world today is not the cure of cancer, the discovery of a new source of energy, or any other specific achievement. It is: How can science maintain its freedom, *and* . . . help preserve a peaceful and effective civilization?" Scientists, in larger numbers than ever before, are wrestling with both parts of the question, knowing perhaps that if they wish to answer one of these, they must answer both together.

12

Metaphors in science and education

All reflection, thought and criticism began in comparison, analogy and metaphor. Faust was wrong: in the beginning was not the act. St. John was right: in the beginning was the word. We are concerned with man, and the world can only exist for man as man knows or imagines it. Metaphor is the route of reason, science and art.[1]

Analogy: . . . a form of reasoning that is particularly liable to yield false conclusions from true premises.[2]

In addressing myself to the metaphors in science and education, I hope I may sidestep the battles about the theories of metaphor from Aristotle to our day, and also not go again over the ground of fine distinctions between metaphor, model, analogy, simile, and all the other tools for performing imitative magic. Rather, I wish to center on praxis, and take off from illustrative examples of the roles metaphor can have for good or ill, first of all in the actual work of scientists. When we know whether and how metaphor lives within the laboratory, and diffuses from there into the wider world, we might begin to discern what the educator must do about it. Even if I limit myself chiefly to physical sciences and slight the others, I still find myself with a vast topic for which I can sketch out here only some of the interesting problems.

To the first question then: Does modern science, properly conducted, really have anything profound to do with metaphor? It will not be universally granted that it has. Ever since Francis Bacon, the use of metaphor has tended to be an embarrassment to some scientists and philosophers. Bacon allowed that metaphors might be "anticipations of nature," but on the whole he dismissed them as serving the "idols" and our natural penchant for fantasy.

Today, those who consider themselves to be the last defenders of the age of reason and inheritors of the battle flags of the old positivism would rather stress scientific discourse in the form of protocol sentences. On this view, good scientific concepts are operational, fairly unambiguously shared by the worldwide community of scientists. Their meaning is as clear as human language can get. Scientists, they will say, differ from humanistic scholars by keeping fundamental decisions free from essentially psychological (esthetic or intuitive) or external (sociological) influences, and let themselves be guided only by the empirical data and logical machinery. On the other hand, metaphors by definition are flexible, subject to a variety of personal interpretations, and often the results of an overburdened imagery. At best, they might be used informally in the classroom or, with caution, in the textbooks. They are part of the natural armamentarium of the artist, poet, or critic, but not of the working scientist.

And it is not only those watchdogs of proper rationality in science who express alarm. Even Colin Turbayne, in his useful *The myth of metaphor*, reserved his first and longest case to an exposure of the great harm he considers to have been done to Western thought by the false use of metaphors in the thinking of great scientists. Mechanism is, he said,

> a case of being victimized by metaphor. I choose Descartes and Newton as excellent examples of metaphysicians of mechanism *malgré eux*, that is to say, as unconscious victims of the metaphor of the great machine. Together they have founded a church, more powerful than that founded by Peter and Paul, whose dogmas are now so entrenched that anyone who tries to reallocate the facts is guilty of more than heresy. . . .[3]

But the implied dichotomy between good metaphor and good science, while widespread, is a vast oversimplification. Comparative linguists have amply demonstrated that our store of metaphors and other imaginative devices determines to a large extent what we can think in any field. Further evidence comes from the findings of historians of science. Their work has shown that fundamentally thematic decisions, even though usually made unconsciously, frequently map out the shape of theories within which scientists progress.

Indeed, it is not too much to say that modern science began with a quarrel over a metaphor. Nicolaus Copernicus, in the first sentence of the introduction to Book I of *De revolutionibus*, stated his belief that "The strongest affection and utmost zeal" is reserved for the promotion of "the studies concerned with the most beautiful objects." Those are the proper subjects of astronomy, "the discipline which deals with the universe's God-like circular movements, and which explains its whole appearance. What indeed is more beautiful than heaven, which of course contains all things of beauty?" This vision provided him, Copernicus says, "extraordinary intellectual pleasure," even as "the Godly psalmist . . . rejoiced in the works of His hands."

But in this realm of divine Beauty, there had appeared a Beast, and Copernicus saw it as his high task to chase it out. As he explains in the preface:

> Those who devised the eccentrics [for modeling the motion of
> planets] seem thereby in large measure to have solved the
> problem of the apparent motions with appropriate calcu-
> lations. But meanwhile they introduced a good many ideas
> [such as the equant] which apparently contradict the first
> principles of uniform motion. Nor could they elicit or deduce
> from the eccentrics the principal consideration, that is, the
> structure of the universe and the true symmetry of its parts.
> On the contrary, their experience was just like someone taking
> from various places hands, feet, a head, and other pieces, very
> well depicted it may be, but not for the representation of a
> single person; since these fragments would not belong to one
> another at all, *a monster rather than a man* would be put
> together from them.[4]

There is some disagreement among historians about the precise nature of the monster, but the main point is clear: The hand of God is not the hand of Dr. Frankenstein who, in Mary Shelley's romance, assembles his monster also out of incongruous parts. To reassert the reign of beauty, Copernicus goes back to what he had called "the first principles of uni-form motion." He rejects non-uniformities and inconstancies of motion – his "mind shudders" at the very consideration of them – and even at the cost of setting the earth into motion, he arrives at a system that has all the earmarks of divine handicraft: the equants are gone, the phenomena are saved; the whole system has symmetry, parsimony, necessity. Indeed, contemplating it as if he himself were viewing it from above, Copernicus is moved to exclaim, at the end of chapter 10, in a moment of uncontrolled enthusiasm: "So vast, without any question, is the divine handiwork of the most excellent Almighty." It was one of the few sentences which the censor of the Inquisition in 1616 insisted on striking out.

But it was too late. The device of uniform motion in a circle was not forced by the data; and as Kepler's ellipses showed later, it was not even the most functional device from the mathematical point of view. Yet the metaphor of uniform circular motion as the divine key for solving the problems posed by the phenomena – even as in antiquity – had infected the thinking from which the scientific revolution of the seventeenth century came. At the very least, the case shows that the function of a metaphor need not be merely, as in Aristotle, a transfer of meaning, but can be "a restructuring of the world," in the words of Sir Ernst Gombrich.[5]

In trying to account for Galileo's irrational refusal to accept Keplerian ellipses (and with them, the additional proofs of the heliocentric system which Galileo was so passionately defending), Alexandre Koyré coined the memorable phrase *hantise de la circularité*; and Erwin Panofsky[6] made the case that this spell of circularity on Galileo was equally at work in his physics and in his esthetic judgments in the arts. In what follows we shall have to come back to at least a few points that have here only been touched on lightly. But the vanquishing of the monster and the triumph of the divine circle remind us that metaphors do not have to be casual indulgences: they can help to make, and defend, a world view.

Lest it be thought that such examples characterized science only in its early stages, let me draw attention to more recent ones. In Thomas Young's 1804 "Reply to the animadversions of the Edinburgh reviewers," he announced the idea for which he is best known, namely that light is fundamentally a wave phenomenon. As he put it, "light is a propagation of an impulse communicated to [the] ether by luminous bodies." He reminds his reader that "It has already been conjectured by Euler, that [contrary to Newton] the colours of light consist in the different frequency of the vibrations of the luminous ether." But this had been only a speculation: Young says, "It does not appear that he has supported this opinion by any argument." And now follows, in a half sentence, the first statement of Thomas Young's striking proposal: "but it [the idea 'that light is a propagation of an impulse to the ether'] is strongly confirmed, by the analogy between the colours of a thin plate and the sounds of a series of organ pipes."[7]

Even before we look at the details of this analogy, we feel the almost breathtaking daring of this transference of meaning between what were previously two entirely separate phenomena. Indeed, the courage for such a jump seemed so ill-advised to George Peacock, devoted friend and editor of Thomas Young's *Works*, when he assembled the volume in 1855, twenty-six years after the death of Young and long after the firm establishment of the wave theory, that he felt he must save the reader from some dreadful mistake; and so George Peacock, D.D., "Dean of Ely, Lowndean Professor of Astronomy in the University of Cambridge, and formerly Fellow and Tutor of Trinity College, FRS, FGS, FRAS," adds, as a stern footnote that is perhaps unique in the literature: "This analogy is fanciful and altogether unfounded. Note by the Editor."

The good Dean, having been taught in the meanwhile by Arago and Fresnel that light waves are transverse rather than longitudinal, perhaps saw more clearly the differences than the similarities. But if we return to the historical situation, as we can with the aid of the details given by Thomas Young, we are struck by the genius that allowed him to make the jump. During his early years at Cambridge, Young had done experiments toward an understanding of the human voice, and for that purpose had studied the way sound is produced in organ pipes. Strangely enough, the subject was in a rather confused state, and it was in fact Thomas Young who proposed the law of superposition which allows one to understand the action of organ pipes in terms of the interference between sound waves traveling in opposite directions within the pipe. He had noted that "The same sound" – for example middle C – could be obtained "from organ pipes which are different multiples of the same length." If you stood in front of a series of such pipes, whose lengths are as 1 to 2 to 3 to 4, and so on, and if some skilful person blows the different pipes in turn and produces the same note, you would know that the mechanism by which organ pipes work, no matter what the details may be, is very likely to be a wave phenomenon; for it will conjure up in your mind the traditional distributions of nodes and anti-nodes on the model of interfering waves, whether they be longitudinal or transverse.

Now, to go to the other part of the offensive half-sentence, that is, from "the sounds of a series of organ pipes" to "the colours of a thin plate." Thomas Young had found in Newton's *Opticks* the beautiful description of the experiments on thin plates and Newton's rings. You will recall that if two thin plates of glass are set up at a slight angle from each other, so that there is a gradually increasing wedge of air between them, and if light of a given color is allowed to fall on the arrangement, the eye placed above the plates observes equally spaced bands of color return to it from the wedge. The height of the air wedge formed by the two glass plates, at the point from which light is returned, is as 1 to 2 to 3 to 4, and so on. As Thomas Young puts it, "The same color recurs whenever the thickness answers to the terms of an arithmetical progression." And he immediately tells us the point on which the metaphor depends: "Now this is precisely similar to the production of the same sound, by means of a uniform blast, from organ pipes which are different multiples of the same length."[8]

We can almost see him standing over the thin plates in the optical experiment, looking at the light displayed at equal distances in the same color, and exclaiming – as one would have to do now if one had not thought of it before – that regardless of detailed mechanism, this must be a wave phenomenon.

Of course he has all the details wrong. Not only is light transverse, which would make little difference to his argument, but the colors seen in the thin plates, even though an interference phenomenon, are not due to a standing wave, as for sound in the organ pipes. Also, there are no pipes in the case of light. But all this is quite secondary and irrelevant. The main thing was his ability to perceive "the analogy between the colours of a thin plate and the sounds of a series of organ pipes." Contrary to George Peacock, you know that if a student, previously ignorant of all these matters, comes to you with the discovery of such analogy, you would be delighted and put him at once to work on your lab team.

I regard the case of Thomas Young as an exemplar of the creative function of metaphor in the nascent phase of the scientific imagination. Anyone who has known or studied scientists at the highest level of achievement knows of this mechanism. Faraday's notebooks are full of it, as is Maxwell's work. For Fermi it was part of his scientific credo to use and reuse the same ideas in quite different settings; for example, a year and a half after a paper dealing with the effect of slow electrons, Fermi was in the unique position of thinking about the effect of slow *neutrons* when his team, evidently by accident, came upon the artificial radioactivity of silver (produced by scattered neutrons).

But of all such examples my favorite is still an autobiographical passage I found some years ago, in Einstein's own hand, in the copy of an unpublished manuscript dating from about 1919 or shortly afterwards, located at the Einstein Archives in Princeton. The title of the document (in translation) is "Fundamental ideas and methods of relativity theory, presented in their development." There, in the middle of a technical paper, Einstein writes suddenly in a personal way about what he called "the happiest thought of my life," which came to him in 1907 and opened the way to go from special to general relativity. He explained in 1919: "As with the electric field produced by electro-magnetic induction [1905], the gravitational field has similarly only a relative existence. For if one considers an observer in free fall, e.g., from the roof of a house, there exists for him during his fall no gravitational field – at least in his immediate vicinity."[9] If we analyze this passage (which is quite coherent with other reports Einstein gave of his thoughts in the matter), we see a number of parallels with the case of Thomas Young. Now the gulf to be negotiated by a metaphoric transference is between the electric and magnetic fields on the one hand, and the gravitation field on the other. The "happiest thought" in 1907, which led indeed to a restructuring of our world picture, was really a simple extension of the analogy from uniform motion in a magnetic field to accelerated motion in a gravitational field – although from the point of view of the physics of the time the analogy might indeed also have been called "fanciful and altogether unfounded."

An earlier passage in the same Einstein manuscript brings us to a point that also has to be discussed, namely, the possible motivation for risking the metaphor. Such risks (which Einstein took repeatedly in his scientific papers) are not made playfully or by accident. The urges to find analogies, and thereby to simplify and unify the various branches of a science, are actively at work in the background of the research of these explorers. In a famous note written to himself on March 19, 1849, Faraday enters the experimental search for a link between gravity and electric and magnetic phenomena with the sentence, twice underlined, "All this is a dream." The dream comes first. Or a nightmare: Einstein, recalling once more the distaste with which he beheld the physics before 1905, in which the current in a conductor induced by a magnet in motion relative to it was thought to be the result of quite different effects, depending on whether the magnet or the conductor was regarded as "really" at rest, exclaims in his 1919 manuscript: "The thought that one is dealing here with two fundamentally different cases was for me unbearable [*war mir unerträglich*]. The difference between these two cases could not be a real difference, but only a difference in the choice of the reference point."

In this mood, the metaphor is a godsent means for bridging an apparent but "unbearable" gulf.

In stressing this active and creative side of metaphoric thinking, I do not want to be misunderstood as giving a normative analysis of recommended scientific procedure. I would rather look at the record to see what did happen. Also, I know well that the cases of brilliant success do not imply that there have been no failures – either a new metaphor misused, or an old, seemingly "dead" but still powerful, metaphor falsely accepted. A case of that latter sort is surely the inability of a group of experimentalists in the 1920s to come upon parity violation in their own data on the asymmetry of the scattering of electrons. The parity and isotropy observed throughout physics at that time forced itself also upon their imagination in this case, and helped them and all others who looked at the data at the time to find an easier way out.

I shall treat later cases of surprising metaphor excess and its costs. But the essentially constructive role metaphor has usually played in the making of science is clear. Andrew Pickering of Edinburgh writes: "Wherever one looks in the history of particle physics, one sees this magical transmutation, producing new theories from old through a process of analogical recycling." He concludes: "I have been trying to suggest that it would be useful to replace the idea that scientists are the passive *discoverers* of the unproblematic facts of nature with the alternative view that they actively *construct* their world." [10] We see here too a disagreement with the opposing traditions that stress the passive role of metaphor, as expressed for example by Richard Boyd (in the volume *Metaphor and thought*, edited by Ortony): "Neither do we, in any important sense, 'construct' the world when we adopt linguistic or theoretical frameworks. Instead *we* accommodate *our* language to the structure of a theory-independent world." [11]

I have asserted that in the work of the active scientist there are not merely *occasions* for using metaphor, but *necessities* for doing so, as when trying to remove an unbearable gap or monstrous fault. I now turn at least briefly to these necessities, and first of all the necessity built into the process of scientific rationality itself, an epistemological necessity that favors the search for and use of metaphors. It is simply the limitation of induction. Where logic fails, analogic continues. The bridge is now made no longer of steel but of gossamer. It breaks often, but sometimes it carries us across the gulf; and in any case there is nothing else that will. As I indicated in Chapter 2, I have recently been exploring the existence of discontinuities that appear in scientific theory construction, forcing what Einstein repeatedly called the researcher's "widely speculative" or "groping constructive attempt," or even a desperate proposal made when one has given up finding other paths to an overarching axiom system. A second discontinuity, Einstein noted, lies in the fact that we often select concepts without some logical necessity, really "arbitrarily"; "considered logically, the concept is not identical with the totality of sense impression referred to; but it is a free creation of the human (or animal) mind." Indeed, the whole "system of concepts is a creation of man," achieved in a "free play."

For these reasons, Einstein warned strongly against making the mistake of thinking of concepts as being the result of "abstraction" from observation. Eventually, Einstein developed a theory of levels, or "stratification of the scientific system," in which the discontinuities between strata draw attention to the need for some sort of creative groping across the discontinuity that is often helped by resort to metaphor. He held that in the striving for logical unity, the theorist is progressively led from a "first layer" to a "secondary system" and on to higher levels, each characterized by more parsimony in concepts and relations, and particularly in the concepts directly connected with complexes of sense experience. "So one continues until we have arrived at a system of the greatest conceivable unity, and of the greatest conceptual paucity of the logical foundations that is compatible with the nature of what is given to our senses" – even though there is no guarantee that "this greatest of all aims can really be attained to a very high degree."

A further necessity for the resort to metaphor in the nascent phase of science-making is precisely what troubled Francis Bacon: the habits and irresistible play of our imagination, and often the imagination in its blatantly anthropomorphic form. While there are good social covenants for removing all traces in the pedagogical recasting of scientific achievement after the imaginative act is over, the making of science presupposes life-world experience. As Immanuel Kant noted, the imagination does not distinguish between life-world experience [*lebensweltliche Erfahrung*] and scientific experience. This is now not a judgment, but an ethnomethodological fact, derived from observing scientists at work. They can't help themselves. Thus, Millikan was so convinced of the existence of the discrete corpuscle of unitary electric charge that he wrote about it in his autobiography as if he had seen it with his own eyes: "He who has seen that experiment, and hundreds of investigators have observed it [has] in effect SEEN the electron." When the droplet he was observing in the electric field was changing speed, an image came quickly to mind, and he wrote: "One single electron jumped upon the drop. Indeed, we could actually see the exact instant at which it jumped on or off."[12]

In the privacy of the laboratory, and sometimes, less cautiously, even in publication, the "idols," and particularly the idol of the theater, play their necessary role, with all the dangers and benefits that this may imply. In a frank and personal essay, Martin Deutsch, a nuclear physicist at MIT, has discussed "the striking degree to which an experimenter's pre-conceived image of the process which he is investigating determines the outcome of his observations," and particularly the "symbolic anthropo-morphic representation of the basically inconceivable atomic processes":

> The human imagination, including the creative scientific
> imagination, can ultimately function only by evoking potential
> or imagined sense impressions . . . I confess I have never met an
> experimental physicist who does not think of the hydrogen
> atom by evoking a visual image of what he would see if the
> particular atomic model with which he is working existed
> literally on a scale accessible to sense impressions – even while
> realizing that in fact the so-called internal structure of the
> hydrogen atom is *in principle* inaccessible to direct sensory
> perception. This situation has far-reaching consequences for
> the method of experimental investigation.[13]

I shall come back to the role of visualization in what follows, but must stop to comment at this point. On this, again, I turn to Gombrich. He notes that the linear character of language makes it hard to hold in mind some concepts that become quite evident when put in diagrammatic form. "This may be one of the psychological reasons for our instinctive equation between seeing and understanding." And he quotes a passage from Cicero on the special appeal of the immediacy of sight:

> Every metaphor, provided it is a good one, has a direct appeal
> to the senses, especially the sense of sight, which is the keenest:
> for while the rest of the senses apply such metaphors as . . .
> "the softness of a humane spirit," "the roar of the sea" and
> "the sweetness of speech," the metaphors drawn from the sense
> of sight are much more vivid, almost placing before the mind's
> eye what we cannot discern and see.[14]

If we had a categorization of scientific metaphors, analogous to Stith Thompson's classification of themes in myth and folklore, or something like the *Atlas Mnemosyne* of Warburg, we would find that the longest sections would go to visualizable metaphors and to anthropomorphic and folkloric metaphors. Human life, the life cycle, and human relationships pervade, in sometimes only slightly disguised form, the most sophisticated scientific papers and, more so, our textbooks. "Strange" particles decay and, in the spark chamber photograph or similar view field, are shown to give rise to new generations of particles, each with its own characteristic lifetime. We speak of families of radioactive isotopes, consisting of a parent, daughters, granddaughters, and so on. We constantly tell stories of evolution and devolution, of birth, adventure, and death on the atomic, molar, or cosmic scale. The chemist's molecules are metamorphosed to enter the life-history of another chemical species. I have always felt it ironic that particularly the newer sciences such as psychology and sociology tend to borrow their metaphors and other terminology so heavily from the older and more respectable physical sciences when, in fact, those have been deriving them in the first place from the most primitive and familiar experiences.

The metaphors based on the human body are undoubtedly the most numerous and seductive in the sciences because, as Vico put it, man is "buried in the body." Donald MacRae is surely right when he says, "If we are to understand the body as metaphor, and as source of metaphors derivable directly or by transformation rules from it, we must remember that our own experience of our bodies is prescientific."[15] This applies both to the metaphors which work for us and to those that eventually do not. Although he did not intend it, MacRae made this point too, and in a manner that brings our attention with uncanny accuracy back to something I discussed at the beginning: "We make our monsters out of bodily parts."

The fourth and last of the reasons why scientists find themselves forced to risk metaphoric thought has to do simply with the fast metabolism of science, so much faster now than for many other fields of thought and action. Scientific vocabulary and imagery are never stabilized. Scientists seem to be working at the edge of an ever more active volcano that showers them with novelties demanding neologisms at an ever-increasing rate. At the same time, of course, old metaphors decay and are lost, or at least are apt to be misinterpreted. Thus the proton has betrayed its etymology by turning out not to be the first of things – and now it is even flirting with the idea of giving up its immortality, settling for a half-life of a mere 10^{32} years, which will cause diamonds not to be forever. While

Homer's Briseis of the fair cheek will remain tender forever in some circumlocution or other, the metaphors of science seem to have a shorter and shorter half-life of their own. And the conditions under which new ones are fashioned seem to verge now occasionally on the frenzy and informality of a big beer-drinking party. We are now invited to settle for quarks that combine the properties of flavor, charm, and "bareness" with either "truth" or "beauty," or, in cruder terminology, with either "top" or "bottom." Some of the metaphors are hammered together out of odd bits and pieces, without attention to delicacies of taste.

It struck me some time ago that since the turn of the century the terminology entering physical science often had a thematic root contrary to the older (and persisting) themata of hierarchy, continuity, and order. That is, the newer conceptions, perhaps corresponding to the characteristic style of our turbulent age, tend to be characterized by the antithetical thema of disintegration, violence, derangement. Evident examples are: radioactive decay, Principles of Impotency, displacement law, fission, spallation, nuclear disintegration, discontinuity (in energy levels), dislocation, indeterminacy, uncertainty, probabilistic causality, strangeness, quantum number, negative states, forbidden transitions, particle annihilation. It is indeed the terminology of a restless, even violent, world.

This brings us inevitably to consider, at least in a first pass, the student new to our classroom. One can say that all the necessities forcing scientists to use metaphors in their work become handicaps for students, who inherit all the troubles and none of the rewards.

Think of what is in the head of your new student, the "metaphor background" and "metaphor readiness." Indeed, in the student's head there may be a disorderly mixture of new and old metaphoric terms. The Big Dipper, the black hole, the big bang, and the big crunch. The harmony of the spheres, the expanding universe, the clockwork universe, attraction and repulsion, inertia, perhaps Schrödinger's cat, left-handed neutrinos, parity breakdown, colored and flavored quarks, gluons, charm, and God playing (or not playing) dice. Also, the heat death, kingdoms of animals and plants, computers that crash or refuse commands, broken symmetry, families of elements, daughter and granddaughter isotopes in radioactive decay, negative feedback, circulation of blood, the tangled bank, the selfish gene, degenerate quantum states, and "everything is relative." (If students go to Wheaton College near Boston, they may encounter a new course in biology being fashioned on radical feminist lines in which, for example, fertilization will be described as an encounter between an aggressive spermatozoon – an old stereotype – and an "equally aggressive egg" that envelopes it, rather than passively accepting what comes.)

The main trouble with this bouillabaisse is that metaphors do not carry with them clear demarcations of the areas of their legitimacy. They may be effective tools for scientists, but pathetic fallacies for students. For the latter, the problem stems in good part from the sociology of communication. Margaret Mead[16] noted more than twenty years ago that scientists at the frontier, where the terminology and imagery are developed, speak mostly to other scientists at or near their own level of understanding. In this way, scientific language has escaped from the realm of "natural language." This is the fate of "any language taught only by adults to adults – or to children as if they were adults. . . . It serves in the end primarily to separate those who know it from those who do not." Since then, linguists and anthropologists have been reinforcing the point that the cure cannot come from simple "translation" but may lie in recognizing that a difference in languages reflects a difference in world views. Without making the mutual accommodation of these views a prominent part of the agenda, science teaching probably has to remain superficial. I refer here to the work of R. Horton on African traditional thought and Western science, and of J. Jones in Margaret Mead's own New Guinea; both Horton and Jones have studied the ways in which the traditional cultures of the new learners differ from the scientific cultures of the teachers, and how and to what limited degree these differences can be decreased.[17]

The most serious charge one might make is that the negligence of scientists, in their role of teachers, to deal with this cultural difference may not be merely a shirking from an admittedly very difficult task. It may even be *functional*, as I shall note later.

Further examples of the powers and pitfalls of metaphors in science and education are offered by relativity theory, and specifically the way the early researchers conceived of length contraction and time dilation. In Lorentz's and Poincaré's work, these were, of course, not symmetrical, but "real" actions of the ether, that legacy of Greek, Cartesian, and Newtonian physics. With Einstein, ether was dismissed as merely "a substratum of nineteenth-century thought" (in the happy phrase of René Dugas). Possibly with the courage gathered by reading Ernst Mach's attack on absolute time and absolute space, Einstein relegated the ether into the dustbin of nonoperational and therefore harmful conceptions (at least in the first decade of his work), and instead drew attention to the reality of the transformation equations themselves, including their implication of complete symmetry. But new metaphors rush in to take the place of old. Thus Minkowski announced in 1908 that "from this hour on," space by itself and time by itself would become mere shadows, and only a sort of union of the two will maintain existence. What he called with great flamboyance "the World" was now a kind of unchanging, crystalline structure in which all past and future processes are represented by lines and their intersections.

At about the same time, more metaphors grew around relativity theory, for example, at the hand of Langevin, who introduced the "twin paradox" in more or less its popular form. Thereby what had slumbered in neutral-looking transformation equations became the widely interesting matter of relative aging, counting of heartbeats, and the like. Some of the literature of the times, both in scientific meetings (e.g., those of the Swiss Physics Society) and in the popular press, show how suddenly and vigorously interest was aroused by this extension of the metaphor to human life.

Another version of the same "metaphor excess" had to do with the discussion given by Einstein in his popular book, and much used elsewhere, of length contraction, for example of the famous and easily visualizable train dashing through that railway station. The "contraction" which previously had been thought to be real within the world view of ether physics was now thought to be directly visible from any other coordinate system. Textbooks, popular articles, and even films began to represent the contraction. Both laymen and scientists seemed to see the contraction in simple and similar ways whenever they wanted to. For about forty-five years after Einstein's first publication, a cubical object, for example, moving at high speed across your line of sight was thought of as presenting itself as a rectangle, its shorter side in the direction of motion "shrunk" by virtue of the length's contraction. But then, through the work of Terrell, Penrose, and others in 1958–59, it became clear, to everyone's surprise, that an observer of a relativistic moving object would see not a distortion; rather, the object would appear not distorted but only rotated. The rotation, or rather "remapping" upon the observer's screen, is itself a metaphor with rather unexplored limits.

A second example of an important metaphor in physical science (already mentioned in passing) is the circular, mandala-like construct as the solution for a great variety of problems. It often seems one might say in the physical sciences: "The circle, or some modification of it, is the answer. What is your question?" This would take us from a common starting point into various directions. One is the Platonic–Aristotelian–Ptolemaic–Copernican–Tychonic . . . sequence. Then would follow the often painful modifications of the circle as a fundamental explanatory metaphor: Kepler's struggle with the ellipse; or Galileo's dismay on discovering that Saturn, far from being a round object, has ear-like appendages (that later were understood to be Saturn's rings, seen at a slowly changing angle through a poor telescope).

As it turned out, motion in a ring around Saturn was not unique in the solar system. Planetary rings were found to be quite commonplace, varying from substantial ones as for the main rings of Saturn or the rings of Uranus, to the more ephemeral dusty ring of Jupiter. Even now, a large and expensive project is trying to find rings around Neptune, on the assumption that the failures to find them previously were simply not credible. A self-respecting planet is now *supposed* to have rings.

The observation of two-dimensional rings around three-dimensional objects reminds us how strangely little resistance was offered to two-dimensional projections and indeed to two-dimensional models, for the purpose both of representation and of theoretical modeling, even in the discussion of the origins of the solar system. The constraints of the two-dimensional surface on which to draw or print must, after all, affect our thinking constantly. Depictions of Laplace's nebula hypothesis, or of T. C. Chamberlin's theory of planet production by the accretion of planetesimals in circular orbits, or of the patterns of motion in the solar nebula according to Weizsäcker's theory, and the like show how much at home we are with two-dimensional thinking for three-dimensional problems.

The penchant for two-dimensional thinking appears even stronger when we turn from the macroscopic mandala of the solar system to its microcosmic equivalent. Ringed Saturn shrank to atomic size when Nagaoka proposed it as a model of the atom in 1903, after having read Maxwell's paper on the stability of motion in Saturn's rings. Rutherford referred to it favorably in a letter and publication of 1911, about the time when he was being joined at his laboratory by Niels Bohr. The model contained a strongly charged central object, with a number of negative electrons of equal mass, arranged in a circle at equal angular intervals.

A rival model of the atom, also influential on physicists from 1911 on, was also frankly two-dimensional and indeed made its way into the physics literature through a lecture demonstration of magnet needles floating in a bowl of water placed below a large cylindrical magnet. Depending on their number, the magnet needles arrange themselves sooner or later in regular two-dimensional patterns. Now a rather familiar demonstration, it was new when published by the American physicist Alfred M. Mayer,[18] and the arrangement of the magnets, projected on a screen, made a great impression on students and popular audiences. They caught the attention of J. J. Thomson. Even the original manuscript pages of (largely two-dimensional) calculations and subsequent publications from Bohr through Sommerfeld show the power of the Saturnian model, suitably but not essentially metamorphosed.

In retrospect it is rather astonishing how well these models worked, from explaining the Zeeman effect and some chemistry to the fine structure of spectral lines – astonishing, because of course it was, from the present point of view, completely wrong at the very point where the metaphor then used was most convincing: the axially symmetrical, easily visualizable playground of equally visualizable events. As noted in Chapter 7, between 1913 and 1927, the old, almost axiomatic *Anschaulichkeit* at the base of models, analogies, and other metaphors in the physical sciences had to be given up.

Though you cannot tell it from the representations in our introductory physics books of today, what Heisenberg achieved over a half a century ago was "complete freedom from planetary orbits." The principal problem with the sensual Kantian notion of *Anschaulichkeit* in the old quantum theory was that the *Anschauungen* were memories or abstractions from the world of perceptions, and consequently were encumbered with pictures that made sense only in terms of classical causality and conservation laws. It implied that matter is infinitely divisible in principle, and hence did not do justice in its very essence to the atomic regime with its discontinuities.

Other solutions to the puzzle of the atom were of course possible in principle. Thus Whitehead[19] suggested that the atom and molecule are to be considered analogous to biological entities, organisms rather than classical particles. And it is also worth noting that chemistry made do with linear or planar models until the 1870s. Structural theory in chemistry did not seriously use three-dimensional space until the "tetra-hedronal carbon atom" was introduced by van't Hoff and le Bell.

The last of the three cases that deserves detailed analysis concerns the metaphoric descriptions of the *task* of the sciences and of what it is supposed to achieve when all is said and done. Immediately our imagination resonates to such images and metaphors as "revolutions," voyages into the unknown, the voyeurist drawing aside of the veil covering Mother Nature, the great mystery story view, the jigsaw puzzle view, the endless-horizon pursuit, and many others.

But the imagery which seems to be the most powerful and motivating one is that of the mountaineer gradually ascending, and thereby gaining not merely the conquest of the peak, but the esthetic, largely visual thrill of an *overview*, encompassing the whole circular area below, from horizon to horizon, and, in the unearthly stillness at that high altitude, seeing at a glance the way the details of the landscape below fit together in one meaningful picture. It matches Bradley's famous definition of metaphysics as "the effort to comprehend the universe not simply piecemeal or by fragments, but somehow as a whole."

You will perceive that this image connects in various ways with matters I have discussed before: the planar and quasi-circular area of action, the importance of visualization, Einstein's "layer theory" of higher, and more and more encompassing levels of theory perfection, with the attendant lengthening of the distance between the concepts and the "facts" of the plane on which there is crowded the "multiplicity of immediate sense experiences" – that plane, above which the system of axioms may be found in painful search. Einstein used that metaphor several times, with all the visual imagery typical of his writings. As he put it in his 1918 *Motiv des Forschens*, the pursuit of science is motivated first negatively, by the

> flight from the everyday life with its painful harshness and wretchedness, and from the fetters of one's own shifting desires . . . the longing that irresistibly pulls a town-dweller toward the silent, high mountains, where the eye ranges freely through the still, pure air and traces the calm contours that seem to be made for eternity.

A second, positive part of the motivation is that "man seeks to form for himself, in whatever manner is suitable for him, a simplified and comprehensive picture of the world." The achievement of such a *Weltbild* is in fact "the supreme task of the physicist." "The longing to behold that preestablished harmony between the world of experience and the theoretical system is the source of the inexhaustible perseverance and patience" of the researcher.[20]

We can trace the majestic metaphor of the *Weltbild* from Goethe,

Schleiermacher, and Alexander von Humboldt. Cassirer[21] put it succinctly: a characteristic and typical *Weltbild* is formed when the chaos of
sense impression is arranged into a cosmos. During the first two decades
of the twentieth century, the fight over the "unity" of the physical *Weltbild* was intense among German scientists. We have become more modest
(or more pedestrian) in our descriptive language, even though our colleagues are getting Nobel Prizes for their inexorable climb toward the
unified theory of all the forces of Nature. Current discussions about the
ultimate attainment of a complete and coherent world picture in the
physical sciences do allow us to discern echoes and overtones from those
earlier debates when the context of a near-sacred mission was never entirely hidden. Scientists used to lapse into poetry, as when Boltzmann
(and after him, Sommerfeld), on contemplating the great synthesizing
power of Maxwell's equations that produced a panoramic unification of
the fields of heat radiation, light, and electric and magnetic fields, turned
to Goethe's *Faust* and quoted the line "Was it a God who designed this
hieroglyph?" [*War es ein Gott, der diese Zeichen schrieb . . . ?*]

Faust's exclamation came as he opened the Book of Nostradamus and
saw the Sign of the Macrocosmos. A fuller quotation goes somewhat like
this:

> Was it a God designed this hieroglyph to calm
> The storm which but now raged inside me
> To pour upon my heart such balm
> And by some secret urge to guide me
> Where all the powers of Nature stand unveiled around me?
> Am I a God? It grows so light!
> . . . [He contemplates the sign.]
> Into one whole how all things blend
> Function and live within each other.

Gombrich uses these passages in his discussion of "The paradox and
the transcendence of language" in *Symbolic images*, and introduces it
with this telling sentence: "It is this effort to transcend the limitations of
discursive speech which links the metaphor with the paradox and thus
paves the way for a mystical interpretation of the enigmatic image." This
insight is applicable directly to the metaphor describing the task of the
scientist that I have been developing here. It becomes rather uncanny
when we read his next sentences:

> The relevant doctrine is adumbrated in Pseudo-Dionysius in a
> passage of crucial importance: "The higher we rise, the more

concise our language becomes, for the Intelligibles present themselves in increasingly condensed fashion. Where we shall advance into the Darkness beyond the Intelligible it will no longer be a matter of conciseness, for the words and thought cease altogether."

In the Western tradition, and chiefly in Platonic philosophy, the most exalted aims are associated with a unification and unity. "Thus Marsilio Ficino can describe the ascent of the mind to the apprehension of the Divine as a return of the soul to its original unity. . . . The ascent to unity leads to the apprehension of Beauty as an analog of the Divine." The realms through which the soul has to rise toward God are arranged in a hierarchy – of Matter, Nature, Opinion, Reason, and Intellect – a "hierarchy of analogies . . . The microcosm as well as the macrocosm must be envisaged as a series of concentric circles surrounding the ineffable unity in ever widening distance." [22] In diagrammatic form we may imagine a sector, or pyramid-like slice, taken from this homocentric construction: at the bottom, at the wide part, are the regions of Matter and Nature; at the top there is convergence and unification of all lines in the point of transcendence.

It might seem that we have gone far beyond the homely metaphor of the scientist as mountaineer, clambering up with a mixture of excitement and pain, until he reaches the top, probably in a state of hyperventilation, and in the euphoria of self-induced narcosis generated by endomorphines in his brain. But the metaphor is much too widespread, through history and cultures, to be dismissed that easily. Rather, it is both the embodiment and the exemplification of an ambition, one which it is now not customary to speak about among scientists, but of which we can constantly overhear confessions in terms of fragments of the whole: the willingness to go to extraordinary efforts in the hope of reaching the elevation from which all puzzles before science will be resolved in one simple, coherent *Übersicht*. The exaltation that beckons is one to which a Kepler or Einstein was courageous enough to give voice openly; but it stands at least as a whispered promise before every candidate who takes part in this expedition to high ground.

This internal state of emotion of the scientist so imbued is intense. Einstein dared to say openly, "The state of feeling which makes one capable of such achievements is akin to that of the religious worshipper or of one who is in love; his daily strivings arise from no deliberate decision or programme, but out of immediate necessity." [23] He probably embarrassed even his fellow scientists when he said so. Such passion is in striking contrast with our generally "cool" and commonplace metaphors of the purpose of scientific work which we usually present to the public and to our beginning students. And this contrast may be significant. For a metaphor can have one of at least four purposes: to serve the individual privately; to serve the circle of the indoctrinated; to serve both of these and the more ignorant public; or to serve chiefly only the more ignorant public. Whether by necessity or not, scientists have been reserving one version of the metaphor, in this case and in many others, to themselves and their fellows, while presenting to the public another, baser version.

It is a situation familiar to anthropologists in other contexts. We must therefore add to Margaret Mead's insight on some chief sources of scientific illiteracy of the public the possibility that the wide sharing of the key metaphors is not only difficult but also not particularly encouraged. The Romantic attacks on established science seem to me motivated in part by this perception. The scientists, on the other side, can hardly help responding to the reaction in just the way they do. After all, what are they to do with the proposals of a Goethe to reorganize the study of colors and optics, when he ends his *Farbenlehre* with a section on "Allegorical, symbolic, mystic use of colour"; he proposes that the scientists open their laboratories to other observers, to "all natures . . . women, children"; and ends with the exhortation, the same as on the title page of Francis Bacon's *Novum organum: Multi pertransibunt et augebitur scientia.* The lonely trek to the epiphanous experience at the peak threatens to become transformed into a family outing.[24]

I end with a listing of further problems of interest to me, which I have been unable to sketch here even in sufficient outline.

(a) Categorization of metaphors. Scientific metaphors will surely allow some categorization (those of processes versus those of structure; biological, mechanistic, technological, topological . . .). There would also appear more clearly the predominance at certain stages of historical development of certain kinds of metaphors compared to others. And synchronically, important debates between metaphoric choices will be thrown into more prominent relief.

Thus Panofsky[25] makes the interesting point that Galileo and Kepler differed not only on the primacy of the circle for celestial mechanics. Kepler thought that "all muscles operate according to the principle of rectilinear movement," the shifting and stretching of straight muscles in a straight line. But Galileo came to the opposite conclusion, by attending to the effect rather than the cause, to the positional change rather than muscular action; and therefore he finds himself writing that "All human or animal movements are circular," with apparent straight-line motion being "only secondary movements depending on the primary ones which take place at the joints" and are circular. Panofsky adds the illuminating observation that the basic contrast is really between a kinematic and a dynamic interpretation of movement, which is precisely what separated Galileo's and Kepler's astronomical notions, too.

(b) More specific tasks for the scientist as educator. Some have been noted at several points, but I would add the explicit need for self-examination by the scientist, to become aware of the metaphoric *distance* between himself and his colleagues on the one hand and his students on the other; to be aware of the metaphoric *dissonance* that reverberates strongly, even though unattended, in every classroom. It was again Heisenberg who, in a well-known story, interrupted Felix Bloch's discourse on new ideas on the geometry of space with a remark: "But space is blue, and birds fly in it." The scientist needs above all watchfully to avoid unintended or misleading but appealing metaphors. More often than not I find so-called popularizations of science shot through with the attempt to gain attention or understanding by banalized or cheapened metaphors. That is just as counterproductive with respect to scientific literacy as failing to explain the proper boundaries of the correct metaphors.

For physical scientists and to some degree biological ones, the use of *demonstrations* in the classroom – often an act of heroic effort and good intentions – can be disastrously counterproductive in terms of the transmission of the essential meaning of the phenomena. This merits special attention. Some years ago[26] I became concerned about the question of what is conveyed by visual presentations during science lectures, and noted the almost built-in divergence between the reality of nature's phenomena experienced in the laboratory of the practicing scientists versus their carefully repackaged and media-transformed versions put before the student.

The matter is complicated by the large separation, whether by necessity or design, between experiential reality and didactic reality even when the actual phenomenon itself is shown. But it gets much worse, if as now happens more and more frequently, even the level of didactic reality is abandoned and the presenter descends to lower levels of reality: the depiction of the phenomenon by film or television; the further deterioration by presenting a machine-shop made analogon (model, animation); or, least expensive of all, a presentation of phenomena in terms of condensed coding (graph, equation, verbal narration).

In these considerations I became impressed by the function of the human presence during such demonstrations. A key observation was that many of the more successful demonstrations are actual happenings – if not on the first at least on the second level of reality, in which the *human body* is involved, as when the lecturer mounts a rotating platform to demonstrate angular momentum. To a discernible degree, such an occasion commands much added attention because the lecturer is putting his dignity and perhaps his safety at risk. A personal commitment, shown by an implicit willingness to take certain risks and evidently go to some trouble: this is the element which the participation of a human being brings that is completely lacking in the operation of the surrogates, that is, in the presentation of a packaged and transformed metaphor, as in a film. And at the same time, another service performed by the presence of the human body alongside the actual phenomenon is to provide a *scale*, as for example in the relation of the hand to the apparatus; that, too, is usually lacking in the depiction-translation.

The scientist-educator of course negotiates easily the jump between the ground metaphor and its debased forms that actually come to the eye in the classroom. But the new student may not be able to follow him in making this leap, the more so as usually nothing is said about any necessity to make one.

Here again Gombrich has pointed to a closely related situation in his remark,

> To primitive mentality, distinction between representation and symbol is no doubt a very difficult one. Warburg described as *Denkraumverlust* this tendency of the human mind to confuse the sign with the thing signified, the name and its bearers, the literal and metaphorical, the image and its prototype. . . . Our language, in fact, favours this twilight region between the literal and the metaphorical. Who can always tell where the one begins and the other ends?[27]

The scientist-educator is more likely to avoid such traps, or at least avoid the full toll, if he or she is more conscious of an *active* obligation to create lively new models, analogies, and metaphors that do not sacrifice the essential scientific content in return for easier transmission. Good writing is of course scarce in all fields; in science education it is both more needed and less frequent. As Richard Feynman and Steven Weinberg[28] show us, honest meaning can be preserved by writing in an engaging way, with wit and vivid style for wide audiences.

(c) Metaphor and thema. In what I have said, I have associated myself with both of two competing views of metaphor: I see metaphor acting sometimes as a means for the transfer of meaning across discontinuity, as a bridge or a boat is a means for transferring a person across a river; or, in other cases, as a more active tool of metamorphosis, of a restructuring of a portion of the world view. In either case, the metaphor has explicit or implicit boundaries. Since the metaphor is always contingent on the context, its boundary will also change as the context shifts (as it becomes possible to cut the atom, or as probabilism and indeterminacy enter).

But while the detailed shape and power of a metaphor change, I see a constancy that endures, and that I regard as the thematic center of the metaphor. I need only indicate here in a word or two the differences I see between metaphors and themata. Themata are near-universals of science (as they are in other cultural artifacts). They operate at the level of structure and serve to endow the successive versions of a metaphor, or a sequence of closely related metaphors, with a meaning that permits the retrieval of the inherent intention despite all evident, or even flamboyant, changes. Thus the sequence of circles, eccentrics, ellipses, ellipsoids, precessing ellipsoids, *Ellipsenverein*, and so on, are variations on one thema, namely the efficacy of geometrical explanation. In fact, as often is the case, a particular metaphor may be at the intersection of two or more themata – in this case both the efficacy of geometrical explanation and the thema of direct, centralized perception (*Anschauung*).

Science has, and always had, a mythopoeic function. The metaphor is one of the tools in that service. That does not mean it is its only function or the chief function, least of all that Dionysus is again in the saddle. Rather, it is a sign that scientific activity is, and has to be, part of a larger cultural metabolism. The scientific imagination is, after all, not the result of Special Creation. Prescientific and nonscientific discourse provides the proto-language of the sciences, and is in turn changed by the products of these sciences. As W. H. Letherdale has noted: "After the capital of ordinary language had been invested by metaphor in science, the words were returned to ordinary language with the accrued interest of their scientific associations." [29] And Turbayne, for all his skepticism, ends by recommending a stance of cautious pragmatism: "[Be] aware there are no proper sorts into which the facts must be allocated, but only better pictures or better metaphors." [30]

Our scientists continue their flourishing traffic with metaphors. And our educators must also sing us new and life-sustaining ones. Their respective tasks differ greatly in detail, but they are grounded in the same aim, an aim to which the description of Copernicus still applies: the promotion of "the studies concerned with the most beautiful objects."

13

"A nation at risk" revisited

On April 26, 1983, after nineteen months of intensive work, the National Commission on Excellence in Education released its report to the president and to the public. Cast in the form of an "open letter," this slim booklet, printed in large type, was entitled *A nation at risk*, and carried the self-confident subtitle "The Imperative for Educational Reform." In the weeks and months that followed, a large number of other reports on the same subject were issued by a variety of other groups, differently constituted, with other kinds of expertise. At their core, they all agreed remarkably in their findings and recommendations. Even more surprising, perhaps, was the unprecedented amount of immediate public attention that these documents generated. The ideas in the National Commission's report have found their way into the continuing debates and struggles over the direction of secondary education in the country. Activist groups with opposing agendas intending very different results are using the findings, language, and arguments of the commission. Presenting a personal view, and without claiming to speak for other members of the commission, it may therefore be useful to summarize what this study contributed to the turbulent "year of the reports"; to suggest what the report did not attempt to do; to point out main polarizations in the responses; and to indicate the presuppositions, as well as the likely next moves of the competing camps.

The National Commission on Excellence in Education was set up in August 1981 by the secretary of education, Terrel H. Bell, acting at President Reagan's request. Mr. Bell said at the time that he was acting in response to "what many consider to be a long and continuing decline in the quality of American education." A main goal was not merely to diagnose the problems but to initiate reform on a grand scale: "We want to seek a vast renewal of the education establishment of this country and a turning more and more toward the pursuit of excellence, to the increasing of standards." The charter charged the commission with examining the nation's educational system, with particular attention to teenage youths, and to make practical recommendations for action to public officials, educators, parents, and others who set school policies.

I was initially dubious about accepting the invitation. At any given time, dozens of well-intentioned commissions are at work on reports having to do with topics that are generally forgotten almost as soon as they are published. At that point in history, a few months into the first triumphant year of Mr. Reagan's presidency, there seemed to be little prospect for a substantial audience for a report on education. Except for occasional references to a gap between the achievements registered in schools in the United States and USSR, public interest in the subject was at a low ebb. The last two high points of public concern with education had come some time ago, in 1958, with President Eisenhower's National Defense Education Act, and in 1965, with President Johnson's Elementary and Secondary Education Act, which intended to provide equal educational opportunities for the poor and minority groups. The momentum in 1981 seemed to be going all the other way. Mr. Reagan had promised to abolish the Department of Education; his administration had initiated cutbacks that augured a marked decrease in federal funding for education. Studies from the Carter years, such as *Science and engineering education for the 1980s and beyond*, had all but disappeared from public discussion.

An indicator of the new direction was the action taken by the National Science Foundation. Its charter directed it not only to support scientific research but also "to initiate and support . . . programs to strengthen . . . science education . . . at all levels"; but from 1970 on, the NSF had given an ever smaller part of its support, with only a brief upturn in the Carter years, to programs for schools, teacher training, and new curriculum development. From fiscal year 1981 to fiscal year 1982, the NSF funds for education were cut by 67 percent; another 34 percent cut was impending for the following year. Except for a program of graduate fellowships, the NSF was on the verge of "zeroing out" all its activities in this field; the deletion of the Science Education Directorate from the NSF's operational chart was scheduled for a few months later. Not only was ours the only advanced nation to strip itself in this manner of central governmental support; more significant, perhaps, was that little audible objection was registered from scientists, teachers, business, or Congress. When a reporter asked the president's science adviser for the reasons behind the phase-out, he was quoted as replying, "We don't know what the best thing to do is." For our commission to expect to receive wide attention for its recommendations, particularly if they differed from prevailing policy, seemed an unlikely prospect, despite Secretary Bell's hope for the outcome.

Equally problematic was the composition of the proposed commission. The list of invited members hardly reflected the education establishments, new or old – there was no dean of education, no education trade union official, no pundit from the educational press. Instead, there were two school principals, a superintendent of schools, and a "teacher of the year"; a renowned chemist, an industrialist-statesman, and a publisher of pedagogic literature; college presidents from southern, western, Ivy League, and community colleges; several members connected with school boards and boards of education; a state governor; and a person identified simply as "a parent." Of course, in addition, dedicated staff from the National Institute of Education was promised. Still, the proposed commission seemed to me likely to become eventually a convenient target of groups whose representatives had been left out. It was more a typical group of consumers of a study on education than the producers of such a report. In the end, this turned out to be one of its chief strengths, just as the timing of the appointment, at the bottom of a curve of attention that could only go up, proved to be fortunate.

None of this could have been foreseen; nor could anyone have imagined that on publication of the report, the response would be so intense that in short order some six hundred thousand copies of the booklet would be distributed (mostly to people who sent for it at $4.50 per copy). The report was also reprinted in large-circulation journals, bringing the estimate of its total distribution to well over six million copies.

Like some others on the eighteen-member commission, I thus accepted the appointment reluctantly, with the explicit understanding that there would be few meetings and that a minority report would be allowed if a need for it developed. Because the initial meeting in October 1981 set the stage for much that followed, it is worth remarking on. When a highly placed administration appointee was asked for an explanation of the radical downturn in federal support for science education, he said, quite simply: "There is no national mandate for such support." The meeting was continued at the White House where Mr. Reagan graciously received the commissioners and shared with them a structured and coherent set of themes and presuppositions about education that have since become more familiar. The president indicated that the encouragement of excellence in education, which should have first priority, might be achieved by making visible examples of good educational practice that already existed. However, he suggested it would also be well to remember the story of Farmer Jones, who confessed that he had no use for the new agricultural extension station down the road since he already knew how to farm better than he cared to. Costly programs, Reagan suggested, are wasted in the absence of sufficient motivation.

Reading from his prepared text, the president continued: since the country spent more on education in recent years "only to wind up with less . . . America [should] get back to stressing fundamentals in our schools" – fundamentals of learning as well as of principles. These principles encompass the following five: 1. Education is "the right and the responsibility of every parent," and institutions serve to "assist families in the instruction of their children"; 2. as in our economy, "excellence demands competition among students and among schools"; 3. diversity and pluralism in American education "has always been one of the strengths of our society, and we welcome the recent resurgence of independent schools"; 4. we cannot "restore educational excellence in schools still plagued by drug abuse, crime, and chronic absenteeism"; 5. let us "begin . . . by allowing God back in the classroom."

If the commissioners had been polled at that stage, they would, in my view, have been unanimous on perhaps only one point: that their most fruitful task would be to look carefully for exemplars of good local educational practice for which there might be hope of successful replication elsewhere. Yet, between the first meeting and the final one, a year and a half later, a remarkable transformation took place within the group. Having started from quite different positions and perceptions, the members reached total unanimity on the final report. Moreover, Secretary Bell accepted it publicly in every detail before transmitting it to the president. The member who had worried initially about the need for a minority report found himself commissioned to do the major part of the penultimate draft of the publication.

The facts had taken over. After dozens of meetings and hearings, together with field visits to schools and colleges, after trying to digest nearly forty commissioned reports from scholars and educators as well as a constant flow of background studies from members of the staff, the commission found itself, in the end, pushed by the data to the shared conviction that the state of American secondary-school education demanded systematic reform. Contrary to the main predisposition initially shared by the group, it had become clear that while identification of exemplars of good educational practice would be useful (and a working paper was subsequently published on it), this was by no means sufficient. The Exemplarians had been turned into Systematists, to use rough shorthand terms that hint at the current antithetical approaches to major educational change.

The commission, in my view, made five essential points. The first is simply the finding that the quantitative and qualitative indicators of the state of education, particularly at the high-school level, are on the whole unacceptably poor. To be sure, there are splendid exceptions. The top few percent of America's students seem to be as good as any in the world. Well-functioning schools exist, sometimes even in very depressed areas. There are numbers of effective and dedicated teachers and principals, some working under very difficult conditions. The school system can be proud of its record on access and retention, both of which have improved greatly during the last twenty years. The public's avowed commitment to good education for all children is so strong that opinion polls give education a top priority for additional federal funds – above, for example, health care, welfare, and even military defense. The reading scores of elementary-school children, many of whom benefited from federally financed remedial reading programs, are markedly better than those of their counterparts in the 1960s who had no access to such programs. For similar reasons, the basic skills of reading, writing, and arithmetic of the lowest-quartile pupils in elementary school appear to have risen. Graduates of President Johnson's "Project Head Start," now in high school, are doing significantly better than those with comparable backgrounds who did not get into that program. The much publicized decline in high school seniors' Scholastic Aptitude Test scores was recently halted, at least temporarily, even if the average score on *achievement* tests, which peaked in the post-Sputnik years, continues to decline at the high-school and college levels.

But apart from the oases of encouraging results, the evidence pointed to conditions which, on balance, were arid and parched. The testimony of educational researchers, employers, teachers, administrators, and students portrayed a system riddled with inadequacies. The charter had asked the commission to look most intently at the high-school years; its inquiry had revealed massive dissatisfaction. Indeed, partway through their work, the commissioners had begun to realize that a politically potent reservoir of frustration existed. It spilled over at the time the various reports appeared, and accounted in good part for the great national interest taken in them.

One need not look far to explain the motivation behind this interest. In one way or another, education engages nearly a third of the American population at any given time – as students, teachers, administrators, suppliers – and therefore is a most ecumenical and cross-societal preoccupation. To borrow a phrase from another battle, preoccupation with education, in most households in the United States, is the pro-family and pro-life activity *par excellence*, involving people of every age, ideological persuasion, and class. The stakes are high; anxiety in the United States (as in other countries) is energized by the perception that the young may not be sufficiently fitted with the skills and intellectual tools and attitudes needed to set down safely on their "landing field" a decade or two from now.

The provocative data found by the commission are now well known, and are not generally in dispute. They must be faced, even if in severely abbreviated form. A charitable count of functional illiteracy in this country yields more than twenty million adults; 13 percent of seventeen-year-olds, with 40 percent among minority youths, are in that same condition. Although that may begin to change a little, thirty-five states allow graduation from high school with little or no attendance in academic subjects. Possibly as a result, an ever-smaller fraction of the student body has been selecting courses with academic content. Eighty percent of high-school students take no science or mathematics after the tenth grade; on the average, graduates emerge with elementary geometry as their peak experience in scientific fields. Only one-third are able to solve mathematics problems that require two steps or more – and the news about one-step problem-solving capacity is not cheerful either: the National Assessment of Mathematics discovered that 50 percent of seventeen-year-olds cannot find the area of a square when given the length of one side.

The ability to sit down and do homework is being rarely tested in practice. The U.S. Department of Defense reported that about a quarter of its recent recruits who graduated from high school cannot read at the ninth-grade level, the level required for basic safety instruction. At the other end, among relatively high achievers, there has been a dramatic decline over the past decade in the number of students who score well on the verbal and mathematics section of the Scholastic Aptitude Test. With the number of test-takers fairly constant, the number scoring 650 or higher fell by 23 percent on the mathematics test and by 45 percent on the verbal test. Similar results were found for the top quartile in the National Assessment of Educational Progress. Fewer have registered to take the College Board Achievement Tests in English and history in recent years; moreover, their average scores have consistently declined. As for the low numbers of students taking foreign languages or calculus, the less said the better. And the test results in international comparisons of academic achievements are equally dispiriting.

In certain subjects, the overlap in instruction in successive years is deadening; tests show that a majority of academic-track students know 80 percent of the content of their textbooks in social science subjects before they have even opened the book. Most of the texts developed by national groups of scholars and teachers in the immediate post-Sputnik years have now been phased out. As if with a sigh of relief, publishers have gone back to the old-style texts that are easier to merchandize, prepared by authors who are easier to control, intended for teachers who are not expected to be trained to use the texts. The nominal 180 days of instruction per year in the United States become on the average only 160 when one considers student absences, and less still when one corrects for poor use of class time and incessant interruptions.

Two other indicators must be mentioned here, though they will figure again later. The federal role in shoring up education, which has always been relatively small but crucial, has been decreasing markedly in the last few years. Equally serious, because the damage is most difficult to repair quickly, the effectiveness of the majority of teachers has been severely compromised, however competent some of them may be individually. It is not too much to say that in most schools in America today, teachers are no longer able to act as professionals but are asked to operate on the level of service personnel. Correspondingly, the average teacher's annual pay, in real terms, has been actually shrinking (by 12 percent in the decade since 1972), and by 1982 had become about $17,000 for a teacher with twelve years of service in the school system.

The toll produced by a lack of reward is evident in many ways: the large teacher dropout rate after the first year or two; the steady inflow of more poorly prepared new teachers, with half of the newly employed mathematics, science, and English teachers being unqualified by license or preparation to teach those subjects. As for updating and retraining, which any professional or skilled worker constantly needs: 80 percent of the teachers in science and mathematics have not had even one substantial (ten-hour) study workshop in ten years. By comparison, for workers in industry the national average of the allowance for retraining in the bigger companies is seven hundred dollars per year, and double that for those in high-technology industrial employ.

The figures cited are only a small sample of the data that so disturbed the commissioners. They lead to the second main point of the report. On some models of social expectation such data might be considered sad but tolerable, an offense to intelligence, dignity, or sense of economy, but not to the social order. It might be argued that any system is bound to reveal certain horrors when forty or more million young people of every sort are kept in school, with authority distributed over sixteen thousand fairly autonomous school districts, each run by a school board made up of local volunteers. On this view, it is sufficient that there are, as there always have been, enough successes in the system to keep the country supplied with workers and managers, scholars and scientists, trained in their after-school careers by industry or universities. As long as the system is kept open enough, and the exemplars and incentives beckon, the best by any measure will rise to the top, as before. To increase resources would be uneconomical, because there is no guarantee the system would then work better; it could be awkward, even dangerous, if the system were to become more efficient, turning out too many overqualified people. More-over, the total bill for educational expenses is already high – on the same order as national defense.

Evidently, the commission took a very different view of the data when it came to the alarming perception that the nation itself is "at risk." At bottom it refused even to consider the concept of triage implied in the social-Darwinian view of how educational systems might function. The report embraced instead the opposite principle, which postulates that an attempt must be made to prepare a safe landing field for every child, no matter how much the attempt may be frustrated in individual cases.

An elaboration of the metaphor of the landing field may be useful. In a world shaped by ever more sophisticated knowledge and technology, it is more than likely that a worker or professional will quickly become obsolete if he or she cannot constantly adapt, learn, and renew. This is true from the beginning of a career to the very end. A common source for the worldwide concern with the effectiveness of education is the acceleration of history itself, which is resulting in new ground rules for the viability of nations for the pedagogic preparation of its citizens. Economic viability is only part of a larger concern. The ultimate aim of education in a free society is to help every young person gain intellectual and moral effectiveness and autonomy, so as to achieve personal freedom. It is unacceptable that by the very design of the system, many should be subjected entirely to the play of uncomprehended and uncontrollable events. This is the modern interpretation of the old prescription, *liberi, liberaliter educati* – free persons, liberally educated – that sets forth the functional link between individual liberty and the institution of a balanced curriculum.

To be sure, there is no easy way to demonstrate that this view (or, for that matter, its opposite) is correct beyond challenge. Most modern educators, however, particularly those ready to be known as reformers, are inspired by this presupposition, which functions for them as a necessary thematic hypothesis. There is also no proof that jobs in the future will demand more brains and more skills from the average citizen, or that jobs in the high-tech sector will exist in sufficient numbers for them. But even those who suggest that industry will be satisfied with the most basic skills at entry level and will teach the rest on the job for those lucky enough to have one, recognize that movement into a career or up the employment ladder will demand knowledge, skills, and habits at a level high enough for individuals to profit from continuing-education programs. The future is not likely to resemble the present. For a large fraction of children in school today, their lives as wage earners may well have as little to do with today's computers and workplaces as today's desirable jobs have to do with yesterday's steam locomotives and printing presses.

Thus, to prepare students for their future landing field means preparing them for a lifetime of learning new competences and of exercising critical thought in ever-changing circumstances; for without such autonomy, the young person is likely soon to be left behind, bewildered by rapid changes in the conditions of work, confused about the issues on which citizens must act. On this model, proper schooling will not only do its share toward securing a tolerable life for the individual and the family, but will also help to avoid a split within our society between the small, technically competent elite on one side and the large mass of the people on the other. That threat to a functioning democracy has preoccupied political thinkers since the time of Jefferson; it is now more compelling than ever. A growing fraction of the decisions made at federal, state, or local levels involves technical judgments, whether they touch economic policy, environmental issues, arms control, or other major concerns. Citizens must be able to participate at the very least by asking persistently the right kinds of questions. The franchise is devalued when major debates among experts are in principle inaccessible to the electorate.

If one reads the report of the commission with these points in mind, it becomes evident why the group thought that America "is at risk," why "educational reform should focus on the goal of creating a Learning Society." The concept of formal education as the preparation for a lifetime of continued self-education in a rapidly changing world is a key to understanding the report's recommendations, though this is rarely noted in the voluminous commentary on the report.

The commission's first specific recommendation, then, was that schools seriously aim to prepare every student, *whether college-bound or not*, to enter and participate in the Learning Society through tested competence in the "new academic basics," so that he or she would have the confidence and motivation to use these tools for further learning. The redefined basics, to be taken during the four high-school years, consist of four years of English, three years of mathematics, three years of science, three years of social studies, half a year of computer science, and, for the college-bound, at least two years of a foreign language, building on earlier courses taken in elementary school. Complementary studies were also recommended in the fine arts, the performing arts, and other subjects, provided the basics are accommodated.

While any such list, in the nature of things, can only be a quantitative rendering of the intended quality of learning, it is a necessary beginning. The disappearance of just such requirements produced the proliferation of nonacademic electives that compromised the standards of achieved competence. Difficult though it may be, it is the task of the school to design the pace and level of instruction to match the student's own best abilities; there is no suggestion that all students should be exposed to one master curriculum.

The difficulties of carrying out such a mandate are at once evident. It cannot be done without a greater number of able school teachers, without greater financial resources allocated to the schools, without a better use of research now available on learning, or for that matter, without the better use of time in the present school day, especially if one wishes to avoid lengthening the school day or school year. These obstinate realities, like those concerning different rates and levels of learning, define the size of the task, but they do not detract from the main purpose. The commission's first recommendation coincides with the finding of a National Academy of Sciences panel of business and education leaders, released in May 1984, that at the school level "virtually the same" basic academic skills need to be mastered by students who will start to work after high school as by those who are college-bound. The NAS report confirmed also that employers regard the major asset of a high-school graduate to be not the mastery of a particular set of job skills, but "the ability to learn and to adapt to changes in the workplace. . . . Those who enter the work force after earning a high school diploma need virtually the same competences as those going on to college, but have less opportunity or time to acquire them. Therefore, the core competences must always come first during the high school years."

Essentially the same conclusion concerning the primacy of academic basics can be reached from a quite different direction. The commission's survey of schools confirmed what had long been noted by educational experts: schools can be found in very disadvantaged settings that do provide a good education for a large fraction of their students. What such schools have in common is their stress on academic basics. The teaching goals are made very clear. and the expectations of student performance are kept high. The climate for learning is set by the school's sense of mission. From this derives an implicit respect for achievement and continuing evaluation; acceptance of strong administrative leadership especially on the part of the principal, and the involvement of both teachers and parents in the operation of the school. Through such procedures, a whole range of problems, from keeping discipline to preparing for employment, is made much simpler.

A third basic realization concerns the awful distance between hopes and realities. The idea of preparing students for autonomy through an academically substantial program could not have come at a worse time, given the new burdens such a reform would place on present schools and teachers. The system is already staggering from nearly two decades of adding important social missions, with greater access being given to high schools for special-need students, including the handicapped, a greater concern with remediation and nutrition, and the continuing efforts at integration. There is need for all these things to continue; still, they have redefined and made more difficult the role and functioning of both teacher and school.

Also, as these missions have been strengthened, there has been a corresponding weakening of the commitment to keep up the school's basic academic purposes. In addition to indicators already mentioned, such as the winding-down of NSF support, one may cite the flagging interest of colleges to provide schoolteachers; the decreasing commitment of many of the great universities to their graduate schools of education; and, above all, the flight of a big fraction of students from academic courses in high school. During the last two decades, the proportion of students enrolled in a meaningless "general track" program, one that is neither academic nor vocational, has gone up by a factor of three, from 12 percent to 42 percent of the student body. These ominous figures measure a betrayal of the hopes of many for quality education. While the school now retains a larger fraction of our young people, they have been drifting in ever-larger numbers into a storage area that prepares them only for a life of menial work or of consumerism. It is an explicit renunciation of the idea of a Learning Society.

Fourth, in common with all other reports on education, it was found that life has become by and large unacceptable for the kind of teacher one most wishes to have in our public schools. The best prepared are the most likely to leave for better jobs outside public school teaching; those who stay on are required to be heroes even when their qualifications to teach the assigned courses are weak, and they have difficulty identifying and challenging the more gifted and talented of their students.

The salary and retraining opportunities teachers have would be quite unacceptable in any other group that has a claim to professionalism. Most teachers receive no help with the administrative and low-skill chores heaped on them. Most have little or no influence in such critical professional matters as the preparation of texts or the selection of books, the keeping of classes to manageable size, the peer review of performance, promotion, and tenure of colleagues. Indeed, few know what subjects they will be required to teach a week hence. Recognition for a teacher's meritorious performance is rare; it may even be prevented entirely by barriers specifically set up for that purpose.

One can understand why a group that through no fault of its own has been so battered reacts with alarm to certain of the specific proposals made to improve the preparation of teachers, to make teaching both a more rewarding and more respected profession. The opposition of some teacher groups does not extend, however, to most of the specific recommendations of the report that colleges and universities do a far better job in pre-service and in-service training of subject matter; that teachers be tested on the mastery of their subjects before they are hired; that salaries be increased; that school boards adopt a twelve-month contract for teachers to ensure time for curriculum and professional development; that the class day be free of administrative interruptions; that so long as the current shortage of science and mathematics teachers continues suitable people from industry and academic life be recruited and trained; that museums and other centers become involved in education; that loans, grants, and other incentives be made available to attract outstanding new recruits to the teaching profession.

Nor are teachers averse to other public-school reforms even when they require further sacrifice and extra effort. As the recent nationwide poll of teachers by Louis Harris and the Metropolitan Life Insurance Company shows, teachers approved by huge majorities a new emphasis in their classes on the study of serious academic subjects, more rigorous graduation requirements, and more homework. They favored also, overwhelmingly, the career-ladder concept, the principle of periodic evaluation, and other changes intended to identify and remove incompetent teachers.

If there is criticism by certain teacher groups, it focuses chiefly on the recommendation to make "salaries professionally competitive, market sensitive, and performance based." The fear is fundamentally one of seeing too great a differentiation in pay and status simply on the basis of "merit," as defined by others. It is understandable that "merit pay" has become something of a red flag. Unless there is a marked upgrading of salaries across the board, with the present low level in many cities and states corrected, any policy concentrating on differentiation of pay must mean, in practice, that most teachers will experience cutbacks when the inadequate pool is reapportioned, often by criteria that the teachers themselves do not trust. Yet the cost of the continuing focus on this one issue, to the exclusion of all the others – especially on one phrase in one of forty recommendations covering some ten pages of the report – may serve only to maximize the political problem of bringing about any change, by deflecting attention from all the other findings and recommendations.

The report's most politically sensitive recommendation, the most revealing certainly in terms of its underlying presupposition, is the last, which was published under the heading "Leadership and fiscal support." It begins with the admonition that "citizens across the nation hold educators and elected officials responsible for providing the leadership necessary to achieve these reforms, and that citizens provide the fiscal support and stability required to bring about the reforms we propose." To be sure, it is recognized that state and local officials have "the primary responsibility for financing and governing the schools"; but there are three specific, detailed implementing recommendations that involve the federal government. First, the federal government should help meet the needs of key groups of students, such as the gifted and talented, the disadvantaged, minorities, and the handicapped. Next, its "role includes several functions of national consequence that states and localities alone are unlikely to be able to meet: protecting constitutional and civil rights for schools and school personnel; collecting data, statistics, and information about education generally; supporting curriculum improvement and research on teaching, learning, and the management of schools; supporting teacher training in areas of critical shortage or key national needs; and providing students financial assistance in research and graduate training." And then, to make certain that the reader not be left in any doubt about the matter despite the report's parentage, a separate recommendation was added:

> The Federal Government has the *primary responsibility* to
> identify the national interest in education. It should also help
> fund and support efforts to protect and promote that interest.
> It must provide the national leadership to ensure that the
> Nation's public and private resources are marshalled to address
> the issues discussed in this Report. (italics in original)

In my opinion, the eventual fate of educational reform in this country depends as much on the will to implement this paragraph as on any other act. Our decentralized educational system never has been and never will be able to make significant across-the-board changes within a five- to ten-year time frame in response to a national challenge if the leverage of the relatively small but vital federal contribution is not brought to bear, in terms of both planning and financing. Moreover, without strong national leadership, the impulse for reform can easily dissipate; in education, the time required to achieve and make visible striking improvements is much longer than the normal attention span that is common in America for a public issue.

It may be useful to note what the commission report did *not* attempt to do. To begin with, it did not call for additional reports. It avoided both the language and the mode of presentation common to academic communities and bureaucracies; it tried to reach its intended reader in unvarnished language. The effort to attain unanimity in the commission did not, to my knowledge, involve compromise on any very strongly felt point. One exception, perhaps, was the unwillingness to estimate the funds that would realistically be needed to put the recommendations in place. This omission, however, was essentially anticipated almost from the beginning, and was remedied, at least in part, by other reports, including that published by the National Science Board, which both set forth and explained the necessary price tag at least for the science- and engineering-oriented programs.

The report studiously avoided stressing what are commonly seen as the "inexpensive solutions" – the identification of exemplars with assumed transplantability, the encouragement of selected teachers and schools by the issuing of ceremonial awards, and the pursuit of other such activities. There is no objection to any of these so long as they do not displace the more essential solutions that will necessarily cost more.

As a consequence of the decision to make the report relatively brief and easy to read, with a separate release of the many commissioned background papers, the final version necessarily was too brief on a number of points that deserved more attention. Not enough, for example, was said about the educational reform needed in America's colleges, nor about the preparation a child ought to receive in the years before high school. "Science every day" is a necessity in the early years if anything like scientific literacy is to be achieved. Not enough was said about the general absence of libraries, or about the run-down and demoralizing physical conditions of so many of the schools in which both teachers and students are forced to spend a good part of their lives. No attempt was made to dwell on the design of stern codes of discipline, as a preliminary to attempting a larger reform; it is probable, in any case, that the causal chain in this instance runs the other way.

Too little was said about the role and responsibility of the educational publishing industry, which functions in this country somewhat as a ministry of education does. The publishers' output and policies are immensely influential in shaping the curriculum. The final text of the report might have said more on the troublesome state adoption system for selecting texts; on the lack of teaching aids and laboratory apparatus in too many schools; on the fact that the total science curriculum throughout the United States, unlike the situation in other advanced nations, not only suffers from being too little and too late, but is upside down intellectually. Whereas the natural sequence in science instruction is to provide a base in physics for subsequent learning in chemistry and then in biology, with parallel development in mathematics, America's high schools almost uniformly go in the opposite order, making each of the subjects more difficult to learn, less likely to be mastered properly.

With respect to other tasks not done, I will mention only three. The report might have made a strong plea that a Learning Society requires a more structured system of widely accessible adult education; this might include the kind of Open University through television and other electronic means that has been found to work well in other countries. On the other hand, it was probably reasonable for the commission to avoid pronouncing on such long-standing and unresolved philosophical battles as to whether the school ought chiefly to help shape the students to its expectations and standards or whether the school ought to be shaped largely by the students' own interests. Such differences are not likely to be resolved by a statement any commission may choose to make.

For similar reasons, the report did not come to grips with the most ominous question of all – how even an "excellent" school can fulfill its multiple functions for a large and growing fraction of our society that appears to languish in the darkness, below the security provided by any safety net. As one teacher put it recently, who will cure a child's inability to learn when the child comes to class hungry, with broken eyeglasses and rotting teeth, having been kept awake all night by family fights born of desperation, and showing all the psychological scars of an inhumane life beyond school? Or as another teacher said, if you have just become a member of the majority of children under age eighteen in this country that lives in single- or no-parent households, how important is algebra?

The nationwide attention that the report received on its release in April 1983 was undoubtedly triggered in good part by two circumstances. One was the president's willingness to launch the report. As Justice Louis Brandeis once wrote: "Our government is the potent, the omnipotent teacher. For good or for ill, it teaches the whole people by its example."

The other was the fact that the report's release took place during a meeting at the White House in the presence of the press. Having received the document some days earlier, the president began his own presentation by citing some of its findings. But his primary reaction to the report may be indicated by his comment:

> Your report emphasizes that the federal role in education should be limited to specific areas, and any assistance should be provided with a minimum of administrative burdens on our schools, colleges, and teachers. Your call for an end to federal intrusion is consistent with our task of redefining the federal role in education. . . . So, we'll continue to work in the months ahead for passage of tuition tax credits, vouchers, educational savings accounts, voluntary school prayer, and abolishing the Department of Education. Our agenda is to restore quality to education by increasing competition and by strengthening parental choice and local control.*

This list of proposed federal actions, though neither discussed nor endorsed in the report, were strikingly consistent with the desiderata set forth in the president's initial meeting with the newly appointed commission, nineteen months earlier. Most of those views were expressed also at other times: in the State of the Union messages of both January 1983 and January 1984; in presidential radio addresses and news conferences, and in the Republican party platform of August 1984.

I believe that some of the initial attention given to the report by the media stems from the perceived dissonance between the recommendations and the interpretations of them. Be that as it may, education suddenly had become a daily, inescapable feature of newspapers and magazines. The other reports that soon followed, with quite congruent recommendations, only added to the momentum. The rapidity with which America is able to embrace (and later drop) an issue has always amazed commentators. In this case, also, the turnaround from the previously low state of visibility of education was striking. There was little doubt that now an

---

*"Tuition tax credits" is a proposal to give a tax deduction to parents whose children attend private schools. "Educational savings accounts" would be analogous to IRA funds now invested by individuals; they would not be taxable. "Vouchers," as usually proposed, are negotiable certificates of specific value issued by the state or local school authority to parents who may use them to pay for their children's education at the school of their choice, whether public, private, or parochial. The source of the money would be funds from the federal government now furnished to the states under the Aid to Disadvantaged Children program.

operational, national mandate on behalf of quality education was waiting to be exploited, provided the political and educational leadership knew how to use the occasion properly.

What has come of all this? What are the chances for an educational renewal similar to the one that followed Sputnik? At this writing, after a year and a half of debates and planning, the prospects are mixed. As a recent, 229-page compendium ("A nation responds") issued by the Department of Education indicates, a large number of individual attempts at change are being considered by state and local authorities throughout America. For example, increases in the required number of academic courses for high-school graduation are being discussed in almost all the states, and have been approved in thirty-five; upgrading procedures for recruitment, certification, professional development, and compensation of teachers are being debated in practically every state, and a few states have taken action; a growing number of college systems report having raised their admissions standards; joint ventures, involving business and industry cooperating with schools and colleges, are becoming more common. Much of this was initiated before the publication of the various reports; given the slow response characteristic of many educational systems, more work will be needed in the long run to improve the curricula and the texts, to make more money available to recruit more good teachers. But, all in all, the products of the "year of the reports" have clearly been used to speed some actions by policymakers at both the state and local levels. Ironically, the public's confidence in what its schools are doing has risen substantially during this same period. One polling firm attributes this to the series of well-publicized studies, contributing to the belief that "many schools have heeded the criticisms made in the reports and have instituted the reforms."

Promising pilot programs already exist that take scientists and engineers from industry, including those who wish to start a new career or take early retirement from a previous one, and train them for service as teachers in areas where there are great staff shortages. Public-awareness programs have sprung up to select good teachers and to recognize them with citations. Professional and scholarly societies have become more active; for example, the American Association for the Advancement of Science is forming a consortium to bring together all the interest groups involved – scientists, engineers, business people, government officials, and educators – to assess the progress being made toward quality science education and to sustain the common national interest.

But the biggest remaining problem, of course, is that of funding, in order to repair past neglect and to achieve higher educational standards. Federal spending for aid to elementary and secondary schools is still significantly lower, in real dollars, than the 4.8 billion dollars appropriated in the last year of the Carter administration. Most states and localities do not have additional resources to meet their educational hopes. Some private foundations and corporations are turning their attention to the problem, but their available funds are not even potentially of the right order of magnitude. Moreover, even if some of the narrowly targeted federal support plans for schools now debated in Congress come to fruition, they will require a 50 percent match with nontax monies.

Apart from financing, there are other substantial obstacles to a lasting, wide-ranging reform. One of these obstacles is the disunity among those who could help in the implementation of common elements in the various proposals. While the recommendations of the commissions were widely welcomed by school boards, professional societies, individual teachers, and citizen groups, they received as little initial endorsement from some prominent members of the traditional educational establishment as they did from the president of the United States – although for diametrically opposite reasons. The basic findings and most of the recommendations were not challenged, but there was substantial opposition nevertheless. A former commissioner of education set the tone; on the release of the report he expressed doubts about the need for serious reform. What ailed the schools, he said, was a case of flu, not of pneumonia. For many months, the senior education columnist of the paper most widely read by educational policymakers devoted his space to repeated attacks, often on essentially semantic points, ending the year with a characterization of this and the other, related reform movements as simply wanting to "get tough" on schools and students, rather than wanting to "get better schools." The difference, he held, is "between clamping down and opening up," between causing "panic" and providing "opportunities." The authors of two of the last reports to reach the public devoted much press commentary to doubts about the recommendations of the earlier reports even while offering proposals of their own that closely paralleled them. One educational expert dismissed all the reports as essentially useless: "All such commissions are ill-equipped for the task of policy analysis." Publishers also expressed their unhappiness with the findings. So did many teacher-union officials, with the striking exception of Albert Shanker. The future student of these discordant notes may well be reminded of the Byzantine disputes of the factions inside beleaguered Constantinople, while Sultan Mehmed II and his troops were literally battering down the walls of the city. The inability of people of allied fates to focus constructive attention on the task at hand turned out to be as damaging as the large, new gun of the sultan.

The average reader of the daily newspaper might even now think that the most burning problem with education today has chiefly to do with whether the tide of mediocrity, in the now-famous metaphor, is still rising or whether it crested some time ago; whether it would be better to add some days to the school year, or make better use of the time now available; what title to bestow on the good teachers who are desperately needed now, but whom we do not presently have; and whether we ought not all to keep silent until some new philosopher of education, a Pestalozzi or a John Dewey, arrives to give us our marching orders.

One immediate casualty of this displacement of attention has been the lost opportunity to reach the citizenry in a more united and hence credible way, to explain how much time, manpower, and money it will really take to upgrade education, to what degree those funds are investments rather than mere expenses, and what the balance of funding sources should be. As John Gardner put it in the new edition of his classic book *Excellence*: "Americans care deeply about the schools; but their minds wander. The basic requirement for effective functioning of the schools is that the public be concerned and involved."

There is as much confusion out there as there is good will. Thus, the public tells the pollsters it wants more serious subjects in the school curriculum – over half want more academic subjects taught than the commission recommended; indeed, more than there is now time in school to do. The public is content to leave primary responsibility for improving and governing the schools principally in local hands, as is both traditional and appropriate; but these ambitions do not translate into votes for sufficient local tax monies (although the pollsters report a substantial majority willing to agree to higher federal taxes if these were channeled to better education).

An opportunity has been lost to consider carefully the future viability of the public-school system if the program of tuition tax credits, vouchers, and educational savings accounts is enacted. When enough parents are dissatisfied with the quality of public schooling, when these proposed devices make an exit from public schools easy, and when the melting-pot myth has given way to a splintering along both ethnic and religious lines, the exodus from a troubled system, with its unmanaged obligations, could be catastrophic. The theory that public schools would be strengthened by a challenge to compete has to be set off against the probability that they would be weakened by such flight, with the attendant loss of community support for the long-established, ecumenical mediator helping to negotiate local or individual differences. It was this last concern that led John Adams to say, "the whole people must take upon themselves the education of the whole people, and must be willing to bear the expense of it."

The alternative to the notion of the common school, commonly supported – which is at the heart of the public educational system – was succinctly put by William F. Buckley, Jr., in a column entitled "The way to excellence," which appeared shortly after the release of the report. The commission "didn't say what is most needed. That, of course, is tuition tax credits. . . . The only obstacle that stands in the way of the substantial privatization of the school system is ideology. Private schools, in situations in which the public schools are not doing the job, are doing what comes naturally in America."

The move toward a substantial privatization of the school system takes yet another form. A number of major corporations are reported to be considering the creation of proprietary elementary and high schools, which they expect to run at a profit. This, of course, is a long way from the useful and well-working partnerships between local businesses and school districts that have become somewhat more common in recent years. Halfway between these extremes is the kind of involvement of the private corporate sector recommended by William Norris of Control Data Corporation in May 1982 at a convocation at the National Academy of Sciences. He said:

> The missing ingredient is nationwide broad-based partnerships
> in which business addresses improvements in education as
> profitable business opportunities in cooperation with other
> sectors. In order to realize the full advantage of the use of
> advanced technology in the educational process, the
> management of schools themselves, or schools within schools,
> should be contracted out to business, which has the expertise
> to use advanced technology efficiently.

On the occasion of receiving the report, the president remarked that "we spent more on education than any country in the world. But what have we bought with all that spending?" During the past two decades, students' scores have declined – "decades during which the federal presence in education grew and grew." This linkage was made even more explicit in the president's radio address four days later: "Bigger budgets are not the answer. Federal spending for education increased seventeen-fold during the same twenty years that marked such dramatic decline in quality." At a press conference eighteen days later (May 17, 1983) a still more detailed causal connection was offered: During the last ten years, he said, scores declined even as "we went from $760 million federal aid to education to about $14.9 billion, and that's a two-thousand percent increase."

Clearly, the commission's report had failed to give sufficient prominence to facts that lead to a quite different conclusion. By far the largest part of the federal funds to education at the elementary and secondary level went not to programs designed to increase scores, but to nonacademic, mandated, social-agenda programs such as nutrition, or to successful efforts to reduce the dropout rate through programs in vocational and bilingual education and help for handicapped children. The exceptions were the federal funds targeted for basic skills, which in fact paid off handsomely in terms of better test scores (as for the nine-year-olds in Title I programs). Even so, the growth of federal funding in the last decade was not so large considering the addition of major new nonacademic programs. The budget of the U.S. Office of Education in fiscal year 1972 was in fact not $760 million but $5.1 billion; allowing for inflation during the ensuing decade, the real increase was thus not 2000 percent, but more nearly 50 percent.

Moreover, the report could have dwelled on the historical fact that from the Morrill Act of 1862 to the GI bill, and in the years that followed, federal aid to education was a bipartisan matter. Many main programs were launched in the Nixon years, and were supported by President Ford. Conservative lawmakers, such as Representative Ashbrook of Ohio and Senator Stafford of Vermont, staunchly defended the key sections of the Elementary and Secondary Education Act of 1965.

The figures, even if explained and corrected, however, would not by themselves be decisive in shaping educational policy today. What counts most is that there appears to be a change from the set of fundamental presuppositions shared in varying degrees by Presidents Eisenhower,

Johnson, Nixon, and Ford, concerning the need to stimulate quality education. The newly operative presuppositions are indicated in the remarks of Secretary of Education Bell; in April 1982 he said that the long-range strategy of the federal government should be to enhance the capacity of others to carry out the responsibility to education. "Ultimate success," he noted, would be "the termination of federal funding." We thus stand on wholly new ground in the battle between competing theories of how to bring about desired social change.

There have been times when effective educational reforms in the traditional sense became preoccupying issues for the nation. This happened in the North and Midwest before the Civil War; in the crusades for improved education in Southern states between 1902 and 1910. At other times, as in the 1960s and 1970s, the movement for school reform attached itself to and merged with other reform movements, important and meaningful in themselves, without putting primary emphasis on quality education as traditionally understood. We are now, I think, at a fork in the road. The potential for strong and long-lasting action exists, but which of two quite different directions we will take is not at all clear. On one side, the main motivation appears to be the removal of the federal presence, the introduction of religious and other private value concerns, and an increasing localization and privatization of pre-college education. On the other side, in a variety of overlapping visions, there is a set of more conventional goals. According to the first view, great hopes are placed in exemplars that are expected to challenge and motivate enough individuals to succeed in our society, because there is already a sufficient degree of opportunity and reason to believe that more growth is coming. On the other view, the system is seen as far too imperfect to be left alone as we move into an uncertain future; a relatively small but crucial infusion of energy, leadership, planning, and funding must come from the national government; and there is a greater sense of urgency to accomplish relatively large improvements.

As things stand today, if one had to bet on the likely outcome of this contest, one might well choose to place one's money on the success of the Exemplarians rather than the Systematists. The former are probably smaller in number, but they are well-placed and more united. The latter are severely divided: incrementalists against revolutionaries; those who see schools as the proper, perhaps even primary vehicle for necessary social development against those who think schools have never known how to do anything well, except, occasionally, to teach academic subjects.

The Exemplarians do not have to worry very much about the resistances in the system. As Chester E. Finn, Jr., explains in a recent article entitled, significantly, "The excellence backlash," changes of the sort the systemic reformers advocate are likely to fall victim to a "full-fledged backlash" that has already started among the various segments of the "education establishment." Principled and high-minded though it may be, the backlash is taking any one of ten forms that Finn develops in substantial detail. Even if the "excellence movement" survives for several more years, Finn sees three almost unresolvable problems built into the system: the sorry state of teacher quality, the politics of enforcing standards by which more children fail to pass and are made to repeat subjects, and the lack of institutional mechanisms to maintain public pressure for better schools.

One must hope that I shall lose that bet, just as I hope that Finn will be proved wrong on many of his somber points. But on his last it is hard to be optimistic. As I look back, three years after first becoming involved in the commission's work, one lesson stands out above all others. There is a basic structural defect in American education that greatly contributes to the confusion and fragility of the national mandate. Despite all the studies and emotion lavished on education, in one important respect it is not too much to say that our young people remain the most neglected fraction of the population. Our country's basic federal laws are so structured that any attention Congress and the administration gives to the needs of the young flows from the temporary triumph of good impulses or the sudden excitement of an easily comprehended challenge, not from the continuing necessity of law. If the Constitution and the Tenth Amendment are interpreted narrowly, as is now the fashion, one cannot be surprised by the movement to phase out most or all federal responsibility for education. Education is not mentioned in the Federalist Papers or in the Constitution. The whole issue was not seen as central to the life and destiny of a people, although one ought to remember that Thomas Jefferson, in asking Congress for a remedy, said: "An amendment of our Constitution must here come in aid of the public education. The influence on government must be shared by all the people."

The omission was natural two centuries ago. We may properly be thankful that we have not been burdened with a solution that would now be outdated. Two hundred years ago, what one learned in the first twelve years of life would, for most, provide sufficient orientation for the rest of their existence. This was appropriate for a rural America, with a relatively small population, surrounded by boundless resources and protecting oceans. But history produces new conditions for personal and national survival. We are living elbow to elbow among determined and strongly motivated competitors – competitors not only in seeking economic or territorial advantage, but in providing models for modern society itself. Under this condition, access to quality education for every young person cannot be built simply on the transient goodwill of local, state, and federal legislators and agencies.

Thus I have come to view that in the long run the health of our schools will not depend at any given time chiefly on how many more hours of history or mathematics are taught, or whether higher scores are attained in literacy tests, or even on the number of excellent teachers recruited. For all this could soon collapse again. Without a device that encourages cumulative improvement over the long haul, without a built-in mandate to identify and promote the national interest in education as well as to "help fund and support efforts to protect and promote that interest" – in the clumsily phrased but critical section of the commission report – we shall go to sleep again between the challenges of a Sputnik and a Honda. If the reform movement does survive, the question to be addressed is how to use the present momentum to put into place, during the next few years, something like a moral equivalent to a "right-to-education" clause, a commitment that will continue to live in the periods between those sporadic and exhausting battles that are a necessary condition of political life in a democracy.

Figure 14.1 President Thomas Jefferson; from an engraving by
Cornelius Thiebout, 1801. Courtesy of the Library of Congress.

14

"The advancement of science, and its burdens": the Jefferson Lecture

As the first of the Jefferson Lecturers to have been brought up in science, one of my themes must be the nature of scientific understanding; and as the first speaker in the series to have come to this country in search of a haven of liberty, my other main theme necessarily concerns the conditions that strengthen or threaten democracy – the more so as the award directs that the occasion be "in the service of the general public interest," bringing to bear the speaker's experiences "upon aspects of contemporary culture and matters of broad public concern."

Thomas Jefferson himself, I believe, would have had no difficulty discerning the connection between the two themes, between the power of science and the condition of society. He saw a double purpose for the pursuit of science: the advancement both of knowledge and of "the freedom and happiness of man." From the many examples that would serve as practical illustrations of Jefferson's twin goals, let me select his protracted study of the design for the humble plough.

In 1788, while acting as Minister in France but still a farmer at heart, he noted the unwieldy plough used there. Little seems to have escaped Jefferson, certainly not any inefficiency that might impede the development of his own young country. His curiosity went to work. He wrote: "The awkward figure of their Mould boards leads one to consider what should be its form." He sketched how the wedge-shaped mouldboard of the plough should raise the soil so that it would topple properly by its own weight.

Three years later he described the progress he had made: "I have

A somewhat expanded version of the Jefferson Lecture, given in two parts, one in Washington, D.C., and one in Boston, in May 1981. The lectureship is further described in the Preface.

imagined and executed a Mould board which may be mathematically demonstrated to be perfect, as far as perfection depends on mathematical principles, and one great circumstance in its favour is that it may be made by the most bungling carpenter and cannot possibly vary a hair's breadth in its form but by gross negligence."

He made a model of it. To improve it further, in 1798, Jefferson, then Vice President of the United States, took from his shelf Emerson's *Treatise on fluxions*, his old college book on the calculus, and re-derived the shape of the mouldboard and ploughshare that would offer least resistance to the soil. In 1805, while President, Jefferson made yet another model, writing to a friend "it took half a day . . . which I could not spare till very lately." He submitted the model to the American Philosophical Society, and he made plans to have the ploughshare cast in iron for wide distribution. He was awarded a gold medal for it by the Society of Agriculture, and published a scientific paper on the design. By May 1808, Jefferson had read in the Memoirs of the French Agricultural Society of the Seine District that an improved plough existed in France. He arranged to have one sent to him, despite the embargo his administration was enforcing on France. By October of that year he had received a French dynamometer and made experiments on the resistance of the plough to show that his was better than the best in France. Last but not least, he advocated the horizontal or contour method of ploughing, "instead of straight furrows." This ecologically sound method, he wrote, "has really saved this hilly country. It was running off into the valleys with every rain, but by this process we now scarcely lose an ounce of soil."

Jefferson is indeed an exemplar of the whole linked sequence of responsible research in science and technology: a concern, at the initial stage, for a technical solution to a socially significant problem; the innovation using, where necessary, the application of mathematics; the experimental tests; the disclosure by publication; the further development with a view to mass production; and, most significantly, the demand that the invention be used to benefit both the people and the land. His was a coherent course of thought and action.

Jefferson persuaded Congress to back the Lewis and Clark expedition as a venture with commercial potential; but to the Spanish authorities, through whose territory they were to pass, he described it as a purely scientific mission. That was shrewd, but also right. The mixture of both motivations was correct. Jefferson regarded the expedition as a scientific survey of the place to which the nation's destiny was likely to lead. This dual-purpose style of research – basic scientific study which had no certain payoff in the short term but was targeted at an area of national importance – deserves a name. We would do justice to call it the "Jeffersonian Research Program."

My intention is not to paint Jefferson as a scientist. On the contrary, my main point is that he and others of his period, while unique in history as political thinkers, were typical of their time in that they considered it as natural to share in the current scientific world picture as in the classics of literature, philosophy, and statecraft. If we turn our attention to this century's intellectuals, we find a situation very different from the comfortable incorporation of science into the world view that characterized the most prominent of the eighteenth-century models. Somewhere on the way, our civilization passed through a discontinuity. I must develop this point now, not because of a nostalgic longing for a return to the eighteenth-century euphoric hopes (voiced in Voltaire's projection that "reason and industry will progress more and more, that the useful arts will be improved, that all the evils that have afflicted man . . . will gradually disappear"); not because a current mutual incomprehension impairs discourse in the Common Rooms of some English colleges; but for a far more important reason – because we are witnessing a disjunction in the political and social life of the nation that curtails a main legacy of Jefferson's America, namely, our power of self-government.

My first witness is appropriately the writer, literary critic, and historian, Lionel Trilling, who in 1972 gave the first Jefferson Lecture, entitled "Mind in the Modern World." He began by reminding his audience of the loss of confidence in the power of mind, a confidence that had been characteristic of Jefferson's time. The old credo was that "if mind were cleared of its inherited illusions and prejudices, . . . what has long been accepted as the inevitable rule of harsh necessity, might be overthrown, and mankind will achieve a felicity which was its immemorial dream and its clear evolutionary destiny." That was once true. But Trilling thought the title of the last book by H.G. Wells was a more appropriate summary of our present condition: *Mind at the end of its tether*. Trilling pointed to two causes for the weakening of what had been a master belief for Jefferson and his time. One reason was our diminishing awareness of the past, what he called our "disaffection from history." But if the loss of history in our time has been, in a sense, a self-inflicted and voluntary disability, such was not the case for the second, equally disorienting one. For, Trilling continued, "the old humanistic faith conceived science, together with mathematics, to be almost as readily accessible to understanding and interest as literature and history." Jefferson, of course, could take that for granted; but this faith has been lost. Science, in our day, Trilling explained, "lies beyond the intellectual grasp of most men. . . . Its operative conceptions are alien to the mass of educated persons. They generate no cosmic speculations, they do not engage emotion or challenge imagination. Our poets are indifferent to them."

Having despaired of the hope Wordsworth expressed in the 1802/ 1805 Preface to the *Lyrical ballads*, that the remotest discoveries of the scientists "will be as proper objects of the Poet's art as any upon which can be employed" if these things "shall be familiar to us," Trilling added a paragraph that captures the anguish of the modern intellectual, to whom nothing is more devastating than the irredeemable remoteness of a whole continent of fundamental knowledge:

> This exclusion of most of us from the mode of thought which is habitually said to be the characteristic achievement of the modern age is bound to be experienced as a wound given to our intellectual self-esteem. About this humiliation, we all agree to be silent; but can we doubt that it has its consequences, that it introduces into the life of the mind a significant element of dubiety and alienation which must be taken into account in any estimate that is being made of the present fortunes of mind?

Trilling's chillingly accurate assessment has implications far beyond the concerns of a few academics. It would be a great error to dismiss it as the parochial anguish of intellectuals, unmindful of the dangers in the real world, a world where over half a billion people cower at the edge of starvation; where some fifty thousand nuclear warheads stand ready; where numerous states subjugate their people under totalitarian rule; and where even the majority of citizens in our democracy regularly fails to exercise its right to select who will govern. On the contrary, there is a direct link between those problems and Trilling's observation. As he noted at the end of his lecture, "this falling off [of the confidence of the power of mind] must be felt as a diminution of national possibility, as a lessening of the social hope. . . . *Mind at the present time draws back from its own freedom and power.*"

Trilling reveals both the intellectual and the political consequences of being denied knowledge without which some of our best thinkers cannot be sure they have a proper hold on the world. This sense of impotence in the face of the social and political decisions being made for us in this scientific age mirrors a fact that has even been documented in opinion polls: It used to be the case that the more educated the individual, the more he approved of science. Distrust was greatest among the least educated. Now there has been a complete reversal; the distrust is least among the most naive. From among the others we are beginning to hear the sentiment with which Bertolt Brecht ended his play *The life of Galileo*: "In time you may discover all there is to discover – but your progress will only be progress away from mankind. The gulf between you and the people will become so great that one fine day you will cry out in jubilation over a new achievement – and be greeted by a cry of universal horror." William Faulkner, in his acceptance speech for the Nobel Prize, said simply: "There are no longer problems of the spirit. There is only the question: When will I be blown up?"

I have been pointing to a polarization that is pitting two groups, once allies, against each other; one, chiefly scientists and their followers, including technology enthusiasts, is cheered by the promise of ever-greater scientific advances; the other is depressed and fearful in the wake of these very advances. How did this separation come about?

We must look first *into the intellectual machinery of scientific advance itself*, to see what is causing these antipodal trends. Is there one common reason for both the euphoria and the despondency? How can an activity be seen as the isolated, esoteric pursuit of a few souls engaged in rapturous contemplation, and also as the dominant, perhaps uncontrollable force directing the very course of society? What is there about modern science that makes it so successful in the hands of relatively few, and so difficult or ominous for the rest?

A key to the answer will be to see that the progress of science and of science-based technology follows a pattern that is built in, and rather different from the progress of other activities. This order conflicts with the layman's impressions, produced by the daily stream of cheerful announcements of new findings and new products emanating from university laboratories and technology-based industry. Their sheer quantity and lack of coherence must be as disorienting to the average person as the cacophony of events that buffeted Tolstoy's Pierre while he stumbled through Napoleon's battlefield. But despite their appearance of randomness, scientific and technical advances reflect a development akin to tree-like growth, a motion with two components, one vertical, toward a higher, more abstract state of scientific specialty, the second horizontal, branching toward the sides, where applications and repercussions radiate into other fields, from engineering to literature. This two-dimensional unfolding can be traced over history. Some components of the development indicate a tendency toward discernible goals – more so for the physical sciences, to which I shall necessarily restrict my specific examples, and less so for some of the other sciences, although in this respect they are all qualitatively similar. We shall find that the model of tree-like growth helps us understand the two main benefits of the advancement of science, and also the two main burdens.

◦~◦~◦~◦~◦~◦

First, consider the order associated with the upward thrust of science. It is motivated by a constant search for unities and simplicities behind nature's spectacle of variety, a reaching for ever higher, more general conceptions that allow one to see common features among the phenomena. I must insert here a disclaimer: Like all who have struggled with this question, I find it utterly mysterious that this search should be so successful – why, for example, the mathematical equation that applies to acoustic waves should be applicable again and again in widely different contexts, from the spreading of temperature profiles to the motion of elementary particles; or how ideas developed in the study of macroscopic phenomena such as superconductivity should turn out to be well adapted to deal with submicroscopic problems concerning the structure of nuclei. I recall a physicist saying, "Why is quantum theory possible at all? Why should a theory which you can write down essentially on one page apply correctly to untold billions of individual cases?" It is fitting to recall Albert Einstein's confession in his letter to Maurice Solovine concerning "the high degree of order in the objective world": "One has no justification to expect it *a priori*. Here lies the sense of 'wonder' which increases ever more with the development of knowledge."

Leaving aside the incomprehensible roots of its success, we do know what stimulates the search for unities and simplicities at ever higher levels of abstraction. While it takes seemingly different forms in different sciences, the upward thrust has been motivated in the physical sciences by the frank desire for "*the complete [intellectual] mastery of the world of sensations.*" With these words, Max Planck, inventor of the early quantum ideas and author of the book *Where is science going?* (1932), answered the question posed by his title. That aim – namely omniscience concerning the world accessible to positive science, and hence the construction of a complete scientific world picture – and the way to attain it, might be called the "Newtonian Program," because Newton, while not the first or last to embrace it, adhered so explicitly to it.

In the Preface to the *Principia*, Newton described his aim and procedure: The observable phenomena (e.g., the fall of objects to earth; some celestial motions) led him to postulate the existence of one general force of gravity by which all bodies attract one another, and from this he was able to deduce in detail "the motion of the planets, the comets, the moon, and the seas." But no sooner had he acknowledged this stupendous achievement than he added, with a hint of his deep disappointment: "I wish we could derive the rest of the phenomena of Nature by the same kind of reasoning from mechanical principles." These would include optics, chemistry, the operation of the human senses. There was the Holy Grail: the mastery of the whole world of experience, by subsuming it ultimately under one unified theoretical structure. Einstein similarly taught that "the noblest aim of science" was the attempt to grasp the "totality of empirical facts," leaving out not "a single datum of experience." Newton hoped even to go beyond that, to an understanding of the Godhead itself, as well as to the laws of social behavior. As he wrote in the *Opticks*:

> And if natural Philosophy in all its Parts, by pursuing this
> Method, shall at length be perfected, the Bounds of Moral
> Philosophy will also be enlarged. For so far as we can know by
> natural Philosophy what is the first Cause, what Power he has
> over us, and what Benefits we receive from him, so far our
> Duty toward him, as well as that toward one another, will
> appear to us by the light of Nature.

Even without this additional aim, and similar ones proposed by the physiocrats of the eighteenth century and the technocrats of the twentieth, the ambition of the scientists' Newtonian Program is breathtaking. To understand better its dynamics, consider the pursuit of the Newtonian Program in terms of a simple diagram. Imagine a horizontal plane on which each point corresponds to a datum of experience, an observation, or a sensation. For example, in the left corner of this plane are points that represent observations of regular planetary motions, from Hipparchus to Tycho Brahe and onward. Next to it, to the right, are points representing the trajectory of projectiles and the falling of an apple; the data of motions of comets; the motions of the moon; the behavior of the tides; the shape of our spinning earth; the variation observed in the time kept by pendulum clocks at different parts of the globe; the propagation of sound and motion in fluids generally. Further toward the right side of this vast plane of experience are the data obtained from various phenomena of optics, from rainbows to diffraction. Beyond that is the curious behavior of charged amber, of currents, magnets, chemical reactions, and so on: the whole accumulation of the "facts" of nature that, at first glance, have little in common with one another.

One way to understand what I have called the Newtonian Program is to see what Newton did with this material. Observation of the phenomena at the left side of the plane led him to propose the existence of a universal force of gravity that diminishes with the square of the distance between bodies. To this postulated force Newton added what he called the three "axioms or laws of motion," some definitions (of mass, momentum, and the like), and a slew of mathematical propositions which nowadays could be greatly condensed. Together, the postulated force, the laws of motion, and the propositions formed Newton's axiom system, suspended above the plane of experience. The reason for this construction was to make logical deductions that were inherent in the axiom system, and to see whether these deductions coincided, or at least were not incompatible, with the known facts, that is, the "points" on a good portion of the plane of experience below.

The Newtonian synthesis was an overwhelming success because it allowed a wealth of apparently disparate phenomena to be related in a very economical way, by seeing that they coincided with deductions from one simple, mechanical axiom system. Moreover, as new phenomena came into view, such as the motion of Uranus and Neptune, the rotation of the galaxies, and other previously uncharted extensions of the plane of experience, they too fell into place under the same "pyramid" formed by the initial axiom at the peak and the deductions radiating down from it.

The next major advance of that sort was Hans Christian Oersted's achievement in 1820, showing that two other, similar pyramidal structures – one accounting for static and current electricity, and the other for the phenomena of magnetism – could be regarded as subsets within a larger and higher pyramid. Oersted found that a magnet needle could be deflected by an electric current, that currents surround themselves with magnetic lines of force. In this way, magnetism and electricity were fused into one conception, the first such unification of apparently separate forces of nature. Oersted's publication provoked an explosion of scientific work, including the discoveries of Ampère and Faraday. Technical applications followed quickly, starting with an electric telegraph.

Half a century later, the plane of experience had become crowded with many more discoveries, but the urge to achieve synthesis had succeeded again, and raised further the height of the axiom system embodied in the structure of physical theory. Maxwell was able to show that the phenomena involved in the propagation of light could be brought under the same roof with the unification of electricity and magnetism that Oersted had discovered. It was another achievement on the scale of Newton's, showing that phenomena that appeared diverse were really closely related.

Advances in the physical sciences have continued in this same style. The experiential horizontal plane has become more extended, and the upward thrust remains powerful. One set of phenomena, such as the emission of beta rays in radioactivity, has been interpreted as the action of a force called the weak nuclear force that acts among leptons (of which the electron is an example). Another large set of nuclear phenomena is now understood in terms of the strong nuclear force acting among quarks.

Einstein had nourished the lifelong hope that a theory could be found to subsume the phenomena involving gravitation and those of electromagnetism under one greater axiom system, incorporating Newton's and Maxwell's pyramids into a yet higher pyramid, a unified field theory. It turned out, however, that this upward movement was achieved in a different way, by uniting the Maxwellian and the weak-force structures under one axiom system, characterized by the action of the "electroweak" force.

Our attention is now focused on the next candidate for further unification, at yet a higher level of new axioms. The name "grand unified theory" has been coined for it. When the fog lifts, the two peaks symbolizing the electro-weak force and the strong force should turn out to have been just the lower portions of a still greater mountain range. If that is true, leptons and quarks would be related, and a lepton-quark conversion, indicated by the decay of protons, should be observable. This further widening of the plane of experience involves exacting experiments to bring into view phenomena that had not even been thought about previously. And beyond that? The attempt will surely be to subsume all four forces – gravity, the electromagnetic, the weak and the strong force – in one theory, a tentative version of which is referred to simply as "supergravity."

In this schema of the Newtonian Program much has been left out, not least the intense analytic activity in the lower foothills without which these upward synthetic thrusts would not be possible. As in the building of the Rocky Mountains, a vast amount of energy has to be supplied from the base. Here it takes the form of painstaking experiments to test the flexibility and limits of provisional theories. Nor have I dwelt on a curious consequence of this mode of advance, namely that it constantly cannibalizes earlier achievements, incorporating the useful parts within the newer setting. The labors of scientists are not really cumulative; rather the new growth makes the old invisible by absorbing it. This is, of course, one of the chief differences from the arts and literature, in which each work retains in principle its individual claim to respect from later practitioners or the public, and perhaps also maintains continued vitality as a source of new inspiration.

Despite its limits, the model I have proposed does illustrate that science, in the words of the philosopher W. V. Quine, is not an assemblage of "isolated bits of belief, but an interconnected system which is adjusted as a whole to the deliverances of experience." The schema also indicates why this audacious method of construction does *not* result in a Tower of Babel: It is at the bottom of the schema that the various languages are spoken, whereas at the top they all become one. In addition, it helps to understand the historic fact that science seems to follow what Alexandre Koyré called "an inherent and autonomous" development. Finally, this outline sketch may explain why the success of the enterprise can on occasion release high expectations and imperialistic ambitions. The physicist Leon Rosenfeld, writing about the 1920s, reported: "It is difficult to imagine the enthusiasm, nay the presumptuousness, which filled our hearts in those days. . . . A friend of mine expressed his view of our future prospects: 'In a couple of years,' he said, 'we shall have cleared up electrodynamics; another couple of years for the nuclei, and physics will be finished. Then we shall turn to biology.' "

A single-minded, optimistic spirit grips the community of scientists engaged in today's high-metabolism, high-stakes enterprise. They seem to be caught up in the promise that perhaps in our own lifetime we shall really know why electrons, protons, and neutrons exist; what the life cycle of the universe is, from the initial Big Bang to the final Big Crunch;

and what the connecting links are between the various states of organized matter, from atoms to living things to societies. Few scientists doubt that the Newtonian Program can eventually be completed, and few could think of a better way to invest their lives than in that grand project. Measured by its own criteria, scientific research has never been of higher quality, nor the prospects greater that the best is just over the horizon.

One answer to the common, plaintive question "Where is science going?" is therefore the practical response of most scientists: If you give us the means, and a flow of spirited young people to assist us at the frontiers of inquiry, and if the sciences are allowed to follow their own manifest destinies, there is virtually no limit to their continued perfection. The haunting promise is that we may achieve a panoramic view of the whole physical world, one that will allow the mind's eye to obtain, at a single glance, a unified perception of all physical events, causes, and effects. In Einstein's words: "The confident belief that this ultimate goal . . . may be reached, is the chief source of the passionate devotion of the researchers." Even if that great project proves endless, it now provides science with a purposeful drive toward a goal, a *télos*. Even though there are always many more "diversifiers" than "unifiers," the vision of the latter helps organize the smaller local tasks; it tides us over the inevitable periods of stagnation and failure; and it will continue to stimulate the community long after our current achievements have come to look small from the new heights.

In sciences other than the physical ones, the upward thrust toward a more encompassing view has been organized on different lines and around other key conceptions. Such differences are of immense interest to the historian of science, but are not significant here, because what will concern us shortly, as one set of consequences of the mode of advance I have sketched, are the *costs* of this method of achieving a scientific world picture – and these costs are much the same for all scientific fields. Yet, before we turn to that, I must refer briefly to the second process in the unfolding of science, the outward movement to the right and left. Here we shall be concerned not with the building of higher conceptual structures in the pursuit of omniscience, but with the primarily instrumental aim of increasing human power through "mission-directed" research.

◦•⌒•⌒•⌒•⌒

For the purpose of illustration, it will be useful to refer to a specific case, such as the targeted research project in 1946–48 that led to the transistor, and for which the 1956 Nobel Prize for Physics was awarded to John Bardeen, William Brittain, and William Shockley of the Bell Telephone Laboratories. Unlike the classic "pure" or "basic" research which generally starts and ends in the minds of scientists, this development quickly triggered a proliferating set of devices, from heart pacemakers to high-speed computers – devices that have begun to change civilization itself, including, indeed, how basic research can be done effectively in the sciences and in some areas of the humanities.

At the ceremony presenting the three Nobelists to the King of Sweden, E.G. Rudberg of the Royal Academy in Stockholm compared them to the "small party of ardent climbers" who have reached the "summit of Everest," even though they had started their assault from a high-altitude camp that "more than a generation of mountaineers had toiled to establish." Such an image applies to the Newtonian Program of autonomous, basic research, but it is not really appropriate for this case or others of the type. A historical study would quickly show that many factors essential to the achievement of these scientists were absent from the typical project we discussed earlier – most important, a mission orientation, embodied in the institutional policy of the Bell Laboratories, announced in 1945, to obtain "new knowledge that can be used in the development of completely new . . . elements of communication systems." The pressure during World War II to develop rectifiers acting as crystal detectors for radar provided a strong impetus, as did an old hope of making amplifiers that avoid the power drain of vacuum tubes, and switching devices without the usual problems of corrosion and slow response. The work was "problem focused"; it depended as much on the expectation of relatively short-term payoff as it did on the quantum physics of the 1920s and 1930s.

When the three scientists were later interviewed, they revealed that much of their success depended also on "extraneous factors," for example, the particular style of work of the laboratory – a pragmatic experimentalism nourished by the rather American mix of science, engineering, and improvisation – and on the mobility of young scientists seeking to train under skilled teachers. In short, the three who made it to the top of Everest were much helped by other mountaineers, by Sherpa porters, by supporting organizations near and far, and by a mandate for useful results.

The mission-directed style of research usually leads to both intended and unintended changes in current technology, engineering, medicine, or agriculture, and so affects the social matrix much faster than do the results of basic research. This penetration may be represented by a diffusion of effects from the realm of scientific concepts into the polity, a lateral outreach that is not a mere by-product but a chief aim of mission-directed research. In essence, the motivation here is instrumental. This style of work might be termed the "Baconian Program," for it was Francis Bacon who urged the use of science in the service of omnipotence, "the enlarging of the bounds of human empire, to the affecting of all things possible."

Whether the title is quite appropriate is unimportant. The significant fact is that in our time a remarkable symbiosis has established itself between the Newtonian and Baconian Programs. There have always been occasional cross influences – historians of science are still debating whether Galileo learned more from the workers in the Venetian shipyards or they from him – but now these two approaches have come to interpenetrate each other thoroughly, just like two mechanical systems that exchange energy more rapidly and vigorously while they resonate with each other. The "pure" and the mission-oriented versions of research send their respective products back and forth, increasing their own rate of advance as they do so – tightening their coupling further, and also blurring the distinction between them.

The marriage of science and technology is undoubtedly permanent and beneficial to each. The experimental side of science has become more technological (the sealing-wax-and-string era of laboratory work ended when World War II technology was declassified), and engineering depends increasingly on a scientific base. Indeed, the laboratory experiments that confirmed the theory of unification of the electro-weak force depended chiefly on products of modern electronic engineering. Conversely, the design of industrial products of biotechnology follows by only a few years or even months the latest results of basic research in genetics and molecular biology.

The first main consequence of the advancement of science in our time is one which we may be sure Jefferson himself would have been delighted to see. The accelerating process of unfolding has brought us to a fairly coherent overview of nature and the design and production of powerful technical devices, while a cybernetic process of positive feedback allows the continued, simultaneous escalation of scientific as well as engineering advance. Another consequence of this process, insufficiently noted, would perhaps have appealed to Jefferson even more. It is not merely a question of the commonplace observation that our physical burdens have lightened, farm life has been transformed, and medicine improved, but of the direct and indirect influence that the results and attitudes of scientific/technological advance have had in extending the very conceptions of *human rights*. Since Thomas Hobbes, to whom the essence of such rights was merely the freedom to eat and be eaten, moral and legal rights have increased greatly. Scientific findings in biology, anthropology, psychology, and other branches are largely responsible for making us certain, as Jefferson's time was not, of the essential homogeneity of the human race – a fact from which springs the entitlement to protection against discrimination. Biomedical advances have made it possible to respond to the clamor for access to medical care, and to support the right of women to assert choice in family planning. Such examples can be greatly extended.

D. D. Raphael has observed that the various twentieth-century declarations of human rights incorporate as necessities what would, in previous centuries, have been luxuries, and that material advance stimulated the continuous expansion of the definition of human needs. J. E. S. Fawcett, surveying the whole issue of protecting human rights, concluded that "fundamental to all effective methods of implementation of human rights" is the development of scientific rationality itself – "independent and objective fact-finding" – as well as the modern technology of rapid communication and publicity. One fact the discussions on modern theories of rights agree on is, in the words of C. B. Macpherson, that the "theory [of human rights] finds itself in need of development to meet changed conditions," and these changed conditions are to a large extent due to the influence of science and technology on the operations and values of modern society.

Increases in productivity have, for example, quite undermined old assumptions of permanent contentiousness and possessive individualism that were based on the notion of the inevitability of severe scarcities of resources. (One of the early, unexpected results of the spread of railways was that local famine became avoidable, and therefore intolerable.) The very increase in the number of charters and debates on human rights in our time may reflect the recognition that central portions of the Hobbes–Lockean "liberal-individualist" tradition, as well as the Rousseau–Marx "anti-bourgeois" traditions, are no longer relevant or sufficient, leaving us with the task of redefining, in the modern context of a technologically driven society, a workable framework for human rights that extends those derived from the Declaration of Independence.

I shall now turn to the two main consequences of the advancement and symbiosis of science/technology that are of a less happy kind: the new burdens. The first and most obvious of these is loss of contact. Despite popularization by the mass media, the concepts and methods of science have become largely inaccessible to all but its practitioners and a small circle of attentive onlookers. On the basis of an extensive anthropological study, Margaret Mead noted more than two decades ago that both school children and adults in our society had, over the previous fifty years, "come to feel that science is something deficient and alien, a discipline that they neither can nor care to understand." While Americans embrace with delight the products of technological industry, the conceptual base of their operations remains opaque to most who use them, unlike the earlier tools and devices that were relatively transparent and comprehensible.

The intellectual brilliance of our sciences, the ingenuity of our technologies, may be characteristic of our culture, but from the point of view of the public and most intellectuals, the thought processes and operations of both have moved behind a dark curtain. There they have taken on a new form of autonomy – isolated from the active participation or real intellectual contact of all but the highly trained. Contrary to eighteenth-century expectations, the scientists are losing what should be their most discerning audience, their wisest and most humane critics.

This burden is, ironically, caused by the very success of the method of scientific/technical advance. As the building of a unified scientific world view proceeded, as the conceptual structure grew more elevated, the chains between the axioms at the top and the empirical base at the bottom had to extend further. The logical structure reaching down from the axioms became more formal and parsimonious; the bridge to ordinary intuitions and to the rationality of everyday thinking was first stretched, and then was lost or became an actual handicap. While in Jefferson's day it was possible for those in his circle to value and comprehend much of the science not only of Newton's time but of their own, this ceased to be the case for interested onlookers a century or so ago. By now, the instructive metaphor is essentially unavailable, and the barriers of communication are virtually insurmountable even for the interested nonscientist –

despite Einstein's warning: "I can think of nothing more objectionable than the idea of science [only] for scientists. It is almost as bad as art for the artists, and religion for the priests."

As the plane of experience expands through the use of specialized or high-technology observational devices, the public progressively loses access to the phenomena of nature. The connection between phenomena and theory, the theory itself, and the way it is constructed, confirmed, and elaborated are, and have to be, fully controlled by the scientific community, and understanding them comes only with long immersion. We can see more clearly one of the chief causes of Trilling's perception. Precisely as science progresses toward its declared goal, and as the rate of its new triumphs increases, the larger yawns the unnegotiated intellectual separation from those standing on the sidelines – all those who feel the shame and sorrow of being excluded "from the mode of thought which is habitually said to be the characteristic achievement of the modern age."

I must now turn to the second cost, which is also a by-product of the dazzling advancement of the modern science-*cum*-technology complex. It is a cost not to the intellectual power and integrity of our cultural trustees, of which Trilling spoke, but to the political effectiveness of our citizenry. The source of this burden lies chiefly in the fact that the metabolism of the science/technology complex has progressively increased since Jefferson's days. As it continues to do so, it enlarges the scale of technological intervention in our lives at the same time as it inhibits our opportunity to evaluate and respond to the ethical and value impacts of these interventions. The old (and sometimes blind) faith in the benign efficacy of technological progress has been waning, even yielding to its very opposite – the fear of an autonomous technology, and the rise of an ideology of *limits*.

The relative modernity of this concern can be illustrated by studying the outcome of the seemingly humble scientific discovery made over a century and a half ago, in October 1831. When Michael Faraday in London put his hand on a copper disk and made it rotate between the poles of a magnet, a steady electric current was induced in the disk. He had discovered (or invented) the electric generator, or dynamo. The simple gadget was a by-product of Faraday's lifelong search for his true and quite "Newtonian" goal – to find through the study of diverse scientific phenomena the unity of all of nature's forces. Faraday's dynamo took the energy his hand gave to the disk and made that mechanical energy reappear as electrical energy which could be drawn off in the wires connected to the disk.

It might have remained a scientific curiosity and a mirror image of the motor effect Faraday had found eleven years earlier when pursuing Oersted's findings. But Faraday's new dynamo (and the version Joseph Henry invented at almost the same time) was a source of electricity more effective and cheaper than the battery that Volta had discovered in 1800; it was soon redesigned to use mechanical energy from any source – wind, water, steam, coal, wood, gas, or oil – and to issue forth electrical energy for immediate transmission to distant points, as far as cables could reach. There, motors reversed the process, yielding back mechanical energy to do work; or resistors could convert the electrical energy to heat and light.

The physics involved is not sophisticated, but the effect on society was. At last, the source of energy did not have to be in the same place as the ultimate user, a crucial difference from the steam engine, burdened down by its fire box, boiler, and fuel. The dynamo could be anywhere and produce its effect with high efficiency at a great distance; its action was almost instantaneous, its capacity in principle virtually limitless, the raw material used for the prime mover relatively cheap, and the sanctions against disposing of the effluents practically absent.

As a result, the total system was immensely potent, indeed irresistible. The dynamo, in hundreds of forms and uses, became the keystone that completed the industrial revolution and the modernization of day-to-day life. The modern city could not have risen without the development of the dynamo, for on it depends essential transportation such as elevators and subways, and all electrical lighting, signaling, communication, and control. Life in the factory, in the workshop, and eventually on the farm was transformed by the new generators, as were industrial processes; the electrochemical industry, which had initially not been considered a beneficiary when the great Niagara Falls power plant was built, by the turn of the century was using more electrical power than lighting and motors together.

What interests us here is that the introduction of electric power into society and the landscape, while not without difficulties, was a remarkable success. One reason was simply that enough time had elapsed to allow a degree of intellectual and political accommodation. The design of commercially viable generators took nearly fifty years after Faraday's first model, and brought together a great variety of interests – scientific, technical, financial, managerial – in many countries. The various problems of "electrification" engaged citizens at every level, from Parliament and Congress to individual towns and villages. The system that finally evolved was a compromise between divergent interests and alternatives that had been widely debated and considered at length.

Such public involvement is just what is becoming less and less possible as the rate increases at which science-based technology is injected into the life of the nation. Thus, the rapid initial deployment in the United States of nuclear reactors for power generation turned out to be a disaster for that industry itself. Concern about the long-range consequences has also been raised by the rapid computerization of the workplace, by the introduction of the robotic factory, and by the vast new weapons systems now on the drawing board and in production. These and similar decisions no longer grow out of an organic adaptation among all the relevant interests, a process of evolution in which sufficient time can elapse to allow technical details to be learned and open debate to take place among an informed citizenry. According to a recent estimate, nearly half the Bills that come before the U. S. Congress have a substantial science/technology component. Yet few in Congress or on its staff have, or can have, the training and background to be really familiar with the problems. We are reminded of Lord Snow's well-known prediction that in such a climate the decision-making process may be usurped by the "scientific overlords," and of Eisenhower's Farewell Address (January 1961), warning not only of "the acquisition of unwarranted influence . . . by the military–industrial complex," but also of the "danger that public policy could itself become the captive of a scientific-technological elite."

In a recent essay, "Coping with technology through the legal process," David L. Bazelon, Chief Judge of the U. S. Court of Appeals for the District of Columbia Circuit, wrote that "some two-thirds of the D. C. Circuit's caseload now involves review of action by federal administrative agencies, and more and more of such cases relate to matters on the frontiers of technology." In discussing how "society can come to terms with science and learn to cope with technological process," Judge Bazelon concentrates on the central issue that virtually every technological innovation introduced to solve a societal difficulty will also have unwanted consequences – hence will require that some painful value choices and difficult policy decisions be made.

Scientists and engineers do have a role to play, but it is a limited one; for as expert witnesses or advisers they are likely to disagree about some basic facts in the case, and to disagree all the more about value-laden inferences from these facts. These diverse opinions, together with all relevant information, must be put on the table – not in a courtroom, where the judge is unlikely to have even a "speaking familiarity" with the issues, but in a public setting: before the decision makers directly responsible to the citizenry; in forums that allow public input and participation; before regulatory agencies; and above all before elected legislatures which, in a democracy, are precisely the bodies intended to make the value choices. Judge Bazelon concludes that this may be a far slower and more cumbersome process than the tempting alternative, where the decision is made by a very few outside the public arena; but that alternative is exactly what must be avoided. He quotes John Stuart Mill:

> Even if the received opinion be not only true, but the whole truth; unless it is suffered to be, and actually is, vigorously and earnestly contested, it will, by most of those who receive it, be held in the manner of a prejudice, with little comprehension or feeling of its rational grounds.

We come here to Thomas Jefferson's lifelong insistence, and the very kernel of his teachings – that citizens are the only safe guardians of their liberty and pursuit of happiness: Not even with the best will in the world can the judges and the courts, the legislators in their ill-attended sessions, or the scientists, engineers, and other specialists advising decision makers play this role for them. All these together are only part of the framework within which citizens can make their franchise effective. To do that, however, requires an adequate level of knowledge, which must be achieved through a "systematic plan of general education" in schools and colleges and later through continued self-study. To James Madison he wrote in December 1787, while discussing the proposed Constitution: "Above all things I hope the education of the common people will be attended to; convinced that on their good we may rely with the most security for the preservation of a due degree of liberty."

For those who are able, Jefferson repeatedly proposed – for the same ultimate purpose – courses of study that were not far from his own at William and Mary, and which put to shame much of what now passes for a typical curriculum. To one of his correspondents (William Green Munford, 1799), who asked about a good reading list for mathematics,

Jefferson recommended some Euclid, some Archimedes, trigonometry ("There is scarcely a day in which [every person] will not resort to it for some of the purposes of common life"), the science of calculation through cube roots, quadratic equations, logarithms, and "fluxions" [calculus] as an "added luxury." He continued, "There are other branches of science, however, worth the attention of every man" and necessary for "our character as well as comfort": "Astronomy, botany, chemistry, natural philosophy, natural history, anatomy, [at least] to possess their general principles and outlines, so that we may be able to amuse ourselves and inform ourselves further in any of them as we proceed through life and have occasion for them." His use of the term "science" shows that he had in mind the balance of the major branches of knowledge; in addressing Joseph Priestley (in January 1800) his list of sciences includes "Politics, Commerce, History, Ethics, Law, Arts, Finearts."

Not for Jefferson the image of the mind at the end of its tether! On the contrary, he confesses freely to believe, with Condorcet, that man's "mind is perfectible to a degree of which we cannot as yet form any conception." To think otherwise, to harbor the "cowardly idea" that the human mind is "incapable of further advances" was to embrace a doctrine that fitted well the purposes of the "present despots of the earth, and their friends." For in Jefferson's view, the main objective, "the freedom of the human mind," is secured only if that mind is educated or engaged in self-improvement. It is in this sense that (in a letter of 1795 to François D'Ivernois) he identified freedom as "the first-born daughter of science"; conversely, as he told Joseph Priestley in 1810, "ignorance puts everything into the hands of power and priestcraft."

Thus the primary purpose of improving knowledge and understanding was, for Jefferson, not to produce a cadre of scientists and literati; not to design better "protection against foreign power"; not even its necessity "for our character as well as comfort," important as all these were. The foremost purpose, as he put it, lay in the importance of knowledge and informed debate "to the preservation of our republican government."

We would do well to remind ourselves today of Jefferson's threefold preoccupation: that the highest good is the exercise by individuals of their natural rights; that the inevitable excesses of power, if unchecked, lead to oppressive regimes; and that the chief safeguard against this course of degeneracy is the wide dissemination of liberal education. He stated this credo most clearly in his courageous "Bill for the More General Diffusion of Knowledge" (1778). It begins with a manifesto that deserves to be inscribed over the doors of our schools:

> Experience hath shewn, that even under the best forms [of government], those entrusted with power have, in time, and by slow operations, perverted it into tyranny; and it is believed that the most effectual means of preventing this would be to illuminate, as far as practicable, the minds of the people at large, and more especially to give them knowledge of those facts, which history exhibiteth, that, possessed thereby of the experience of other ages and countries, they may be enabled to know ambition under all its shapes, and prompt to exert their natural powers to defeat its purposes. . . . Whence it becomes expedient for promoting the publick happiness that those persons, whom nature hath endowed with genius and virtue, should be rendered by liberal education worthy to receive, and to be able to guard the sacred deposit of the rights and liberties of their fellow citizens. . . .

One may quarrel about the details of a realistic education that could safeguard the mind's freedom and effectiveness, now that the nation has grown vastly from the three million inhabitants of Jefferson's predominantly rural, isolated America. Even allowing for the sophistication of modern knowledge, the complexities of modern problems, and the potentially more calamitous results of bad decisions, one may consider his curriculum "worthy of the attention of every man" to be unreasonably demanding. However, national policy does respond to its leaders' expectations.

Yet the evidence points to a catastrophic reduction of the normal expectation and preparation in the education of our future citizens. The findings are chilling in every field, and those for the sciences merely typical. A carefully assembled study by the National Academy of Sciences for the President in 1980 concluded that the current trend is a tailspin "toward virtual scientific and technological illiteracy." Moreover, as if to thumb its nose at Fate, the next administration began in 1981 to phase out all federal support for science education at the college and precollege levels – the only industrial country to do so.

The need for an informed and confident citizenry to assure that the processes of democracy continue to work has not diminished. The very opposite is the case. In addition to adopting the role of alert watchdog of national and local policy, we all bear the burden of making personal decisions whose soundness depends on our ability at least to ask questions about their technical context. Everyone is caught in this vortex: the engineer who discovers in midcareer that, while he knows how to design a new plant, as he was trained to do, he now has to learn how to evaluate the ecological impact; the administrator forced to deal with ambiguous scientific data or probabilistic risk–benefit analyses in some economic dispute; the board chairman as well as the greenest worker, when the old office is turned over to electronic machinery and the shop floor is automated; the parents having to decide what processed food to feed their child; the doctor or clergyman when faced with ethical problems raised by new technologies that allow us to initiate, prevent, or end a life at will.

If one has no base on which to formulate probing questions, can one actually give informed consent to planned surgery? How can jobs be designed for people who are not skilled or numerate? How is the issue of privacy affected when bank accounts are absorbed into the electronic fund transfer system? How does one react in an informed way – neither hysterically nor techno-enthusiastically – to a plan for siting a power plant nearby, whether nuclear or not?

There are a few inspiring examples of collaborations between scientists and the public, bringing enlightened solutions to complex technical problems that face the nation. One is the cessation in the early 1960s of nuclear weapons testing in the atmosphere, chiefly as the result of a major educational campaign mounted by scientists and others concerned about the health dangers of radioactive fallout. (One Congressman was reported to say: "When parents wrote me that Strontium-90 from fallout was getting into their children's teeth and bones, I knew I had to pay attention: they were spelling Strontium correctly.") A second, more recent example was the debate about the safety of laboratory experimentation with recombinant DNA that was finally resolved in a manner satisfactory to practically all involved. Both these and a few other such instances had in common patient educational efforts and protracted debate among scientists and groups of laymen.

For just that reason, these are the exceptions. As a rule, there is not enough time or talent to build up successful communication between groups of concerned scientists and voters to help form a national decision agreeable to each. Technical details are (or at least seem) so complex to the layman, and their initial base of knowledge is so frail, that serious effort is discouraged. This is most evident today in the difficulty of conducting a meaningful debate on weapons policy. The arms race is full of absurdities, not the least being that it is a grotesque offspring of a perverse coupling, in each country, between international, unifying science and parochial, aggressive nationalism. The worst of these absurdities is that the arms race, with all its costs and dangers, has long since ceased to be reasonable on grounds of national security. Larger and more expensive systems are being designed and deployed that at best will have only marginal value in terms of security, and are far more likely to divert resources and destabilize the fragile geopolitical equilibrium.

The purely technical facts, widely accepted among competent evaluators, and accessible with some effort to any seriously interested citizen, point to the wisdom of calling a halt, especially to the accumulation of nuclear arms. Indeed, beyond one hundred or so weapons on each side, deterrence is saturated. What drives the race now, in both superpowers, are political and perceptual rather than technical factors, perceptions of power rather than physical realities. It would be a triumph if this fact could be made clear to the citizens of the nuclear powers. One of the main obstacles is surely the widespread inability or unwillingness to handle key concepts of a technical kind. Even political leaders with the right intentions and good scientific advice have been disabled by the fear of backlash. We now learn from those close to President Kennedy that in July 1963 he felt unable to follow the recommendation of his scientific-technical civilian advisers to sign an agreement on a complete nuclear weapons test ban with the Soviets, including an end to tests of significant size not only in the air, sea, and outer space, but also underground. Such a comprehensive agreement – which Eisenhower had also considered essential – could have changed world history by containing the nuclear weapons race at that point, and incidentally would have preserved the technological advantage the United States had then. The Soviets, in a notable departure from past positions, were prepared to accept two or three annual inspections on their soil for monitoring suspected cheating. Technically this was sufficient, for it provided a high probability of being able to discover really dangerous new developments (involving a long series of tests, rather than one or two that might escape notice). Moreover, better seismographic monitoring at a distance without on-site inspection could be, and soon was, achieved to provide further deterrence against breaking the offered agreement.

Yet Kennedy did not dare ask Congress for such a treaty. He felt, probably correctly, that despite all its advantages, the proposal would be rejected because it was based on concepts of probability – one of the notorious blanks in modern education. As the ensuing debate showed, a majority of citizens, unfamiliar with the rudiments of probability, required a higher (and to the Soviets, unacceptable) number of on-site inspections, in order to feel that they knew "for certain" – even though the only certainty was that the lack of a comprehensive ban would assure the dangerous and costly continuation of the nuclear menace, the inevitable increase in the size, variety, and proliferation of those weapons, and the introduction of additional systems.

The Academy report I quoted earlier warned that "important national decisions involving scientific knowledge will be made increasingly on the basis of ignorance and misunderstanding." As the divergence widens between those who make policy and citizens who lack the knowledge to assess their proper interests, the threat increases: the nation is in danger of being torn in two. The wound already felt by sensitive humanists such as Trilling must sooner or later become a traumatic separation – the most ironic cost of the advance of the modern science/engineering complex. On one side of the gulf will be a relatively small, technically trained elite, consisting chiefly of scientists, engineers, technicians, and other highly skilled individuals, amounting to a few percent of the population. As an increasing proportion of major decisions have a scientific/technical component, they will supply the new potentials as well as advice on how to direct and use them. On the other side will be the huge majority of the people, without sufficient language, tools, or methods to reason or to argue with the experts, to check on the options they present, or to counter either their technical enthusiasms or their doomsday warnings. That majority will effectively place itself in the hands of the elite, perhaps sinking quietly into the comforts and amusements which technology has helped provide. Some cynics may even welcome such a state of affairs, for the ignorant tend to be easier to govern and to divert into militant philistinism.

These new illiterates will be slaves with respect to the key issue of self-governance, a possibility expressed by Cardinal Newman in the *Idea of a university*: "Not to know the relative disposition of things is the state of slaves." For them, in a grotesque reversal of the Enlightenment dream, the answer to the question, "Where is the modern advance of knowledge and power taking us?" will be, "Into a new slavery."

I have chosen that word advisedly. The historian John Hope Franklin, in his 1976 Jefferson Lecture, reminded his audience of the conditions that helped keep slavery intact in early America. One of these was the denial in the eighteenth century that slaves could *reason.* Jefferson himself once expressed the opinion, although "hazarded with great diffidence," that among the slaves "one could scarcely be found capable of . . . comprehending the investigation of Euclid." That was his touchstone. For many eighteenth-century thinkers, slaves would not come under the Law of Nature which, as John Locke had said, is Reason, and from which law Locke derived the natural rights. In Locke's *Second treatise on government*, which Jefferson claimed to know almost by heart as he drafted the Declaration "without the help of books," Locke taught that man has not the liberty to renounce liberty, to renounce being a naturally free being, and to make himself a slave of others. To be a slave, in Locke's words, is to be "degraded from the common state of rational creatures," hence a species apart.

The slaves of the eighteenth century were not given the choice, were not responsible for their shackles. Our new slaves will be different in this respect; for drawing one's mind back from one's own freedom and powers amounts to a willful self-estrangement, a voluntary renunciation of self-government on the hard issues that determine the fate of a people.

I have outlined the long-range consequences of the intellectual and political costs facing us, largely by-products of the accelerating advancement of knowledge against the background of citizens unable to understand enough to take command of their own destiny. It is a stark picture, and the practical obstacles for the more obvious remedies are formidable. Yet I would not have addressed this issue had I thought the present course uncorrectable. I have referred to a few of the tools for establishing a new equilibrium, ranging from educational programs to the relocation of at least some basic research into areas where there is some motivation to pursue more Jeffersonian Research Programs, alongside the current Newtonian and Baconian ones. There are others; the nation does not lack good ideas. Rather, it is the scale and seriousness of our current efforts which are inadequate. An assertion of national will and leadership is sorely needed to learn how to live in the modern age while preserving one's dignity and self-governance.

The idea of the "Learning Society" has been the mainspring of every period of civilization, from Athens and the Renaissance to the founding of our republic; it must now be broadened to encompass not only Jefferson's person of "genius and virtue," but the whole population. In this way, the hope of the Enlightenment will not be dissipated, and the energies characterizing both the people and contemporary scholarly and scientific advance will be directed to the benefit of each.

◆·◆·◆·◆·◆

In 1812, Jefferson, then seventy, replied from Monticello to his friend John Adams. Downcast at that dark point in history, he wrote:

As for France and England, with all their preeminence in
science, one is the den of robbers, and the other of pirates.
And if science produces no better fruits than tyranny, murder,
rapine and destitution of national morality, I would rather
wish our country to be ignorant, honest and estimable as our
neighboring savages are.

But wither is senile garrulity leading me? Into politics, of
which I have taken final leave. I think little of them, and say
less. I have given up newspapers in exchange for Tacitus and
Thucydides, for Newton and Euclid; and I find myself much
the happier.

But Jefferson's native optimism prevails all the same, an optimism we can share if we do not neglect the tasks before us. He tells John Adams:

[Your letter] carries me back to the times when, beset with
difficulties, we were fellow laborers . . . , struggling for what is
most valuable to man – his right of self-government. . . . We
rode through the storm with heart in hand, and made a happy
port. Still we do not expect to be without rubs and difficulties;
and we have had them. . . . And so we have gone on, and so we
shall go on, puzzled and prospering beyond example in the
history of man. And I do believe we shall continue to grow,
to multiply and prosper, until we exhibit an association,
powerful, wise and happy, beyond what has yet been seen
by man.

NOTES

Chapter 1. Thematic presuppositions and the direction of scientific advance

1 Philipp Frank, *Einstein: His life and times* (New York: Knopf, 1947), p. 217. As his correspondence with Frederick Lindemann (kept at the Einstein Archives) shows, Einstein was "particularly pleased" to enter into what he hoped would be "regular contact" with Oxford, and he seems to have considered this lecture as part of that process. Indeed, Einstein added to the prefatory sentence cited above: "May I say that the invitation makes me feel that the links between this University and myself are becoming professionally stronger?" At that time, Einstein had made up his mind not to return to Germany. But he had not yet decided, among various possibilities, where to settle.

2 It is of some importance to note here the publication history of Einstein's Herbert Spencer Lecture – a confusing history, although in that respect by no means different from that of many of Einstein's important essays. Einstein read his lecture in English, apparently the first time he had dared to do so at Oxford. As we know from his correspondence and diary of that time, he was studying English, but felt that he had a quite incomplete mastery of the language. The original manuscript of Einstein's lecture was in German, and has been published in his collection *Mein Weltbild* (Frankfurt am Main: Ullstein Verlag, 1977), pp. 113–19, under the title "Zur Methodik [*not* Methode] der theoretischen Physik." In the English version, as actually delivered, Einstein acknowledged his "thanks to my colleagues at Christ Church, Mr. Ryle, Mr. Page, and Dr. Hurst, who helped me – and perhaps a few of you – by translating into the English the lecture which I wrote in German."

Unfortunately, the English translation, as published as a small booklet by Oxford University Press (1933), left a good deal to be desired. Key portions of the original manuscript were rendered quite freely. Perhaps for this reason, a different English translation was prepared (by Sonja Bargmann) when Einstein later published a collection of his essays under the title *Ideas and opinions* (hereafter referred to as *I.O.*) (New York: Dell, 1954), pp. 270–6. In quoting from Einstein's Spencer Lecture, and indeed from his other publications, I have gone back to the corresponding original German essays and prepared my own translations where necessary.

305

3 Filmer S. C. Northrop, "Einstein's conception of science," in *Albert Einstein: Philosopher-scientist*, Paul A. Schilpp, ed. (Evanston, IL: Library of Living Philosophers, 1949), p. 407.

4 Quoted in Richard K. Gehrenbeck, *C. J. Davisson, L. H. Germer, and the discovery of electron diffraction*, Ph.D. dissertation, University of Minnesota, 1973, pp. 343–4.

5 Gerald Holton, *Thematic origins of scientific thought: Kepler to Einstein*, hereafter referred to as *Thematic origins* (Cambridge, MA: Harvard University Press, 1973), pp. 190, 234–5.

6 *Albert Einstein, Hedwig und Max Born, Briefwechsel, 1916–1955*, Kommentiert von Max Born (München: Nymphenburger Verlagshandlung, 1969), pp. 257–8; *The Born–Einstein letters*, Correspondence between Albert Einstein and Max and Hedwig Born from 1916 to 1955 with commentaries by Max Born, trans. by Irene Born (New York: Walker, 1971), p. 192.

7 After the observation of the bending of light, published in November 1919, Einstein amended this sentence for the edition printed in 1920: now there remained only one consequence drawn from the theory which had not been observed (the red shift of spectral lines); but he added, "I do not doubt at all that this consequence of the theory will also find its confirmation soon." Albert Einstein, *Über die spezielle und die allgemeine Relativitätstheorie*, 7th ed. (Braunschweig: Vieweg, 1920), p. 70.

It is in this way that we must understand those other isolated places in Einstein's writings where he seems to advise that we must throw away the whole hard-won theory when there appears a single disproof of one of its predictions. A similar example is in an article on relativity written at the request of *The Times* of London in November 1919, at the height of the astounding euphoria about his work, when the British eclipse expedition had just produced its successful experimental results. Toward the end he listed the three observable consequences then expected from his theory, and added (*I.O.*, p. 232): "The chief attraction of the theory lies in its logical completeness. If a single one of the conclusions drawn from it proves wrong [*unzutreffend*], it must be given up; to modify it without destroying the whole structure seems to be impossible."

However, one of his three predictions, concerning the value of the displacement of spectral lines toward the red end of the spectrum in the light from stars with large mass, was by no means easy to test, and the shift was found to be neither systematic nor of the predicted amount. Yet Einstein persisted.

8 I analyzed this case in "Subelectrons, presuppositions, and the Millikan–Ehrenhaft dispute," Chapter 2 of my book, *The scientific imagination: Case studies* (Cambridge: Cambridge University Press, 1978), pp. 25–84.

9 R. A. Millikan, "On the elementary electrical charge and the Avogadro constant," *Physical Review*, 2 (1913): 109–43, completed June 2, 1913. See also Millikan's *The electron: Its isolation and measurement and the determination of some of its properties* (Chicago: University of Chicago Press, 1917), p. 106.

10 Lest it be thought that Millikan was only lucky in guessing which of the data were really usable, I hasten to point out that he continued to exhibit his skill under much more difficult circumstances immediately after this work on the electron. He re-

sumed his experiments on the photoelectric effect, for which he became best known. For ten years he worked with a wrong presupposition that light did not exhibit the quantization of energy. But in the end, he proved the quantum hypothesis experimentally – as he said in his Nobel Prize address, "contrary to my own expectation."

11   For example, A. Einstein, *I.O.*, op. cit. (n. 2), p. 272. The quotations in the next six paragraphs are from the same source, pp. 273–276.

12   In the essay "*Physik und Realität*," 1936, trans. as "Physics and reality," in Einstein, *I.O.*, op. cit. (n. 2), p. 294.

13   In Pierre Speziali, ed., *Albert Einstein, Michele Besso, correspondence 1903–1955* (Paris: Hermann, 1972), p. 527.

14   Isaiah Berlin, *Concepts and categories* (New York: Viking Press, 1979), p. 159.

15   Albert Einstein, "Autobiographical notes," in *Albert Einstein: Philosopher-scientist*, op. cit. (n. 3), p. 53.

16   Albert Einstein, "Induktion und Deduktion in der Physik," *Berliner Tageblatt* (Supplement), December 25, 1919.

17   Albert Einstein, "Reply to criticisms," in *Albert Einstein: Philosopher-scientist*, op. cit. (n. 3), pp. 673–4. See also p. 678: "categories are necessary as indispensable elements of thinking."

18   A brief survey of thematic analysis is provided in the Introduction and Chapter 1 of Gerald Holton, *The scientific imagination: Case studies*, op. cit. (n. 8).

19   Nicholas Copernicus, *On the revolutions*, Jerzey Dobrzycki, ed., Edward Rosen, trans. (Baltimore: Johns Hopkins University Press, 1978), p. 4.

20   Max Planck, "Verhältnis der Theorien zueinander," in *Die Kultur der Gegenwart*, Paul Hinneberg, ed., Part III, vol. 1 (Leipzig: B. G. Teubner, 1915), p. 737.

21   J. T. Merz, *A history of European thought in the nineteenth century* (London: Blackwood, 1904), I, pp. 251–2.

22   Ludwig Büchner, *Kraft und Stoff: Empirisch-naturphilosophische Studien*, 9th ed. (Leipzig: Theodor Thomas, 1867), p. 89.

23   Ernst Mach, *Die Mechanik in ihrer Entwicklung, historisch-kritisch dargestellt*, 2nd ed. (Leipzig: F. A. Brockhaus, 1889), pp. 437–8.

24   Compare "Aufruf," *Physikalische Zeitschrift*, 13 (1912): 735–6; and Friedrich Herneck, "Albert Einstein und der philosophische Materialismus," *Forschungen und Fortschritte*, 32 (1958): 206. I thank Dr. Herneck for kindly making available to me a copy of the original typescript of the *Manifesto*.

25   Albert Einstein, "Motiv des Forschens"; a rather loose English translation was published in Einstein, *I.O.*, op. cit. (n. 2), pp. 224–7.

26   A. Einstein, "Autobiographical notes." in *Albert Einstein: Philosopher-scientist*, op. cit. (n. 3), pp. 59–61, 81. Emphases in original.

  In the Spencer Lecture, Einstein raises this whole problem only gently and at the end, by saying: "Meanwhile the great stumbling block for a field theory of this kind lies in the conception of the atomic structure of matter and energy. For the theory is fundamentally non- atomic insofar as it operates exclusively with continuous functions of space," unlike classical mechanics which, by introducing as its most important element the material point, does justice to an atomic structure of matter. He does see a way out: "For instance, to account for the atomic character of electricity the field equations need only lead to the following conclusion: The region of three-dimensional space at whose boundary electrical density vanishes everywhere always contains a total electrical charge whose size is represented by a whole number. In the continuum theory, atomic characteristics would be satisfactorily expressed by integral laws without localization of the entities which constitute the atomic structure." In referring to the total electric charge whose size is represented by a whole number, he points of course to the result of R. A. Millikan's work.

27 Einstein, in *I.O.*, op. cit. (n. 2), p. 272.
28 Einstein, "Physik und Realität" ["Physics and reality," in Einstein, *I.O.*, op. cit. (n. 2), p. 318.]
29 Indeed, Joseph Needham may well be right that the development of the concept of a unified natural science depended on the preparation of the ground through monotheism, so that one can understand more easily the reason that modern science rose in seventeenth-century Europe rather than, say, in China.
30 The case is quite general. Thus, Kepler's world was constructed of three overlapping thematic structures, two ancient and one new: the universe as theological order, the universe as mathematical harmony, and the universe as physical machine. Newton's scientific world picture clearly retained animistic and theological elements. Lorentz's predominantly electromagnetic world view was really a mixture of Newtonian mechanics, as applied to point masses, determining the motion of electrons, and Maxwell's continuous-field physics. Ernest Rutherford, writing to his new protégé, Niels Bohr, on March 20, 1913, gently scolds him: "Your ideas as to the mode of origin of spectra in hydrogen are very ingenious and seem to work out well: but the mixture of Planck's ideas [quantization] with the old mechanics make it very difficult to form a physical idea of what is the basis of it." In fact, of course, Bohr's progress toward the new quantum mechanics via the correspondence principle was a conscious attempt to find his way stepwise from the classical basis.
31 Einstein, "On the theory of relativity," in *I.O.*, op. cit. (n. 2), p. 246.

Chapter 2. Einstein's model for constructing a scientific theory

1 Albert Einstein, "Ernst Mach," *Physikalische Zeitschrift*, *17*, (1916): 101–4.
2 Albert Einstein, "Reply to criticisms," in *Albert Einstein: Philosopher-scientist*, Paul A. Schilpp, ed. (Evanston, IL: The Library of Living Philosophers, 1949), pp. 683–4. This book includes (pp. 3–94) Einstein's "Autobiographical notes"; a number of essays on Einstein's work by scientists and philosophers; and (pp. 665–88) a set of supplementary comments by Einstein.
3 Albert Einstein, *Ideas and opinions*, new translations and revisions by Sonja Bargmann (New York: Crown, 1954, and, with somewhat different pagination, New York: Dell, 1954). These essays are based on *Mein Weltbild*, a collection of Einstein's essays, edited by Carl Seelig, and other sources.

    Ideas and opinions will hereafter be referred to as *I.O.* in the text. Among the essays cited are the following (all in the Crown edition):

        "Remarks on Bertrand Russell's Theory of Knowledge," 1944 (pp. 18–24)
        "A mathematician's mind," 1945 (Letter to J. S. Hadamard) (pp. 25–6)
        "Principles of theoretical physics," 1914 (pp. 220–23)

"Principles of research," 1918 (better: "Motive of research") (pp. 224–7)

"What is the theory of relativity?," 1919 (pp. 227–32)

"Geometry and experience," 1921 (pp. 232–46)

"On the method of theoretical physics," 1933 (Herbert Spencer Lecture) (pp. 270–6)

"The problem of space, ether, and the field in physics," 1930–1934 (pp. 276–85)

"Physics and reality," 1936 (pp. 290–323)

"The fundaments of theoretical physics," 1940 (pp. 323–35)

"On the generalized theory of gravitation," 1950 (pp. 341–56)

4   Einstein wrote the essay in 1949 as the opening article for the book *Albert Einstein: Philosopher-scientist*, op. cit. (n. 2). I have analyzed the early portion of the "Autobiographical notes" in G. Holton, "What, precisely, is 'thinking'? Einstein's answer," *Einstein: A centenary volume*, in A. P. French, ed. (London: Heinemann. Cambridge, MA: Harvard University Press, 1979).

5   Throughout this paper, parentheses within quotations are given as they appear in the original, and square brackets identify editorial additions or explanations.

6   Einstein, "Autobiographical notes," in *Albert Einstein: Philosopher-scientist*, op. cit. (n. 2), pp. 11–12, and often elsewhere.

7   *I.O.*, pp. 307, 309; cf. letter to Besso, August 28, 1918, quoted in Gerald Holton, *Thematic origins of scientific thought: Kepler to Einstein* (Cambridge, MA: Harvard University Press, 1973), p. 229.

8   Ibid., pp. 254, 286.

9   Einstein, "Autobiographical notes," in *Albert Einstein: Philosopher-scientist*, op. cit. (n. 2), p. 21.

10 Quoted in Holton, op. cit. (n. 7), p. 377.

11 Einstein, "Autobiographical notes," in *Albert Einstein: Philosopher-scientist*, op. cit. (n. 2), p. 13.

12 Ibid., p. 674.

13 Ibid., pp. 13, 7.

14 However, Mach, with W. Ostwald, is scolded by Einstein (ibid., p. 49) for his "positivistic philosophical attitude" which misled them into opposing atomic theory. They were victims of "philosophical prejudice," chiefly "the faith that facts by themselves can and should yield scientific knowledge without free conceptual construction."

15 Einstein, "Ernst Mach," op. cit. (n. 2), pp. 101–4.

16 Einstein, "Autobiographical notes," in *Albert Einstein: Philosopher-scientist*, op. cit. (n. 2), p. 13.

17 Ibid., pp. 9, 11.

18 For a discussion of the concept of "suspension of disbelief," see Gerald Holton, *The scientific imagination: Case studies* (Cambridge: Cambridge University Press, 1978), pp. 71–2. Even the "dean" of logical positivists of his day, Hans Reichenbach, might have agreed, for he said "The physicist who is looking for new discoveries must not be too critical; in the initial stages he is dependent on guessing, and he will find his way only if he is carried along by a certain faith which serves as a directive for his guesses," etc. Hans Reichenbach, "The philosophical significance

of the theory of relativity," in *Albert Einstein: Philosopher-scientist*, op. cit. (n. 2), p. 292. But he went on to deny that such mechanisms can or should be of interest to "the philosopher of science."

19   *Stud. Hist. Phil. Sci. 8* (1977): 49–60. See also P. Mittlestaedt, "Conventionalism in special relativity," *Foundations of physics, 7* (1977): 573–83.

20   W. Heisenberg, *Physics and beyond* (New York: Harper and Row, 1971), p. 63.

21   Einstein, "Autobiographical notes," in *Albert Einstein: Philosopher-scientist*, op. cit. (n. 2), p. 22.

22   Ibid., pp. 20, 22, 21. Wherever possible I have checked the published English-language translation of material Einstein published first in German, and where necessary have corrected the translation, as in this case.

23   Ibid., p. 23.

24   Ibid., pp. 21–3.

25   Ibid., pp. 23–5.

26   The commonly agreed-upon structure of writing scientific papers for publication, which makes it seem that the gathering of data and induction from them formed the beginning of scientific work, has prompted P. B. Medawar to call the scientific paper a "fraud" and a "travesty of the nature of scientific thought." P. B. Medawar, "Is the scientific paper a fraud?," *The listener* (1963): 377–8.

27   A. Einstein, "Considerations concerning the fundamentals of theoretical physics," *Science, 91* (1940): 487, as translated from the original essay, "Das Fundament der Physik" (1940), p. 106 of the revised collection of Einstein's essays, *Aus meinen späten Jahren* (Stuttgart: Deutsche Verlags-Anstalt, 1979). The latter gives the original German-language text of this and other essays that had previously been available only in an English translation. It should also be noted here that the earlier (German) edition of this book is to be used with great caution, since it contained German-language *re*translations from the English-language publications.

28   A. Einstein, "Induktion und Deduktion in der Physik," *Berliner Tageblatt*, December 25, 1919.

29   Einstein, "Reply to criticisms," in *Albert Einstein: Philosopher-scientist*, op. cit. (n. 2), pp. 673–4.

30   For example, refer to Holton, *Thematic origins*, op. cit. (n. 7); Holton, *The scientific imagination*, op. cit. (n. 18); and Chapter 1 of this book.

31   M. Planck, "Prinzip der Relativität," discussion at German Physical Society, March 23, 1906.

32   Einstein, "Autobiographical notes," in *Albert Einstein: Philosopher-scientist*, op. cit. (n. 2), p. 7.

Chapter 3. Einstein's scientific program: the formative years

1   See, for example, the collection *Mein Weltbild*, edited by C. Seelig (Frankfurt: Verlag Ullstein, 1977) and A. Einstein, *Aus meinen späten Jahren* (Stuttgart: Deutsche Verlags-Anstalt, 1979).

2   I shall refer to the first nine papers of Einstein:
     a) "Folgerungen aus den Kapillaritätserscheinungen," *Annalen der Physik*, ser. 4, 4 (1901): 513–23.

b) "Über die thermodynamische Theorie der Potentialdifferenz zwischen Metallen und vollständig dissoziierten Lösungen ihrer Salze, und eine elektrische Methode zur Erforschung der Molekularkräfte," *Annalen der Physik*, ser. 4, *8* (1902): 798–814.

c) "Kinetische Theorie des Wärmegleichgewichtes und des zweiten Hauptsatzes der Thermodynamik," *Annalen der Physik*, ser. 4, *9* (1902): 417–33.

d) "Eine Theorie der Grundlagen der Thermodynamik," *Annalen der Physik*, ser. 4, *11* (1903): 170–87.

e) "Zur allgemeinen molekularen Theorie der Wärme," *Annalen der Physik*, ser. 4, *14* (1904): 354–62.

f) "Eine neue Bestimmung der Moleküldimensionen," (Bern: Wyss, 1905), 21 pp.

g) "Über einen die Erzeugung und Verwandlung des Lichtes betreffenden heuristischen Gesichtspunkt," *Annalen der Physik*, ser. 4, *17* (1905): 132–48.

h) "Die von der molekularkinetischen Theorie der Wärme geforderte Bewegung von in ruhenden Flüssigkeiten suspendierten Teilchen," *Annalen der Physik*, ser. 4, *17* (1905): 549–60.

i) "Zur Elektrodynamik bewegter Körper," *Annalen der Physik*, ser. 4, *17* (1905): 891–921.

3   Quoted by M. Besso in *Albert Einstein, Michele Besso, correspondence 1903–1955*, edited by Pierre Speziali (Paris: Hermann, 1972), p. 550.

4   Not to speak of twenty-two newly discovered abstract reviews of books and papers that Einstein published on these subjects in the *Beiblätter* of the *Annalen der Physik* between 1905 and 1907. See M. J. Klein and A. Needell, "Some unnoticed publications by Einstein," *ISIS 68* (1977): 601–4.

5   Quoted in Carl Seelig, *Albert Einstein, eine dokumentarische Biographie* (Zurich: Europa Verlag, 1954), pp. 61–2.

6   A. Einstein, "Autobiographical notes," in *Albert Einstein: Philosopher-scientist*, Paul A. Schilpp, ed. (Evanston, IL: Library of Living Philosophers, 1949), p. 27. Where necessary I have provided corrected translations of quotations from Einstein's original German essays.

7   *Einstein, Besso, correspondence*, op. cit. (n. 3), pp. 3–4.

8   Ibid.

9   Seelig, op. cit. (n. 1), p. 74.

10 A. Einstein, "Eine neue elektrostatische Methode zur Messung kleiner Elektrizitätsmengen," *Physikalische Zeitschrift*, 9 (1908): 216–17. He had laid the theoretical base for the method in *Annalen der Physik*, 22 (1907): 569–72.

11 The sixth paper, using the sequence given in Margaret Shields' bibliography in *Albert Einstein: Philosopher-scientist*, op. cit. (n. 6). With a brief *Nachtrag*, this paper was published in *Annalen der Physik*, 19 (1906): 289–306.

12 In *Albert Einstein: Philosopher-scientist*, op. cit. (n. 6), p. 47. Even in the *Annalen der Physik*, in which Einstein published all his early papers, there had been an article by F. M. Exner in 1900 that showed that microscopic particles move with greater average speed at higher temperature.

13 *Einstein, Besso, correspondence*, op. cit. (n. 3), p. 14.

14 A similar attitude underlies Einstein's address at the Salzburg meeting in 1909. See "Über die Entwicklung unserer Anschauungen über das Wesen und der Konstitution der Strahlung," *Physikalische Zeitschrift*, *10* (1909): 817–26.

15   In *Albert Einstein: Philosopher-scientist*, op. cit. (n. 6), p. 47.
16   It was, however, perhaps just for that reason, bound to seem "very revolutionary," as Einstein put it in his high-spirited letter to Conrad Habicht in the spring of 1905; see Seelig, op. cit. (n. 5), pp. 88–9.
17   Quoted in M. Klein, "Einstein, specific heats and the early quantum theory," *Science, 148* (1965): 177.
18   "Über die vom Relativitätsprinzip geforderte Trägheit der Energie," *Annalen der Physik, 23* (1907): 371–2.
19   "Zur Theorie der Brownschen Bewegung," *Annalen der Physik, 19* (1906): 372.
20   E.g., in "Entwurf einer verallgemeinerten Relativitätstheorie und einer Theorie der Gravitation. I. Physikalischer Teil," *Zeitschrift für Mathematik und Physik, 62* (1913): 225, and similarly in *Annalen der Physik, 38* (1912): 1059. In his first review article, written in good part for didactic purposes in 1907 (see n. 30), he presented his work on relativity as "the unification of the Lorentzian *theory* and the relativity *principle*" (italics supplied). His reference at the Salzburg 1909 meeting was similar (see op. cit. in n. 14). In the titles of his papers until 1911, Einstein used the term "Relativitätsprinzip" rather than "Relativitätstheorie."
       Incidentally, in the titles of Einstein's first papers on general relativity he repeatedly used "verallgemeinerte Relativitätstheorie," rather than the later phrase "allgemeine Relativitätstheorie." In his correspondence, Einstein continued to use the earlier phrase; cf. his letter to M. von Laue, January 17, 1952 (in the Einstein Archive at Hebrew University in Jerusalem).
21   He used the term informally, e.g., in correspondence with Besso [*Einstein, Besso, correspondence*, op. cit. (n. 3), p. 526]. He showed his willingness to adhere to the basic invariant space-time element *ds* even when it became clear that it threw in doubt the "physical meaning (measurability-in-principle)" of the individual coordinates; see "Entwurf" op. cit. (n. 20), p. 230, and *Annalen der Physik, 35* (1911): 930 ff. And he confessed that, all things considered, the term would have been preferable; in a letter of September 30, 1921, to E. Zschimmer of Jena in the Archive at Hebrew University, Jerusalem, he writes: "Now to the name relativity theory, I admit that it is unfortunate, and has given occasion to philosophical misunderstandings. The name 'Invarianz-Theorie' would describe the research *method* of the theory but unfortunately not its material content (constancy of light-velocity, essential equivalence of inertia and gravity). Nevertheless, the description you proposed would perhaps be better; but I believe it would cause confusion to change the generally accepted name after all this time."
       J. L. Synge expressed himself similarly: "Much as I dislike the name [relativity theory] (I would much prefer to follow Minkowski, but it is now too late)." [In *Albert Einstein's theory of general relativity*, G. E. Tauber, ed. (New York: Crown, 1979), p. 199.] Synge alludes to H. Minkowski's remark in his 1908 talk "Space and Time": "that the word 'relativity-postulate' for the requirement of an invariance . . . seems to me very feeble [*sehr matt*]." Others wrote in the same vein; e.g. Arnold Sommerfeld, "Philosophie und Physik" [1948], in A. Sommerfeld, *Gesammelte Schriften*, Vol. 4 (Braunschweig: Vieweg, 1968), pp. 640–1.
22   I have discussed this method in detail, based on Einstein's writings, in Chapter 2 "Einstein's model for constructing a scientific theory." Needless to say, Einstein's published papers in their architectural details do not necessarily correspond point

for point with the sequence of his actual thought processes in arriving at his conclusions.

23 As his work proceeded, Einstein became more aware of this feature of his method,
and more daring still. See, for example, his "Autobiographical notes": "I have
learned something else from the theory of gravitation. No ever so inclusive collection of empirical facts can ever lead to the setting up of complicated equations. A
theory can be tested by experience, but there is no way from the experience to the
setting up of a theory. . . . Once one has those sufficiently strong formal conditions,
one requires only little knowledge of facts for the setting up of a theory." [*Albert
Einstein: Philosopher-scientist*, op. cit. (n. 6), p. 89.]

Of course, this does *not* mean that Einstein had no interest in experimental facts
as such; indeed he respected greatly some of the "artists" in the field of experimental physics or astrophysics, and often enjoyed puzzling over new experimental
results or apparatus. Moreover, he insisted that the ultimate goal of theory construction must be the detailed and complete coordination of theory and experience
– e.g., in the remark, "A theoretical system can claim completeness only when the
relations of concepts and experienced facts are laid down firmly and unequivocally.
. . . If one neglects this point of view, one can only attain unrealistic systems."
[Quoted in F. Herneck, *Forschungen und Fortschritte*, 40 (1966): 133.]

24 Planck was listed as special editorial consultant on the masthead of the *Annalen*.
Einstein's relativity paper was received at the *Annalen* on June 30, 1905. The editor,
Paul Drude, famous for his writings on light and ether, had just moved from Giessen University to Berlin. Until just two and one-half months earlier, the *Annalen*
requested that all manuscripts be sent to Drude in Giessen. One can only speculate
what the fate of the manuscript might have been there.

25 Walter Kaufmann, "Über die Konstitution des Elektrons," *Annalen der Physik*, 19
(1906): 487–553. Emphasis in original. A preview was given in W. Kaufmann,
"Über die Konstitution des Elektrons," *Sitzungsberichte der königlichen Academie
der Wissenschaften*, 45 (1905): 949–56.

26 As discussed in G. Holton, *Thematic origins of scientific thought* (Cambridge,
MA: Harvard University Press, 1973), pp. 189–90, and in detail in A. I. Miller's
"On the history of the special relativity theory," in *Albert Einstein, his influence on
physics, philosophy and politics*, P. C. Aichelburg and R. Sexl, eds. (Wiesbaden:
Vieweg, 1979), pp. 89–108.

27 Max Planck, "Die Kaufmannschen Messungen," *Physikalische Zeitschrift*, 7
(1906): 753–61.

28 W. Wien, *Über Elektronen*, 2nd ed. (Leipzig: B. G. Teubner, 1909), p. 32. Lorentz,
who never fully accepted Einstein's relativity, was most generous in acknowledging
its power and originality. But he also put his finger on a widely felt dismay with
Einstein's method when he remarked: "Einstein simply postulates what we have
deduced. . ." [H. A. Lorentz, *The theory of electrons* (1909), p. 230].

29 *Physikalische Zeitschrift*, 9 (1908): 762. In the preface of his first edition (1911) of
the first textbook on the relativity theory, *Das Relativitätsprinzip* (Braunschweig:
Vieweg, 1911), p. i. Max von Laue still could not point to incontrovertible experimental evidence in favor of Einsteinian relativity, but stressed the lack of persuasive falsifications, and the argument from congeniality – the two favorable criteria of a theory that Einstein also approved of. Max von Laue wrote:

In the five and a half years since Einstein's founding of relativity theory this theory has gathered attention in growing measure. To be sure, this attention is not always equal to adherence. Many researchers, including some with well-known names, consider the empirical grounds not sufficiently firm. Worries of this kind can of course be helped only through further experiments; in any case this book strengthens the proof that not a single empirical ground exists against the theory.

More extensive is the number of those who cannot find the intellectual content congenial, and to whom particularly the relativity of time, with those consequences that sometimes really appear to be quite paradoxical, seem unacceptable.

30 A. Einstein, "Über das Relativitätsprinzip und die aus demselben gezogenen Folgerungen," *Jahrbuch der Radioaktivität und Elektronik*, 4, 411–62 (dated 1907, but appeared in 1908). I discussed this response of Einstein to Kaufmann first in Holton, op. cit. (n. 26), pp. 234–6; but I include here this paragraph and the next in response to queries about Einstein's reaction, made in the question period following the public delivery of a briefer version of this paper.

31 Einstein continued to assign probabilities, e.g., in the last sentence of *The meaning of relativity*: "Although such an assumption is logically possible, it is less probable than the assumption that there is a finite density of matter in the universe" [Princeton, NJ: Princeton University Press, (1922, 1945), p. 108].

32 A. Einstein, *Les Prix Nobel en 1921–1922* (Stockholm, 1923), and *Nobel Lectures, 1901–1921* (Amsterdam: Elsevier, 1967), pp. 482–90.

33 The point is put in a personal way in one of Einstein's manuscripts in the Archive, entitled "Fundamental ideas and methods of relativity theory, presented in their development," dating from about 1919 or shortly afterward. Speaking about the fact that prior to relativity theory the theoretical interpretation of induction was quite different depending on whether the magnet or the conductor is considered in motion, he confessed that this produced in him, as perhaps in no one else, a discomfort that had to be removed: "The thought that one is dealing here with two fundamentally different cases was for me unbearable [*war mir unerträglich*]. . . . The phenomenon of electromagnetic induction forced me to postulate the special relativity principle." In this way, "a kind of objective reality could be granted only to the electric and magnetic field together."

34 For example, "Physik und Realität," *Journal of the Franklin Institute*, 221, (1936): 317. Cf. also "Über die vom Relativitätsprinzip geforderte Trägheit der Energie," op. cit. (n. 18).

35 *Albert Einstein: Philosopher-scientist*, op. cit. (n. 6), p. 81.

36 "Physik und Realität," op. cit. (n. 34), p. 317.

37 As listed in the section "The role of thematic presuppositions," Chapter 2 in this book.

38 In *Albert Einstein: Philosopher-scientist*, op. cit. (n. 6), p. 23.

39 At the Jerusalem Einstein Centennial Symposium, P. A. M. Dirac gave on March 20, 1979, a paper on "Unification: Aims and principles," in which he said:

It seems clear that the present quantum mechanics is not in its final form. Some further changes will be needed, just about as drastic as the changes which one made in passing from Bohr's orbits to a quantum mechanics. Some day a new relativistic quantum mechanics will be discovered in which we don't have these infinities occurring at all. It might very well be that the new quantum mechanics will have determinism in the way that Einstein wanted. This determinism will be introduced only at the expense of abandoning some other preconceptions which physicists now hold, and which it is not sensible to try to get at now.

So, under these conditions I think it is very likely, or at any rate quite possible, that in the long run Einstein will turn out to be correct, even though for the time being physicists have to accept the Bohr probability interpretation – especially if they have examinations in front of them.

40   "Physik und Realität," op. cit. (n. 34), p. 315.
41   Einstein, letter to M. Solovine, March 30, 1952, copy in the Einstein Archive at Hebrew University in Jerusalem.

Chapter 4. Einstein's search for the Weltbild

1   A. Einstein, "Physik und Realität," *Journal of the Franklin Institute*, *221*, (1936): 316–317 (in translation). In this paper, I am relying primarily on Einstein's own statements rather than on those of his commentators. Also, I am providing corrected translations of them wherever the otherwise available translations make this necessary.

2   A. Einstein, *Mein Weltbild*, Carl Seelig, ed., (Amsterdam: Querido Verlag, 1934). The letters in the Einstein Archive in Jerusalem show that the publisher used the title at the suggestion of Einstein's biographer and son-in-law, Rudolph Kayser. Einstein himself was understandably displeased with the choice of the title. He considered it "tasteless and misleading." This book was, after all, a collection of separate essays, not the presentation of a coherent proposal that would deserve what Einstein had called elsewhere "the proud term 'Weltbild'."

3   C. Geertz, *The interpretation of cultures* (New York: Basic Books, 1973), p. 141.

4   For example, see the section, "What, precisely, is 'thinking'," in A. Einstein, "Autobiographical notes," *Albert Einstein: Philosopher-scientist*, P. A. Schilpp, ed. (Evanston, IL: Library of Living Philosophers, 1949), p. 7, and Einstein's essay "The laws of ethics and the laws of science," serving as preface to Philipp Frank, *Relativity, A richer truth* (Boston: Beacon Press, 1950), also in *Aus meinen späten Jahren* (Stuttgart: Deutsche Verlags-Anstalt, 1979), pp. 53–5.

5   A. Einstein, "In memoriam Marie Curie" (1935) in *Aus meinen späten Jahren* (n. 4), p. 207 (in translation). In a tribute in memory of Marie Curie, he expressed what multitudes appeared to feel about him, too: "It is the ethical qualities of its leading personalities that are perhaps of even greater significance for a generation and for the course of history than the purely intellectual accomplishments. Even these latter are, to a far greater degree than is commonly credited, dependent on the stature of character." Several of the essays cited in this paper may be found, usually in somewhat freer translation, in [A. Einstein], *Ideas and opinions* (New York: Crown, 1954).

6   Erwin Panofsky, *Meaning in the visual arts* (New York: Doubleday, 1955), p. 24.

7   Einstein's "Autobiographical notes," op. cit. (n. 4), p. 95.

8 The events surrounding Einstein's arrival are detailed in Philipp Frank, *Einstein, sein Leben und seine Zeit* (Munich: Paul List Verlag, 1949; reissued by Friedrich Vieweg, 1979). Also, between April 3 and May 30, 1921, the *New York Times* carried a number of detailed accounts of Einstein's visit. Passages quoted below are from the *New York Times's* accounts of April 3, 4, and 5.

9 *Report of the Seventy-ninth Meeting of the British Association for the Advancement of Science, Winnipeg, 1909* (London: John Murray, 1910), pp. 3–29.

10 Einstein's explicit interest in methodology asserted itself quite early, was repeated almost insistently, and continued to the end. He introduced the distinction between "*induktive Physik*" and "*deduktive Physik*" in 1914, in his Inaugural Address in Berlin ("Prinzipien der theoretischen Physik," in *Mein Weltbild* (n. 2), pp. 110–13). The closely related formulation of "*konstruktive Theorie*" vs. "*Prinziptheorie*" appeared in print first in his *London Times* essay of 1919, "What is the theory of relativity?" (original given in n. 2, pp. 127–31).

11 H. A. Lorentz, "Electromagnetic phenomena in a system moving with any velocity less than that of light," *Proceedings of the Academy of Sciences of Amsterdam*, 6 (1904), as reprinted (incompletely) in *The principle of relativity* (New York: Dover, n.d.), pp. 11–34.

12 Among those examining it from the perspective of the history of science are Stephen Brush, Philipp Frank, Charles Gillispie, Stanley Goldberg, Werner Heisenberg, Tetu Hirosige, Max Jammer, Martin Klein, Russell McCormmach, Arthur I. Miller, Stephen Toulmin, and others.

13 W. Kaufmann, "Entwicklung des Elektronenbegriffs," *Verhandlungen der Gesellschaft deutscher Naturforscher und Ärzte* [held Sept. 1901] (Leipzig: F. C. W. Vogel, 1902), pp. 125–6. Two years later he concluded from his experiments that both β-rays and cathode rays "consist of electrons whose mass is purely electromagnetic in nature"; W. Kaufmann, "Über die 'Elektromagnetische Masse' der Elektronen," *Göttinger Nachrichten* (1903), p. 103.

14 A. Einstein, *Aether und Relativitätstheorie* (Berlin: Julius Springer, 1920).

15 "Autobiographical notes" (n. 4), pp. 21–3.

16 A. Einstein, "Über einen die Erzeugung und Verwandlung des Lichtes betreffenden heuristischen Gesichtspunkt," *Annalen der Physik*, ser. 4, *17* (1905): 132–48. There, Einstein had also other reasons for objecting to current theory, e.g., that it led to the problem later termed by Paul Ehrenfest the "ultraviolet catastrophe."

17 A. Einstein, "Über die vom Relativitätsprinzip geforderte Trägheit der Energie," *Annalen der Physik*, 23 (1907): 372.

18 Quoted in Carl Seelig, *Albert Einstein, eine dokumentarische Biographie* (Zürich: Europa Verlag, 1954), pp. 61–62.

19 "Kinetische Theorie des Wärmegleichgewichtes und des zweiten Hauptsatzes der Thermodynamik," *Annalen der Physik*, ser. 4, 9 (1902): 417–33.

20 Einstein, "Über die vom Relativitätsprinzip," op. cit. (n. 17), p. 372.

21 A. Einstein, "Zur Elektrodynamik bewegter Körper," *Annalen der Physik*, ser. 4, *17* (1905): 891–921, given in (approximate) translation in *The principle of relativity* (n. 11), pp. 37–65.

22 "Autobiographical notes" (n. 4), p. 53.

23 Compare A. Einstein's letter to Carl Seelig (February 19, 1955, in Einstein Archive at Hebrew University, Jerusalem), in which he recalls his first paper on relativity in

1905: "What was new in it was the recognition that the significance of the Lorentz transformations went beyond the connection with the Maxwell equations and concerned the nature of space and time in general. Also new was the insight that the 'Lorentz invariance' is a general condition for every physical theory. This was for me of special importance because I had recognized earlier that Maxwell's theory did not describe the microstructure of radiation and therefore was not generally valid."

The last sentence refers to Einstein's paper earlier in 1905, concerning a quantum-physical approach to light (n. 16).

24 As quoted in Ernst Cassirer, *Substance and function, and Einstein's theory of relativity* (New York: Dover, 1953), p. 371.

25 Note that unlike others such as Poincaré, Einstein regarded the principle of relativity neither as a more or less exact empirical truth nor as a statement to be derived from a future theory. It was truly a postulated axiom, and moreover one that extended not only to one branch of physics but to all.

26 It is worth noting that in setting up later the principle of general relativity, Einstein applied the same postulational method he had used in 1905, proceeding by the generalization of the top-most principles.

27 Cf. T. Hirosige, "The ether problem, the mechanistic world view, and the origin of the theory of relativity," *Historical Studies in the Physical Sciences*, 7 (1976): 3–82.

28 A. Einstein, "Physik und Realität," op. cit. (n. 1), pp. 316–17.

29 A. Einstein, "Considerations concerning the fundamentals of theoretical physics," *Science, 91* (1940): 487, as translated from the original essay, "Das Fundament der Physik" (1940), p. 106 of the recently revised collection of Einstein's essays, *Aus meinen späten Jahren* (n. 4). The latter gives the original German-language text of this and other essays that had previously been available only in an English translation. It should also be noted here that the earlier (German) edition of this book is to be used with great caution, since it contained German-language *re*translations from the English-language publications.

30 A. Einstein, "Motiv des Forschens," in *Mein Weltbild*, (n. 2), pp. 108–9 (in translation). Similar passages occur repeatedly later, e.g., in the essay of 1930, "Das Raum-, Aether- und Feld-Problem der Physik" (ibid., p. 144): "The theory of relativity is a fine example for the fundamental character of the modern development of theory. For the initial hypotheses are becoming ever more abstract, remote from experience. In return, however, one is brought closer to the noblest aim of science which is to cover with a minimum of hypotheses or axioms the maximum number of empirical facts through logical deduction."

31 A. Einstein, *Les Prix Nobel en 1921–1922* (Stockholm, 1923), and *Nobel Lectures, 1901–1921* (Amsterdam: Elsevier, 1967), pp. 482–90.

32 In one of his last essays, "Relativity and the problem of space" (Appendix V in the revised edition of A. Einstein, *Relativity, the special and the general theory*, New York: Crown, 1952, p. 150), Einstein comes back to this point: "It appears therefore more natural to think of physical reality as a four-dimensional existence, instead of, as hitherto, the *evolution* of a three-dimensional existence." (Emphasis in original.)

33 Ibid.

34 "Autobiographical notes" (n. 4), pp. 59–61.

35 "Autobiographical notes" (n. 4), p. 81 (Emphases in original).
36 "Das Raum-, Äther- und Feld-Problem der Physik" (n. 30), p. 147.
37 "On the generalized theory of gravitation" (1950), in *Ideas and opinions* (n. 5), p. 352. See also "Das Fundament der Physik" (1940), in *Aus meinen späten Jahren* (n. 4), pp. 112–13. "To construct a theory . . . one must also have a point of view that sufficiently restricts the unlimited variety of possibilities."
38 Op. cit. (n. 14), p. 14.
39 A. Einstein, "Maxwells Einfluss auf die Entwicklung der Auffassung des Physikalisch-Realen," reprinted in *Mein Weltbild*, pp. 159–62.
40 A. Einstein, "Das Fundament der Physik" (n. 29), pp. 106–21.
41 Einstein to von Laue, January 17, 1952, copy of letter in the Einstein Archive at Hebrew University, Jerusalem.
42 A. Einstein, "Reply to criticisms," in *Albert Einstein: Philosopher-scientist*, op. cit. (n. 4), p. 667.
43 "Relativity and the problem of space" (n. 32), p. 157. (Emphasis supplied.)
44 A. Einstein, "Physik und Realität," in *Aus meinen späten Jahren* (n. 4), p. 96.
45 A. Einstein, "Über die Entwicklung unserer Anschauungen über das Wesen und die Konstitution der Strahlung," *Physikalische Zeitschrift*, *10* (1909): 817–28.
46 A. Einstein, "Zur Quantentheorie der Strahlung," *Mitteilungen der physikalischen Gesellschaft Zürich* (1916): 47–62; also in *Physikalische Zeitschrift*, *18* (1917): 121–8. A good discussion is given by Martin J. Klein, "Einstein and the wave-particle duality," *The Natural Philosopher*, *3* (1964): 1–49.
47 A. Einstein, "Das Fundament der Physik" (n. 29), p. 121.
48 This has been pointed out most recently by Eugene Wigner, "The basic conflict between the concepts of general relativity and of quantum mechanics," address given at the meeting of the American Physical Society, April 25, 1979. Professor Wigner added "It is not unreasonable to say that the general relativists' attention is focussed on macroscopic objects, the quantum uncertainties of the position of which are negligible, whereas the quantum theorists' attention is concentrated on microscopic objects, atoms and particles, the gravitational interaction of which he can neglect when compared with the other interactions."
    To this, Einstein might have responded [with a passage in "Physik und Realität," in *Aus meinen späten Jahren* (n. 4), p. 101]: "The introduction of a spacetime continuum may be considered as contrary to nature, in view of the molecular structure of everything which happens on a small scale. Perhaps the success of Heisenberg's method points to a purely algebraic method of nature-description, to the elimination of continuous functions from physics. But then one must in principle give up the use of the spacetime continuum. It is not unthinkable that human ingenuity will find someday methods which make possible proceeding along this path. For the time being, however, this project appears like an attempt to breathe in empty space."
    From 1905 on, when the introduction of discontinuity, in the form of the light quantum, forced itself on him as a "heuristic," hence not fundamental, point of view, Einstein clung to the continuum as a fundamental, thematic conception. It appeared in what he called his Maxwellian program to fashion a unified field theory; and in almost passionate tones it is defended in his letters. Atomistic discreteness and all it entails was a problem, not a solution.

49 P. A. M. Dirac, "Unification: Aims and principles," address given on March 21, 1979 at the Jerusalem Einstein Centennial Symposium, published as "The early years of relativity," in *Albert Einstein, historical and cultural perspectives*, Gerald Holton and Yehuda Elkana, eds. (Princeton, NJ: Princeton University Press, 1982), p. 85.

50 Cf. G. Holton, *Thematic origins of scientific thought: Kepler to Einstein* (Cambridge, MA: Harvard University Press, 1973), pp. 243–6.

51 Op. cit. (n. 39), p. 159.

52 "Autobiographical notes" (n. 4), p. 5.

53 "Das Fundament der Physik" (n. 29), p. 120.

54 Max Planck, quoted in Ernest Lecher, *Physikalische Weltbilder* (Leipzig: Theodore Thomas Verlag, 1912), p. 84.

55 Op. cit. (n. 18), p. 89.

56 A. Einstein, *Über die spezielle und die allgemeine Relativitätstheorie* (Braunschweig: Friedr. Vieweg, 1917), p. 52.

57 "Physik und Realität," in *Aus meinen späten Jahren* (n. 4), pp. 67–9.

58 Op. cit. (n. 39), p. 159.

59 Major scientists have often found themselves besieged with requests to confess to being scientific "revolutionaries," but most of them have disclaimed the label. Even Heisenberg, who came perhaps closer than any other twentieth-century scientist to deserve the label – and who certainly was at least a true scientific radical (see Chapter 8) – persistently and eloquently opposed the facile notion of science as a succession of revolutionary acts. See for example the record of the lengthy "Discussion with Professor Heisenberg," pp. 556–73, in Owen Gingerich, ed., *The nature of scientific discovery* (Washington, DC: Smithsonian Distribution Press, 1975). Heisenberg was there surrounded by people who were practically begging him to accept the accolade, or at least to agree with a popular opinion regarding the scientific growth through revolutions. He steadfastly refused to do so, and also explicitly opposed this attitude.

60 *New York Times*, April 4, 1921.

61 Ibid., April 16, 1921.

62 A. Einstein, "Über Relativitätstheorie," in *Mein Weltbild* (n. 2), p. 131; similarly in his essay for the *London Times* of November 28, 1919, "Was ist Relativitätstheorie?" (ibid., p. 131).

63 A. Einstein, *Über die spezielle*, op. cit. (n. 56), p. 52.

64 H. A. Lorentz, *The Einstein theory of relativity* (New York: Brentano's, 1920), pp. 23–4. Einstein himself did not hesitate to introduce the term *beauty of a theory* into his scientific papers; e.g., *Sitzungsberichte der preussischen Akademie der Wissenschaften*, (1919): 349–56.

Chapter 5. Einstein and the shaping of our imagination

1 Daniel N. Lapedes, ed., *McGraw-Hill dictionary of scientific and technical terms*, 2d ed. (New York: McGraw-Hill, 1978), pp. 512–13. As another measure in the continuing, albeit sometimes only ritualistic, reference made in the ongoing research literature to Einstein's publications, Eugene Garfield has found that during the period 1961–75 the serious scientific journals *in toto* carried no less than 40 million citations to previously published articles. Of these, 58 cited articles stand

out by virtue of having been published before 1930 *and* cited over 100 times each; and among these 58 enduring classics, ranging from astronomy and physics to biomedicine and psychology, 4 are Einstein's. See E. Garfield, *Current Contents*, 21 (1976): 5–9.

2 Quoted from J. J. Thomson, *Reflections and recollections* (London: G. Bell and Sons, Ltd., 1936), p. 431 [italics in original]. See also Philipp Frank, *Einstein: His life and times* (New York: Knopf, 1947), p. 190. Frank's book is one of the good sources for documentation on the reception and rejection of Einstein's theories by various religions and philosophic and political systems, ranging from Cardinal O'Connell's assessment that "those theories [Einstein's as well as Darwin's] became outmoded because they were mainly materialistic and therefore unable to stand the test of time" (p. 262), to the attempt of a Nazi scientist to overcome his aversion sufficiently to "recommend Einstein's theory of relativity to National Socialists" as a weapon in the fight against "materialistic philosophy" (p. 351).

3 These articles, and excerpts from some other publications dealing with the influence of Einstein's work, have been gathered in L. Pearce Williams, ed., *Relativity theory: Its origins and impact on modern thought* (New York: Wiley, 1968). It is a useful compendium, and I am indebted to it for a number of illustrations to be referred to below.

4 John Passmore, *A hundred years of philosophy*, rev. ed. (New York: Basic Books, 1966), p. 332.

5 As reported by Paul M. LaPorte, "Cubism and relativity, with a letter of Albert Einstein," *Art Journal*, 25, no. 3 (1966): 246.

6 Ibid. See also C. H. Waddington, *Behind appearances: A study of the relations between painting and the natural sciences in this century* (Edinburgh: Edinburgh University Press, 1969; Cambridge, MA: MIT Press, 1970), pp. 9–39. At the Jerusalem Einstein Symposium, (1979), Professor Meyer Schapiro presented an extensive and devastating critique of the frequently proposed relation between modern physics and modern art.

7 *Contact*, 4 (1923): 3. I am indebted to Carol Donley's draft paper, "Einstein, too, demands the muse" for this lead and others in the following paragraphs.

8 *Selected essays of William Carlos Williams* (New York: Random House, 1954), p. 283.

9 Ibid., p. 340.

10 J. P. Sartre, "François Mauriac and freedom," in *Literary and philosophical essays* (New York: Criterion Books, 1955), p. 23.

11 Lawrence Durrell, *Balthazar* (New York: E. P. Dutton, 1958), Author's Note, p. 9.

12 Ibid., p. 142.

13 For a good review of details, to which I am indebted, see Alfred M. Bork, "Durrell and relativity," *Centennial Review*, 7 (1963): 191–203.

14 L. Durrell, *A key to modern British poetry* (Norman: University of Oklahoma Press, 1952), p. 48.

15 Ibid., pp. 25, 26, 29.

16 All quotations are from William Faulkner, *The sound and the fury* (London: Chatto and Windus, 1961), pp. 81–177. I thank Dr. J. M. Johnson for a draft copy of her interesting essay, "Albert Einstein and William Faulkner," and have profited from some passages even while differing with others.

17 The chapter is shot through with references to light, light rays, even to travel "down the long and lonely light rays."

18 In *Les Prix Nobel en 1950*, Stockholm, Imprimierie Royale, 1951, p. 71.

19 Jean Piaget, *The child's conception of time* (New York: Ballantine, 1971), p. vii.

20 Jean Piaget, *Genetic epistemology* (New York: Columbia University Press, 1970), p. 69; see also p. 7.

21 Jean Piaget, *Psychology and epistemology* (New York: Grossman, 1971), p. 82; see also pp. 10, 110. A similar statement is to be found in Piaget's *Six psychological studies* (New York: Vintage Books, 1968), p. 85.

22 For example, Jean Piaget, with Bärbel Inhelder, *The child's conception of space* (New York: W. W. Norton, 1967), pp. 232–3; *The child's conception of time* (London: Routledge and Kegan Paul, 1969), pp. 305–6; *Biology and knowledge* (Chicago: University of Chicago Press, 1971), pp. 308, 337, 341–2. I wish to express my thanks to Dr. Katherine Sopka for help in tracing these references.

23 Werner Heisenberg, "The representation of nature in contemporary physics," *Daedalus* (Summer 1958): 103–5.

24 Lionel Trilling, *Mind in the modern world: The 1972 Jefferson Lecture in the Humanities* (New York: Viking Press, 1972), pp. 13–14.

25 Albert Einstein, *Über die spezielle und die allgemeine Relativitätstheorie, gemeinverständlich* (Braunschweig: Vieweg, 1917). It was often translated and to this day is perhaps his most widely known work.

26 Albert Einstein and Leopold Infeld, *The evolution of physics* (New York, Simon and Schuster, 1938).

27 Albert Einstein, *On the method of theoretical physics* (Oxford, Clarendon Press, 1933); see also Chapter 1.

28 Albert Einstein, "Physics and reality," *Journal of the Franklin Institute*, 221 (1936): 349–82.

29 First published in English in 1933, in his *The modern theme* (New York, W. W. Norton).

30 Ibid., pp. 135–6.

Chapter 6. Physics in America, and Einstein's decision to immigrate

1 See, for example, Philipp Frank, *Einstein, his life and times* (New York, 1947), and Alan Beyerchen, *Scientists under Hitler: Politics and the physics community in the Third Reich* (New Haven, 1977).

2 Letter of Sept. 9, 1920, quoted in Siegfried Grundmann, "Die Auslandsreisen Albert Einsteins," *NTM, Schriftenreihe für Geschichte der Naturwissenschaften, Technik und Medizin* 2, no. 6 (1965):4.

3 Letter of Einstein to the German Embassy in Tokyo; quoted in report of the embassy to its Ministry of Exterior, Berlin, Jan. 3, 1923; quoted in Grundmann, "Die Auslandsreisen Einsteins" (n. 2), p. 9.

4 Charles Weiner, "A new site for the seminar: The refugees and American physics in the 1930s," in Donald H. Fleming and Bernard Bailyn, eds., *The intellectual migration: Europe and America, 1930–1960* (Cambridge, MA, 1969), pp. 190–1.

5 Daniel J. Kevles, *The physicists: The history of a scientific community in modern America* (New York, 1978), p. 221.

6 For example, the scholars listed in nn. 1, 4, and 5 above; and also Armin Hermann, *The new physics: The route into the atomic age* (Munich, 1979), chap. 11, "The mass migration from under"; Laura Fermi, *Illustrious immigrants*, 2d ed. (Chicago, 1971); J. H. Van Vleck, "American physics comes of age," *Physics Today*, 17 (June 1964): 21–6; Katherine Sopka, *Quantum physics in America, 1920–1935* (New York, 1981); Stanley Coben, "Scientific establishment and the transmission of quantum mechanics to the United States, 1919–1932," *American Historical Review*, 76 (1971): 442–66; Stanley Coben, "Foundation officials and fellowships: Innovation in the patronage of science," *Minerva* 14 (Summer 1976): 225–40; Spencer R. Weart, "The physics business in America, 1919–1940: A statistical reconnaissance," in Nathan Reingold, ed., *The sciences in the United States: A bicentennial perspective* (Princeton, 1979). In Kevles's book, see particularly chapters 14–16 and his bibliographical essay on Resources, pp. 450–7, for further primary sources. See also Paul K. Hoch, "The reception of central European refugee physicists of the 1930s: USSR, U.K., U.S.A.," *Annals of Science*, 40 (1983): 217–46; and Roger H. Stuewer, "Nuclear physicists in a new world: The émigrés of the 1930s to America," *Berichte zur Wissenschaftsgeschichte*, 7 (1984): 23–40.

7 J. Robert Oppenheimer, in a filmed interview conducted Nov. 1, 1966, by Charles Weiner for Harvard Project Physics; transcript at American Institute of Physics, Center for the History of Physics, New York; see n. 4, p. 191.

8 "May I say that the invitation [to give the lecture] makes me feel that the links between this University and myself are becoming professionally stronger?" Trans. from Albert Einstein manuscript; cf. *Mein Weltbild* (Frankfurt am Main, 1977), pp. 113–19.

9 Luce Langevin, "Paul Langevin et Albert Einstein d'après une Correspondance et des Documents inédits," *La Pensée*, No. 161 (Feb. 1972): 29.

10 Stanley Goldberg, "The early response to Einstein's special theory of relativity, 1905–1911: A case study in national differences" (Ph.D. diss., Harvard University, 1968).

11 Frank, *Einstein*, op cit. (n. 1), p. 186.

12 Albert Einstein, "My first impressions of the U.S.A.," in *Ideas and opinions* (New York, 1954), pp. 16–19.

13 Frank, *Einstein*, op. cit. (n. 1), p. 178.

14 Fritz Stern, "Einstein's Germany," in Gerald Holton and Yehuda Elkana, eds., *Albert Einstein: Historical and cultural perspectives* (Princeton, 1982).

15 Sopka, *Quantum physics*, op. cit. (n. 6), pp. 3.40–3.42.

16 See Weiner, "A new site," op. cit. (n. 4), p. 226.

17 Sopka, *Quantum physics*, op. cit. (n. 6), pp. A.17–A.26.

18 Ibid., p. A.29.

19 Figures derived from the *Rockefeller Foundation directory of fellowship awards, 1917–1950* (New York, 1951); Myron Rand, "The national research council fellowships," *The Scientific Monthly*, 58, no. 2 (Aug. 1921); and *NRC check list of grants-in-aid for May 1929–December 1937*. I thank Charles Weiner for making many of these data available to me.

20 Werner Heisenberg, *Physics and beyond* (New York, 1971), chap. 8, p. 94.
21 R. A. Millikan, *The autobiography of Robert A. Millikan* (New York, 1950), pp. 215, 217.
22 Ibid., p. 221.
23 Ibid., p. 239.
24 See Gerald Holton, *Thematic origins of scientific thought: Kepler to Einstein* (Cambridge, MA, 1973), p. 320.
25 Letter of Oct. 23, 1931, from the Dean of Christ Church, announcing Einstein's election to a "Research Studentship" in the College. Einstein Archive at Hebrew University, Jerusalem. I thank Professor John Stachel for having drawn my attention to the letter.
26 Reprinted as Document 160 in C. Kirsten and H.-J. Treder, eds., *Albert Einstein in Berlin 1913–1933* [Berlin, 1979], 1:239–40.
27 Travel Diary no. 5, Dec. 1931–Feb. 1932. In Einstein Archive, Jerusalem.
28 R. A. Millikan, "The new opportunities in science," *Science, 50* (1919): 297, quoted in Kevles, *The physicists*, op. cit. (n. 5), p. 169.
29 Victor Weisskopf, quoted in Fleming and Bailyn, *The intellectual migration*, op. cit. (n. 4), p. 222. It is also significant that Einstein's own scientific publications, from 1935 on, refer more frequently than before to work going on in America and to articles published in American scientific journals.
30 Ironically, Einstein was barred (without his knowledge) at the highest level from participating in sensitive U.S. war research – again for fear of his internationalist tendencies. See Bernard T. Feld, "Einstein and the politics of nuclear weapons," in Holton and Elkana, *Albert Einstein*, op. cit. (n. 14).

Chapter 7. *"Success sanctifies the means": Heisenberg, Oppenheimer, and the transition to modern physics*

1 Max Jammer, *The conceptual development of quantum mechanics* (New York: McGraw-Hill, 1966), p. 219.
2 For example, in the *Archives of the sources for the history of quantum physics*, twelve sessions, November 30, 1962–July 12, 1963.
3 *Archives*, op. cit. (n. 2), pp. 5–6.
4 Heisenberg's paper, "Zur Quantentheorie der Linienstruktur und der anomalen Zeemaneffekte," appeared in *Zeitschrift für Physik, 8* (1922): 273–97. For further details, including a good discussion of Heisenberg's "unique style" and the reaction to it, see David C. Cassidy, "Heisenberg's first paper," *Physics Today, 31* (1978): 23–8, and David C. Cassidy, "Heisenberg's first core model of the atom: The formation of a professional style," *Historical Studies in the Physical Sciences, 10* (1979): 187–224. As Cassidy points out, half-integral quantum numbers had been used by A. Landé and others, but other of Heisenberg's conscious deviations from accepted principles such as half-integral momenta and a magnetic core had not. "Not only had Heisenberg introduced real non-integral momenta, but he had also violated [mostly without explicitly drawing attention to it] the Sommerfeld quantum conditions, the angular-momentum selection rules, space quantization, classi-

cal radiation theory, the Larmor precession theorem, and the semi-classical criterion of perceptual clarity (*Anschaulichkeit*) in model interpretation" (ibid., pp. 190–1).

Among other useful articles touching on these points, see Paul Forman, "The doublet riddle and atomic physics *circa* 1924," *Isis*, 59 (1968): 156–74; Edward MacKinnon, "Heisenberg, models, and the rise of matrix mechanics," *Historical Studies in the Physical Sciences*, 8 (1977): 137–88; Daniel Serwer, "*Unmechanischer Zwang:* Pauli, Heisenberg, and the rejection of the mechanical atom, 1923–1925," *Historical Studies in the Physical Sciences*, 8 (1977): 189–256; David Bohm, "Heisenberg's contribution to physics," in W. C. Price and S. S. Chissick, eds., *The uncertainty principle and foundations of quantum mechanics* (New York: Wiley, 1977), pp. 559–63 (on Heisenberg's "radically new mathematical and physical account of the facts as a whole"); and A. Hermann, K. V. Meyenn, and V. F. Weisskopf, eds., *Wolfgang Pauli: Wissenschaftlicher Briefwechsel mit Bohr, Einstein, Heisenberg, u.a.*, vol. I, 1919–29 (New York: Springer-Verlag, 1979).

5 W. Heisenberg, *Physics and beyond* (New York: Harper & Row, 1971), ch. 1; see also ch. 20.

6 W. Heisenberg, *Tradition in science* (New York: Seabury Press, 1978), p. 40; a collection of the essays of the late period. See also his articles, "Development of concepts in the history of quantum theory," *American Journal of Physics*, 43 (1975): 389–94, and "The nature of elementary particles," *Physics Today*, 29 (1976): 32–9.

7 "Discussion with Professor Heisenberg," in O. Gingerich, ed., *The nature of scientific discovery* (Washington, D.C.: Smithsonian Institution Press, 1975), 556–73, esp. pp. 560 and 565.

8 W. Heisenberg, "Über quantentheoretische Umdeutung kinematischer und mechanischer Beziehungen," *Zeitschrift für Physik, 33* (1925): 879–93.

9 W. Heisenberg, "Quantenmechanik," *Die Naturwissenschaften, 14* (1926): 989–94.

10 W. Heisenberg, "Über den anschaulichen Inhalt der quantentheoretischen Kinematik und Mechanik," *Zeitschrift für Physik, 43* (1927): 172–98.

11 See A. I. Miller, "Vizualization lost and regained: The genesis of the quantum theory in the period 1913–1927," in Judith Wechsler, ed., *On aesthetics in science* (Cambridge, MA: MIT Press, 1978), pp. 73–102. Also, A. I. Miller, "Redefining Anschaulichkeit," in *Physics as natural philosophy: essays in honor of Laszlo Tisza on his Seventy-fifth birthday*, Abner Shimony and Herman Feshbach, eds., (Cambridge, MA: MIT Press, 1982), which carefully treats the complex of meanings and transitions associated with the conception of *Anschauung* and *Anschaulichkeit* in German physics research in the 1920s.

12 Max Jammer, *The philosophy of quantum mechanics* (New York: Wiley, 1974), p. 21. Jammer assigns Max Born a relatively secondary role in the further development: "As Max Born soon recognized, the 'sets' in terms of which Heisenberg had solved the problem of the anharmonic oscillator" were entities known from the theory of matrices. "Within a few months Heisenberg's new approach was *elaborated* by Born, Jordan, and Heisenberg himself." A similar estimate is given in Jammer, *Quantum mechanics*, op. cit. (n. 1), p. 197.

13 Heisenberg, "Quantenmechanik," op. cit. (n. 9).

14 See M. Born, *Zeitschrift für Physik*, 37 (1926): 863 and *38* (1926): 803; also Born's letter to Einstein, Nov. 11, 1926.

Heisenberg was upset that Born imposed the notion of probability on the quantum mechanics that Heisenberg thought of as a closed theory; moreover, Born was using wave functions taken from Schrödinger's wave mechanics that purported to be based on a continuum interpretation of the atomic regime, replete with pictures, i.e., *Anschauungen*. Heisenberg immediately submitted a paper in which Born's probabilistic interpretation of the wave function is not mentioned and Schrödinger is roundly criticized [W. Heisenberg, "Schwankungserscheinungen und Quantenmechanik," *Zeitschrift für Physik*, 40 (1926): 501–6]. There Heisenberg demonstrated that a probability interpretation emerges naturally from quantum mechanics and is properly understood in terms of essential discontinuity (quantum jumps).

Later, Heisenberg, distancing himself further from Born, went so far as to declare Born's statistical interpretation of the wave function to be the result of "developing and elaborating an idea previously [1924] expressed by Born, Kramers, and Slater." Heisenberg, in M. Fierz and V. F. Weisskopf, eds., *Theoretical physics in the twentieth century* (New York: Interscience, 1960), pp. 40–7. Moreover, Born's original probabilistic interpretation of 1926, "Zur Quantenmechanik der Stossvorgänge," *Zeitschrift für Physik*, 37 (1926): 863–7, with all its successes, failed when applied to electron diffraction. Again, it was Heisenberg's role here to make a crucial step, by changing the interpretation of the ψ waves, "not to regard them as merely a mathematical fiction but to ascribe to them some kind of physical reality." Jammer, *Philosophy of Quantum Mechanics*, op. cit. (n. 12), p. 44.

15 Heisenberg, *Physics and Beyond*, op. cit. (n. 5).

16 "Discussion with Professor Heisenberg," op. cit. (n. 7). See also the last section of Chapter 4.

17 See, for example: Stanley Coben, "Scientific establishment and the transmission of quantum mechanics to the United States, 1919–1932," *American Historical Review*, 76 (1971); Laura Fermi, *Illustrious immigrants*, 2d ed. (Chicago: University of Chicago Press, 1971); Daniel J. Kevles, *The physicists: The history of a scientific community in modern America* (New York: Knopf, 1978) (see esp. chap. 14–16 and his bibliographical essay on Resources, pp. 450–7 for further primary sources); Katherine R. Sopka, *Quantum physics in America, 1920–1935* (New York: Arno Press, 1980); J. H. Van Vleck, "American physics comes of age," *Physics Today*, 17 (1964); Spencer R. Weart, "The physics business in America, 1919–1940: A statistical reconnaissance," in Nathan Reingold, ed., *The sciences in the United States: A bicentennial perspective* (Princeton, NJ: Princeton University Press, 1979); and Charles Weiner, "A new site for the seminar: the refugees and American physics in the 1930s," in Donald H. Fleming and Bernard Bailyn, eds., *The intellectual migration: Europe and America, 1930–1960* (Cambridge, MA: Harvard University Press, 1969).

18 J. H. Van Vleck, "American physics" op. cit. (n. 17), p. 25.

19 Ibid., p. 24.

20 Quoted in Kevles, *The physicists*, op. cit. (n. 17), p. 169.

21 Of excellent use for the following passages are the newly published letters, in Alice Kimball Smith and Charles Weiner (eds.), *Robert Oppenheimer, letters and recol-*

lections (Cambridge, MA: Harvard University Press, 1980). The quotations that follow are from the letters in that book.

22 M. Born and R. Oppenheimer, "Zur Quantentheorie der Molekeln," *Annalen der Physik*, 84 (1927): 457–84. The first footnote reference in the paper is to a collaborative paper by Born and Heisenberg, published three years earlier.

23 Smith and Weiner, eds., *Robert Oppenheimer*, op. cit. (n. 21), p. 227.

24 E.g., in A. Hermann et al., eds., *Wolfgang Pauli*, op. cit. (n. 4).

25 Born and Oppenheimer, "Zur Quantentheorie der Molekeln," op. cit. (n. 22).

26 Smith and Weiner (eds.), *Robert Oppenheimer*, op. cit. (n. 21).

Chapter 8. Do scientists need a philosophy?

1 Steven Weinberg, "The search for unity: Notes for a history of quantum field theory," *Daedalus* (Fall, 1977): 17–18. A similar point is made about the development of particle physics, including Yukawa's contribution, in *The birth of particle physics*, L. M. Brown and L. Hoddeson, eds. (Cambridge: Cambridge University Press, 1983), pp. 286–292, 294–303.

2 A. Einstein, "Reply to criticisms," in *Albert Einstein: Philosopher-scientist*, Paul A. Schilpp, ed. (Evanston, IL: The Library of Living Philosophers, 1949), pp. 683–4.

3 Also refer to Chapter 7, "Success sanctifies the means."

4 Editor's summary in *Springs of scientific creativity: Essays on founders of modern science* (Minneapolis, MN: University of Minnesota Press, 1983), p. vi.

5 Peter Galison, "Bubble chambers and the experimental workplace," in *Observation, experiments, and hypothesis in modern physical science*, Peter Achinstein and Owen Hannaway, eds. (Cambridge, MA: MIT Press, 1985), pp. 309–72.

6 Julian Schwinger, "Two shakers of physics: Memorial lecture for Sin-itiro Tomonago," in *The birth of particle physics*, op. cit. (n. 1), pp. 364–5.

7 Hilary Putnam, "Philosophers and human understanding," in A. F. Heath, ed., *Scientific explanation* (Oxford: Clarendon Press, 1981), pp. 99–120.

8 In Pierre Speziali, ed., *Albert Einstein, Michele Besso, correspondence 1903–1955* (Paris: Hermann, 1972), pp. 114–15.

9 Henry Harris, "Rationality and science," in A. F. Heath, ed., *Scientific explanation*, op. cit. (n. 7), pp. 36–52.

10 Peter Galison, "How the first neutral current experiments ended," *Reviews of Modern Physics*, 55 (1983): 487–91, 505–6.

11 Sheldon L. Glashow, "Particle symmetries of weak interactions," *Nuclear Physics*, 22 (1961): 579–88.

12 Howard Georgi and Sheldon L. Glashow, "Unity of all elementary-particle forces," *Physical Review Letters*, 32 (January 1974): 438–41.

13 Carolyn Eisele, "Peirce, Charles Sanders," in *The dictionary of scientific biography*, C. C. Gillispie, ed., Vol. 10 (New York: Scribner, 1973), pp. 484–5; see also Christine Ladd-Franklin, "Charles S. Peirce at the Johns Hopkins," in *The Journal of Philosophy, Psychology, and Scientific Method*, 13, no. 26 (1916): 718.

14 Putnam, op. cit. (n. 7), p. 118. The work of the philosopher Stephen Toulmin (e.g., *Human understanding*) also shows great sensitivity to the facts of scientific practice.

15   E.g., Chapters 1 and 2 of this book, and in G. Holton, *Thematic origins of scientific thought: Kepler to Einstein* (Cambridge, MA: Harvard University Press, 1973) and *The scientific imagination: Case studies* (Cambridge: Cambridge University Press, 1978).

16   Also refer to Chapter 4, "Einstein's search for the *Weltbild.*"

Chapter 9. Science, technology, and the fourth discontinuity

1   Mazlish, Bruce. "The fourth discontinuity." *Technology and Culture, 8* (1967): 1–15.

2   Ibid., p. 4.

3   "How basic research reaps unexpected rewards" is, in fact, the title of a pamphlet released by NSF, February 1980.

4   For some reasons why, in the past, "we have not found good ways of encouraging much-needed inquiry, especially in the areas of the environment, the control of population growth, and the conversion of energy," see R. S. Morison, "Introduction" to the book *Limits of scientific inquiry*, R. S. Morison and G. Holton, eds. (New York: W. W. Norton, 1979), p. xviii.

5   As Bruce L. R. Smith and Joseph J. Karlesky [*The state of academic science: The universities in the nation's research effort* (New York: Change Magazine Press, 1977)] have noted, institutional deficiencies, including short cycles of funding by government and industrial sponsors, have also adversely affected the pursuit of "pure" research and resulted in a certain lack of venturesomeness. They identified in academic institutions a "notable shift away from basic research to applied and mission-oriented research" – precisely in the opposite direction from the combined mode – "and from risk-taking to relatively safe and predictable lines of inquiry." In industry, the same problem has appeared, as documented in the biennial *Science indicators*, with a decade-long decrease in basic research expenditure and a change to short-term goals, often to "defensive research" that aims chiefly at the protection of old products against regulations.

6   Press, Frank. "Science and technology: the road ahead." *Science, 200* (1978): 737–41.

7   Tenth Annual Report of the National Science Board, *Basic research in the mission agencies: Agency perspectives on the conduct and support of basic research, 2 August 1978* (Washington, DC: U.S. Government Printing Office, 1978). p. 303.

8   Elsewhere in the NSF report (p. 286) there is a frank discussion about the practical differences of drawing the line between basic and applied science in the mission agencies: "Every agency science administrator is plagued by the mission relevance question, especially in relation to basic research. For example, the Office of Naval Research (ONR) identifies support of pure mathematics as highly relevant to the Navy's mission, but perhaps this would not be so regarded in other sectors." Similarly, "NSF has been plagued since its inception by persons who ask how many of the supported projects can be justified and to what extent they relate to any conceivable national purpose. Scientists within the agencies feel that skepticism is due to a lack of understanding of what basic science is about and how it relates to the national purpose. The science administrator is caught between the scientist, who believes any scientific enquiry is justified, and skeptical citizens or Congressmen,

who wonder how esoteric enquiries can warrant public fund support. As the pressure mounts, the research administrator finds applied research easier to justify than basic" (ibid.).

A more thorough analysis would demand that one face a number of other practical problems: for example, how to deal with differences existing between basic and applied science in different fields, such as mathematics and cultural anthropology, or even physics and biology; how to prevent such political problems as either seeming to promise too much or incurring backlash when a problem turns out to be even more long-range and complex than originally foreseen on the most cautious model; or, for that matter, how to institutionalize the support of research in the combined mode in order to immunize it, at least during the early, vulnerable phase, from the axe of practical-minded budget cutters during a period of general retrenchment. As a top official of the OMB recently said, "Frankly, basic research is necessary in the long run, vulnerable in the short run." A good test case of this sort is the tortuous progress toward the fusion reactor, an effort in which a large component of fundamental research has been involved over a period of two decades without clearly reaching the promised goal of a limitless supply of energy with relatively low risk. The frustration that has been building up was captured in a recent comment by a Congressional aide: "There is a feeling that if you leave it to the scientists all the time, you won't get any energy out of it. It's becoming unacceptable to have an energy problem that costs half a billion dollars a year and doesn't produce anything."

9   See, for example, *Science and Government Report*, 9 (2) for February 1, 1979, and *Hearings before the Subcommittee on Science, Technology, and Space, on Oversight on the Office of Science and Technology Policy*, March 7 and 21, 1979.

10   Two volumes; Government Printing Office, Superintendent of Documents, Washington, DC 20402, Stock Numbers 038-000-00442-5 and 038-000-00441-7.

11   The legislative language requiring the NSF to prepare such volumes specifies that it "identify and describe situations and conditions which warrant special attention within the next five years, involving, (1) current and emerging problems of national significance that are identified through scientific research, or in which scientific or technical considerations are of major significance, and (2) opportunities for, and constraints on, the use of new and existing scientific and technological capabilities which can make a significant contribution to the resolution of problems identified . . . or to the achievement or furthering of program objectives or national goals. . . ."

12   Op. cit. (n. 10), vol. 1, pp. 1–4.

13   Op. cit. (n. 10), vol. 1, p. 31.

14   Op. cit. (n. 10), vol. 2, p. 530.

15   See Elzinga, Aant. "The Swedish science discussion 1965–1975." *Social Indicators Research*, 7 (1980): 379–99. The article is very interesting and, to many U.S. readers, will be disturbing. It also warns that "as pressure of sectorization increases, so does polarization of the scientific community into those who are receptive and those who resist this development." The opposition in fact contains not only the group "arguing from the idea of a free autonomous science based on liberal values," but also other groups, "often with a radical leftish social inclination, who oppose sectorization in its concrete form because the overall goals are determined by monopoly capitalist class interests" (p. 391).

16 The National Program for Science and Technology, 1978–1982." (CONACYT, Mexico, D.F., 1978) and Edmundo Flores, "Mexico's program for science and technology, 1978–82." *Science, 204* (1979): 1279–82.

17 For example, reports by George A. Pake and by Lewis M. Branscomb in *Physics Today,* April 1978 and April 1980, respectively.

18 For representative brief descriptions, see the growth and development of STS education – three examples. *Science, Technology, and Human Values, 5* (Spring 1980): 31–35. The programs described are those at Lehigh University, Stanford University, and MIT.

Chapter 10. The two maps

1 Shortly thereafter, he also added that the action between magnet needle and current loop is reciprocal, a discovery usually associated with Ampère, who independently published it later. For good accounts of Oersted's work and motivation, see L. Pearce Williams' account in the *Dictionary of scientific biography*; Kristine Meyer, "The scientific life and works of H. C. Ørsted," in *H. C. Ørsted, Naturvidenskabelige Skrifter,* edited by K. Meyer (Copenhagen: Host, 1920), Vol. 1, pp. XIII–CLXVI; Robert C. Stauffer, "Speculation and experiment in the background of Oersted's discovery of electromagnetism," *Isis, 48* (1957): 33–50; Bern Dibner, *Oersted and the discovery of electromagnetism* (Norwalk: Burndy, 1961); and Barry Gower, "Speculation in physics: The history and practice of Naturphilosophie," *Studies in history and philosophy of science* (New York: Pergamon, 1973), Vol. 3, pp. 301–56.

2 For a brilliant analysis of the changing tasks of historians, see Judith N. Shklar, "Learning without knowing," *Daedalus, 109* (2) (1980): 53–72, from which the last two quotations are taken.

3 For a faithful description of the system of education in a Gymnasium of Vienna by an exact contemporary, see Egon Schwarz, *Keine Zeit für Eichendorff: Chronik unfreiwilliger Wanderjahre* (Königstein: Athenäum Verlag, 1979), pp. 11–18.

Chapter 11. From the endless frontier to the ideology of limits

1 Vannevar Bush, *Pieces of the action* (New York: Morrow, 1970), p. 64.

2 Ibid., p. 65.

3 Ibid.

4 Vannevar Bush, *Science, the endless frontier* (Washington, DC: National Science Foundation, 1945), p. xxvi.

5 Arthur Kubo, MIT Ph.D. thesis. "Technology assessment of high-level nuclear waste management," p. 181. He refers to the National Academy of Sciences study "Disposal of radioactive wastes on land," 1957, requested by the Atomic Energy Commission.

6 Bush, *Science,* op. cit. (n. 4), pp. 3–4.

7 Ibid., p. 1. One more irony is that when the National Science Foundation was set up as the direct result of Bush's campaign, its mandate *excluded* the "war against disease," the support of medicine and its related sciences that may have been

of greatest personal interest to Roosevelt, as well as to all civilian research on weapons.

8   President Carter's covering letter of August 2, 1978, for the Tenth Annual Report of the National Science Board, and accompanying NSF News Release. Tenth Annual Report of the National Science Board, *Basic research in mission agencies: Agency perspectives on the conduct and support of basic research*, August 2, 1978 (Washington, DC: U.S. Government Printing Office, 1978).

9   Don K. Price, "Endless frontier or bureaucratic morass?," in *The limits of scientific inquiry*, Gerald Holton and Robert S. Morison, eds. (New York: Norton, 1979), p. 85.

10 Harvey Brooks, "The problem of research priorities," in *The limits*, op. cit. (n. 9), pp. 171–90.

11 George Sarton, *The study of the history of science* (1936; rpt. New York: Dover, 1957), p. 5.

12 Imre Lakatos, "Understanding Toulmin," *Minerva*, 14 (1976): 128.

Chapter 12. Metaphors in science and education

1   Donald G. MacRae, "The body and social metaphor," in *The body as a medium of expression*, J. Benthall and T. Polhemus, eds. (London: Allen Lane, 1975), p. 59.

2   Alan Bullock and Oliver Stallybrass, eds., *The Fontana dictionary of modern thought* (London: Collins, 1977), p. 20. (*The Harper dictionary of modern thought*, New York: Harper and Row.)

3   Colin Turbayne, *The myth of metaphor* (New Haven, CT: Yale University Press, 1962), p. 5.

4   Nicolaus Copernicus (c. 1530, trans. E. Rosen), *De Revolutionibus Orbium Coelestium* (Baltimore, MD: Johns Hopkins University Press, 1978), pp. 4–7; my italics.

5   Ernst Gombrich, *Symbolic images: Studies in the art of the renaissance* (London: Phaidon, 1972), p. 166.

6   Erwin Panofsky, "Galileo as a critic of the arts: Aesthetic attitude and scientific thought," *Isis*, 47 (1956): 3–15.

7   Thomas Young, "Outlines of experiments and inquiries respecting sound and light," in *Miscellaneous works of Thomas Young*, Vol. I, George Peacock, ed. (London: Murray, 1855), pp. 80–81.

8   Ibid., p. 82.

9   Gerald Holton, *Thematic origins of scientific thought* (Cambridge, MA: Harvard University Press, 1973), pp. 363–4.

10 Andrew Pickering, "Exemplars and analogies," *Social Studies of Science*, 10 (1980), pp. 497–502.

11 Richard Boyd, "Metaphor and theory change," in *Metaphor and thought*, A. Ortony, ed. (Cambridge: Cambridge University Press, 1979), p. 408.

12 Gerald Holton, *The scientific imagination: Case studies* (Cambridge: Cambridge University Press, 1978), pp. 37–8.

13 Martin Deutsch, "Evidence and inference in nuclear research," *Daedalus*, 87 (Fall 1958): 88–9.

14 Gombrich, op. cit. (n. 5), p. 167.
15 MacRae, op. cit. (n. 1), p. 67.
16 Margaret Mead, "Closing the gap between the scientists and the others," *Daedalus*, 88 (1959): 139–46.
17 P. Logan, "Language and physics," *Physics Education*, 16 (1981): 174–77.
18 Alfred M. Mayer, *Silliman's American journal*, April 1878; reprinted in *Philosophical Magazine*, 5 (1878), 397–8.
19 Alfred N. Whitehead, *Science and the modern world* (New York: New American Library, 1948), p. 105.
20 Holton, *Thematic origins*, op. cit. (n. 9), pp. 376–7. See also Chapter 4.
21 E. Cassirer, *Philosophie der Symbolischen Formen* (Berlin: B. Cassirer, 1925).
22 Gombrich, op. cit. (n. 5), p. 168.
23 Holton, *Thematic origins*, op. cit. (n. 9), p. 37.
24 This is my own translation from J. W. Goethe, *Zur Farbenlehre*, in *Goethes Werke*, Vol. XII (Hamburg: Christian Wegner Verlag, 1955; 4th ed., 1962), pp. 522–3.
25 Panofsky, op. cit. (n. 6).
26 See Gerald Holton, "Conveying science by visual presentation," in *Education of vision*, G. Kepes, ed. (New York: Geo. Braziller, 1965).
27 Gombrich, op. cit. (n. 5), p. 125.
28 Richard Feynman, *The character of physical law* (London: Cox & Wyman, 1965); Steven Weinberg, *The discovery of subatomic particles* (New York and San Francisco: Freeman, 1983).
29 W. H. Letherdale, *The role of analogy, model and metaphor in science* (New York: American Elsevier, 1974), p. 242.
30 Turbayne, op. cit. (n. 3), p. 217.

ACKNOWLEDGMENTS FOR PREVIOUSLY PUBLISHED VERSIONS OF CHAPTERS

Most chapters in this book have been substantially reworked from their original form, as given in the following publications:

Chapter 1, see "Thematic presuppositions and the direction of scientific advance," in A. F. Heath, ed., *Scientific explanation: Papers based on Herbert Spencer Lectures given in the University of Oxford* (New York: Oxford University Press, 1981), pp. 1–27.

Chapter 2, see "Constructing a theory: Einstein's model," *The American Scholar*, 48, no. 3 (Summer, 1979): 309–40.

Chapter 3, see "Einstein's scientific program: The formative years," in Harry Woolf, ed., *Some strangeness in the proportion, a centennial symposium to celebrate the achievements of Albert Einstein* (London: Addison-Wesley, 1980), pp. 49–65.

Chapter 4, see "Einstein's search for the *Weltbild*," *Proceedings of the American Philosophical Society*, 125, 1 (February, 1981): 1–15.

Chapter 5, see "Einstein and the shaping of our imagination," in Gerald Holton and Yehuda Elkana, eds., *Albert Einstein, historical and cultural perspectives: The Centennial Symposium in Jerusalem* (Princeton, NJ: Princeton University Press, 1982), pp. vii–xxxii. Copyright © 1982 by Princeton University Press. Reprinted with permission of Princeton University Press.

Chapter 6, see "The migration of physicists to the United States," in Jarrell C. Jackman and Carla M. Borden, eds., *The muses flee Hitler, cultural transfer and adaptation, 1930–1945* (Washington, DC: Smithsonian Institution Press, 1983), pp. 169–89.

Chapter 7, see "Success sanctifies the means," in Everett Mendelsohn, ed., *Transformation and tradition in the sciences. Essays in honor of I. Bernard Cohen* (Cambridge: Cambridge University Press, 1984), pp. 155–175.

Chapter 8, see *Times Literary Supplement*, No. 4, 257 (November 2, 1984):1231–4.

Chapter 9, see "Science, technology, and the fourth discontinuity," in R. A. Kasschau, R. Lachman, and K. R. Laughery, eds., *Information technology and psychology, prospects for the future* (New York: Praeger, 1982), pp. 1–19. Copyright © 1982 by Praeger Publishers. Reprinted with permission of Praeger Publishers.

Chapter 10, see Oersted Medal Response at the Joint American Physical Society–American Association of Physics Teachers Meeting, Chicago, January 22, 1980 ["The two maps," *American Journal of Physics*, 48, no. 12 (December, 1980):1014–19. Copyright 1980 by AAPT.]

Chapter 11, see "From the endless frontier to the ideology of limits," in Gerald Holton and Robert S. Morison, eds., *Limits of scientific inquiry* (New York: W. W. Norton & Co., 1979), pp. 227–41.

Chapter 12, see "Metaphors in science and education," in William Taylor et al., eds., *Metaphors in education* (London: Heinemann Educational Books, for the Institute of Education, University of London, 1984), pp. 91–113.

Chapter 13, see *Daedalus*, 113, no. 4 (Fall 1984):1–27.

INDEX